machine
shop
operations and setups

The Authors

Orville D. Lascoe

Professor Lascoe is the Director of the Manufacturing Engineering Laboratories in the School of Industrial Engineering at Purdue University, which is the largest and most complete machine tool laboratory in the country. Professor Lascoe has many years of industrial experience in tool engineering and supervision. He has served as a consultant on manufacturing and engineering problems and has organized machine tool laboratories in several vocational schools and universities.

Active in machine tool association programs, Professor Lascoe is a contributing author to publications on machine tool operations. His research activities extend to fields such as Machinability, Plastics, Electro-discharge and Electro-chemical machining processes. Winner of the 1963 A.S.T.M.E. National Education Award, he has represented the U.S. Department of Commerce at international trade fairs and served as a consultant on metalworking equipment in the far east.

Clyde A. Nelson

Mr. Nelson is the Director of Manpower Development and Training Programs for the Chicago Public Schools System. Mr. Nelson brings to this textbook a lifetime of experience with machine tools. Beginning as a machinist apprentice, he became journeyman, machinist, foreman, and finally supervisor of apprentices.

Mr. Nelson completed his undergraduate studies in Industrial Education at Purdue University. He received his M.S. from Chicago State University. He has served as machine shop instructor at several Chicago vocational high schools, and as lecturer in Industrial Education courses for several Illinois colleges. Mr. Nelson was appointed Supervisor of Vocational and Practical Arts, Machine Shop Division, for the entire Chicago Public Schools System, an assignment which led to his present position. He has also been involved as a metalworking consultant in industrial training programs.

Harold W. Porter

Dr. Porter is Dean of Instruction and Acting Chairman of General Education at the State University of New York Agricultural and Technical College in Canton. His varied career has included teaching vocational education in Wisconsin, serving on the Wisconsin State Board of Vocational and Adult Education. He taught at Purdue University for twelve years and headed up the Industrial Education undergraduate program, during which time he received a Ph.D. in Industrial Psychology.

He has also been Chairman of the Industrial Education and Arts Department at Kansas State College in Pittsburgh, and has served as a consultant to the Pakistan government with responsibility for establishing vocational and technical schools, standards and curricula. He has also supervised training in prisons and reformatories, and was a consultant to the Bureau of Prisons.

machineshop

operations and setups

4th edition

lascoe   nelson   porter

AMERICAN TECHNICAL SOCIETY   CHICAGO, 60637

Preface

Machine tool technology has advanced tremendously in the last decade. As in other fields, today's trainee is faced with two tasks: (1) mastering the fundamentals of basic machine tools and (2) keeping up with the new concepts being introduced in machine tool technology. Even now, modern machine tools are flowing from research and development to an industry which is not fully prepared to utilize them. Many times this is due to a lack of trained manpower.

The purpose of this text is, therefore, twofold: (1) to provide basic training in conventional machine tool operation and (2) to provide the trainee with the knowledge he will need to understand the latest machine tool processes and developments.

Even though chip-producing machine tools vary in size, shape and function, the cutting theory of the single-point cutting tool underlies every chip-producing machine tool operation. Amply stressed in this text, this carryover of principles from one machine tool to another reveals to the student basic similarities in an otherwise often bewildering variety of machine tools.

Equally important, this new Fourth Edition prepares the student for such innovations as Numerical Control and the Electrical Energy processes, which have literally revolutionized today's metalworking industry.

TransVision teaching aids, an innovation in machine shop textbooks introduced with the previous edition, have been enthusiastically received and, therefore, retained in this Fourth Edition. TransVisions have proved highly successful in communicating information and techniques that are difficult for the student to visualize, understand, and retain when presented in conventional ways. The subjects chosen for presentation in the three sets of TransVisions were carefully chosen for maximum benefit to the student; they are the micrometer, lathe, and milling machine.

Throughout the text color has been used to indicate forces, movement, and the possible motions of each machine tool. Careful preparation of illustrations has assured that the student will better comprehend machine tool capabilities and movements.

In this Fourth Edition, chapters on Power Saws and Turret Lathes have been added—the latter with a special introduction and an actual production setup—and several chapters have been greatly expanded. This text continues to benefit, however, from the research that went into the original edition. Content and organization are based on that research and on the knowledge and experience of the authors in machine tool technology.

The Publishers

Acknowledgements

The authors wish to express their respect and appreciation to the dedicated training directors and instructors throughout the country whose comments and questions led to improvements or corrections in this text.

We are grateful to the many machine tool builders and to the related companies and associations listed below who have assisted in the preparation of this text by supplying needed illustrations and technical information regarding machine tools, accessories, cutting tools, basic operations, techniques, and processes.

Allegheny Ludlum Steel Corp.
American Tool Works Co.
Atlas Press Co., Clausing Div.
Bay State Products Co.
Black & Decker Manufacturing Co.
Bridgeport Machine Tool Co.
Brown & Sharpe Manufacturing Co.
The Carborundum Co.
Chicago Public Schools, Manpower Training Div.
Charmilles Corp. of America
Cincinnati Lathe & Tool Co.
Cincinnati Milacron Co.
Cincinnati Shaper Co.
Cleveland Twist Drill Co.
Collins Microflat Co.
Delta File Works, Inc.
The DoALL Co.
Electric Hotpack Co., Inc.
General Dynamics Corp., General Atomic Div.
General Electric Co., Machining Development Lab.
General Electric Co., Metallurgical Products Dept.
General Motors Corp.
Giddings & Lewis Machine Tool Co.
Goddard & Goddard Co.
G. A. Gray Co.
Hardinge Bros., Inc.
Harig Products Co.
Jacobs Manufacturing Co.
Kearney & Trecker Corp.
Landis Tool Co.
R. K. LeBlond Machine Tool Co.

Lindberg Hevi-Duty Div., Sola Basic Industries
Lodge & Shipley Co.
Lodge & Shipley Co.
Lufkin Tool Co.
Macklin Co.
Minneapolis-Honeywell Corp.
Modern Machine Shop Magazine
Monarch Machine Tool Co.
Morse Twist Drill & Machine Co.
National Automatic Tool Co., Inc.
National Machine Tool Builders Assoc.
National Safety Council
National Twist Drill and Tool Co.
Nicholson File Co.
Norton Co.
Pratt & Whitney Div., Colt Industries, Inc.
Purdue University
Racine Hydraulics & Machinery, Inc.
Ready Tool Co.
Rockford Machine Tool Co.
Rockwell Manufacturing Co., Delta Power Tool Div.
Sheffield Corp.
Shore Instrument & Manufacturing Co., Inc.
South Bend Lathe, Inc.
Standard Tool Co.
The L. S. Starrett Co.
Taft-Peirce Manufacturing Co.
Warner & Swasey Co.
V. R. Wesson Co.
Westinghouse Corp.
J. H. Williams & Co.
Wilson Co.

Contents

A Handy Guide to Operations, Setups, and Techniques

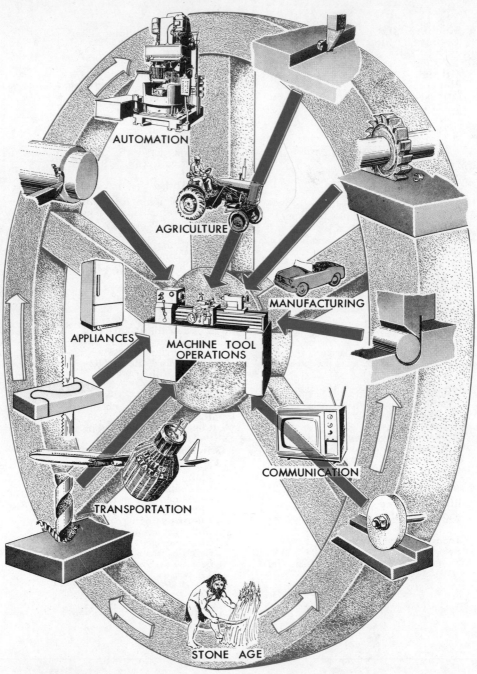

AUTOMATION

AGRICULTURE

MANUFACTURING

APPLIANCES

MACHINE TOOL
OPERATIONS

COMMUNICATION

TRANSPORTATION

STONE AGE

"Machine tools determine how much a nation produces and how well its people live."
–National Machine Tool Builder's Association

Machine Tools:

Measure of Man's Progress

Chapter

1

How much a nation produces determines how well its people live. First, we must produce enough food, clothing and shelter. Only then can we go on to produce the other benefits of civilization. A society's standard of living is in direct proportion to the amount it produces. People know that their own interests are served best when their society produces the greatest good for the greatest number. The greatest good for the greatest number means many things, but first and foremost it means the greatest amount of goods or material things for the greatest number of people. To produce these goods has required a steady increase in the quality of tools.

Tools — The Basis of Industry

Man's progress has been governed by the tools he has developed. See the frontispiece for this textbook. In the earliest times, man was limited by the movements his hands and arms were capable of making. Craftsmen worked metal with muscular effort and hand tools. Today, metal workers use powered machine tools, Fig. 1, to shape and form the myriad parts which, when assembled, comprise our modern world of machinery. Piece parts are assembled into every conceivable product to satisfy our needs and wants. These products or piece parts all have certain standard forms and shapes which may be classified as *solid concentric, flats and flanges, cups or cones, nonconcentric,* and *spiral repetitive*, Fig. 2. All of these forms and shapes require shape refinement to close tolerance and finishes by basic machine tools.

1

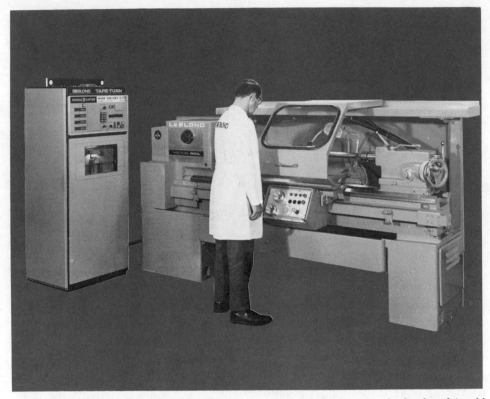

Fig. 1. This numerically controlled lathe is an example of the many tools developed to aid man's progress. (R. K. LeBlond Machine Tool Co.)

Today, every product known — from a paper clip to a space vehicle — is a product of machine tools. If not used directly in the manufacture of the product itself, machine tools are required to produce the machinery and equipment necessary for its processing. Without machine tools, modern man could not exist at his present level of material well-being; without machine tools modern civilization, as we know it, could not exist. Think what the world would still be without the automobile, the huge generators which supply electric power, manufacturing and earth-moving equipment.

Without machine tools, the present population of the world could not even adequately feed or clothe itself. Tractors, cultivators, reapers, trucks and railroad equipment for transporting food to distant markets, refrigeration and processing equipment for meats and other perishables, looms and fabric mills for quantity production of cloth—all of this would be impossible without the

BASIC CONCEPTS OF MATERIAL SHAPING BY MACHINE TOOLS

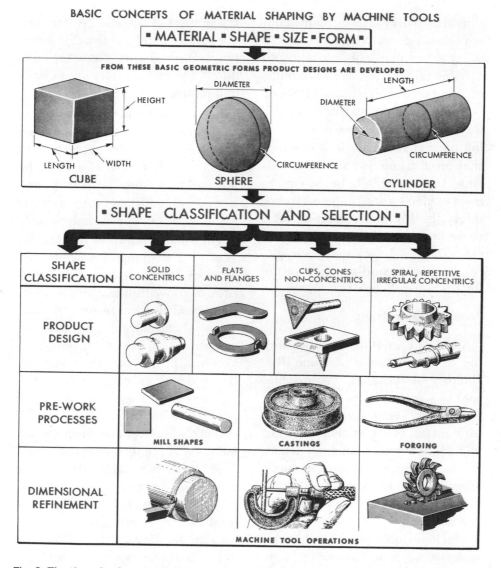

Fig. 2. The three basic geometric forms shown in this chart are used to develop all product designs.

machine tools of civilization.

Consider the combinations of machines required to produce items such as the lock on your front door, the lawnmower, the stove and re-frigerator. The parts for the equipment required to manufacture these items, as well as the machinery used in the production of this book, had first to be formed and precisely machined by machine tools. Many modern conveniences exist only because today's machine tools can cut and form metal to extremely close limits of accuracy at high rates of speed.

Tools always have been and always will be man's means of changing natural resources into products that meet his needs and wants. Machine tools contribute to the enrichment of man's work, his home, travel, safety, health, and recreation. Machine tools, perhaps more than any other factor, make possible a successful present and a secure future for the greatest number.

Basic Machine Tools

What is a machine tool? According to the National Machine Tool Builders Association, "A machine tool is a power-driven machine not portable by hand, used to shape or form metals or materials by cutting, impacting, forming, eroding, deplating, or a combination of these processes." From this definition it is obvious that machine tools vary in type and design.

Machine tools may be classified under three main categories: *Conventional chip producing tools, conventional non-chip producing tools*, and the *new generation of machine tools*, which may be chip or non-chip producing tools.

Chip Producing Tools. Machines of this type shape metal to a size and contour by cutting away the unwanted portions. A chip-producing machine tool usually refines pre-worked shapes that are produced by casting, forging or steel mill rolling.

Non-Chip Producing Tools. This second type of machine shapes metals by shearing, pressing, and drawing to a desired shape. The pre-worked materials normally come from steel mills and producers of granular or powdered materials.

The New Generation of Machine Tools. This group of machines was developed to overcome the limitations of conventional machine tools. Electrical discharge and electrochemical machines, for example, are completely different in construction and in the way they shape or finish metal as compared to conventional machine tools. To make it possible to machine the new, exotic materials developed to meet the demands of space technology, machine tools must now be provided with increased power and more accurate and precise controls. In some cases, machine tools are even provided with automatic gaging devices.

For increased power, the new machine tools are often driven by hydraulic or pneumatic means. To increase the rate of production as well as the preciseness of machined parts, automatic programming has also been added to what we formerly referred to as conventional machine tools.

Training for Advancing Technology

The impact of changing technology requires serious thought in your selection of a career. Industry's need for competent, technically skilled workers is greater today than ever. Skill in machine operation is certainly valuable to high school graduates, apprentices, and technicians. However, employment in today's industry requires some theoretical background.

The technically skilled machinist must be capable of planning for the new machining concepts and calculating setups. He must have training in all phases of material processing as well.

Training through Apprenticeship

Many skilled machinists are developed through apprenticeship training programs. In every factory, large or small, in every machine shop, the apprenticeship program is important.

Apprenticeship means learning a trade on the job with supplemental instruction in classes. The apprentice machinist must learn machine shop technology, machine shop mathematics, blueprint reading, and other related subjects. Public and private schools offer such training through vocational education.

A potential machinist may enter apprenticeship after having completed one or more years of pre-apprenticeship training in a trade or vocational school. In some cases, time is deducted from the apprenticeship term in recognition of this training.

Once selected, the apprentice is given the very best of training, both on the job and in related classes under the vocational programs. There are certain principles, certain standards to be reached, which the apprentice must follow in training to become a highly skilled machinist. These standards benefit both the apprentice and the employer. The employer naturally wants the apprentice to get the preparation that will develop him into a thoroughly skilled worker. The apprentice naturally wants to develop his skill, so that he can command a higher wage. In order that both the employer and the apprentice can gain their objectives, the apprentice must take his apprenticeship seriously, must perform satisfactory work, must be punctual in his related classes and must stay with the apprenticeship and trade.

The Machinist-Technician

Because of the various operations that must be performed in the metal working shop, there are many types of machines. Each is designed to do work of a specific nature. There are some 500 basic types of machine tools, each built in a range of sizes and configurations. And, each type and size has many special workholding devices, tooling features and attachments.

Precision of operation is the most outstanding characteristic of today's machine tools, and accuracy of measurement has simultaneously been the basic factor that has made such tools practical. Dimensional accuracies have already progressed from thousandths of an inch to ten-thousandths, and now approach a millionth. A millionth of an inch (0.000001 in.) is about 1/300th the thickness of a human hair.

The machine tool technician is in great demand in the metalworking industries. He must be familiar with every type of machine tool and must keep abreast of new concepts related to machine tool technology. A properly-trained technician must:

1. possess knowledge of machine tool operations and setups.

2. have a strong background in applied mathematics,

3. have a knowledge of the workability of materials,

4. be able to interpret drawings and prints which describe shape refinements to close tolerances and finishes, and

5. develop attitudes and safety habits as they are applied in industry.

In addition to the skills and knowledges listed above, the trained machinist-technician must be able to master operational sequences so that the information may be programmed on tape or computers.

Opportunities in industries for the individual properly qualified by training are excellent, especially if the individual will take advantage of the many educational training programs available. Machine tools are capital investments. They are expensive, and therefore require a highly intelligent group of machine-tool technicians to direct and operate them to their fullest capacity in producing parts at mass production rates.

Safety in the Training Shop

The Occupational Safety and Health Act of 1970, referred to as "OSHA," provides for the establishment and enforcement of occupational safety and health standards covering the nation's workplaces and school shops. The law authorizes inspections and investigations to determine compliance with its standards and provides for penalties for violations.

OSHA places serious responsibilities on the school shop instructor. Each instructor must become familiar with the safety standards in his state related to this Act and make certain that the school shop layout qualifies under the new safety regulations as a safe place to work. Penalties for failure to correct violations includes fines for each serious violation if not corrected within a reasonable period.

Safety in the Shop Layout

In planning a machine shop layout or revising existing layouts to meet the safety standards, the following areas should be considered by the instructor:

1. Machine tools should be arranged with ample working space so that the operator has ample room to safely operate the machine, perform routine maintenance and check all controls.

2. Aisles and passageways must be wide enough for free access to the machines and in and out of the shop. Aisles should be not less than 3 ft. wide. Lanes for aisles and passageways should be painted in black and/or white.

3. Fire fighting equipment should be properly identified and located where it can be seen from any part of the shop.

4. Provision should be made for first aid cabinets for minor scratches and cuts. Records should be kept of these minor accidents.

5. Shop noise level should comply with OSHA requirements.

Point-of-Operation Guards

All machine tools and portable power tools capable of receiving guards must be so equipped to safeguard the operator during the machining cycle. This requirement extends to lathes, drill presses, milling machines, grinders and all other production tooling machines. Basically, all machine tools parts which undergo straight line or rotary motion to carry the tool or workpiece to the point of operation require guarding. It becomes the instructor's job to identify and provide guards for all such danger points.

Hand tools should not be issued or permitted to be used if they are in unsafe condition. Wrenches should not be used when jaws are sprung to the point where slippage occurs. Impact tools such as drift pins, wedges and chisels should be kept free of mushroomed heads. Wood handles on files, scrapers, etc., should be free of splinters and tight in the tool.

General Housekeeping

The school shop should be a model layout and exhibit the best housekeeping possible. Space should be allocated for collection of chips. Proper chip removal tools should be used for removing chips from the pans of machines. Brushes should be provided for cleaning chips from machines and containers should be available for rags used to wipe down machines.

Material storage areas should be properly identified and safe racks provided for storage with ample space between the storage and machine tool areas.

A safety bulletin board should be located where it can be seen. Current safety bulletins should be posted weekly to keep students alert to potential hazards in the shop.

Safety Rules and Regulations

Safety rules and regulations should be distributed and the course should begin with a safety lecture before the students are permitted to operate machine tools. The following safety rules should be scrupulously adhered to:

1. Wear safety glasses when operating or observing machine tools.

2. Remove rings, jewelry, etc. before running the machine.

3. Never wear loose clothing, such as dangling ties, sweaters, etc. while running a machine.

4. Wear an approved shop apron.

5. Never leave a running machine untended or start and stop a machine for someone else.

6. Never leave a chuck wrench in the chuck of a machine.

7. Check that the safety switch is turned off before cleaning the machine.

8. Always ask questions when you are not sure what to do.

9. Students are not permitted to turn switches on or off on main electrical panels.

10. Use tools from cabinets designated for your machine only.

11. Do not use mallets or hammers to remove lathe centers or drill chucks.

12. Long hair must be covered; use a hair net.

Things to Remember Before Going to Work

1. Proper selection and use of tools for the job is a must.

2. Keep tools in top condition; report damaged tools.

3. Check the machine setup before starting to prevent damage to tools and machines.

4. Place tools in a convenient tray, not on beds and tables of machine tools.

5. Oil the machine before starting to work.

6. Assume a correct posture at the machine.

7. Be mindful of others around you.

8. Cover oil spots on floors with oil soaking compounds to prevent falls.

9. Check point-of-operation guards before starting the machine.

NOTE: Please see review questions at end of book.

Measuring Tools:

Semi-precision and Precision

In machine shop work, every piece must be machined accurately to size within close limits. Careful measuring of the workpiece is necessary to insure proper fit and satisfactory operation of each part when it is assembled with other parts. Inaccurate and careless measurements are worthless. They waste time and materials. A good machinist must be responsible for accurate work. He must, therefore, be able to use with speed and accuracy the measuring tools discussed in this chapter.

Semi-precision Measuring Tools

Linear Measure

The English system of linear measure (measuring in a straight line) is the standard adopted by American industry. The common unit of length is the inch. The inch may be divided into common fractional parts or decimal fractional parts.

The fractional system divides a whole into equal parts. A whole can be divided into two equal parts, or halves. A half can then be divided into two equal quarters, a quarter into eighths, an eighth into sixteenths, and so on. This method of dividing is illustrated in Fig. 1.

The steel rules used in the machine shop are calibrated in divisions of $\frac{1}{2}$, $\frac{1}{4}$, $\frac{1}{8}$, $\frac{1}{16}$, $\frac{1}{32}$, and $\frac{1}{64}$ of an inch. Divisions of $\frac{1}{64}$ of an inch are about as fine as can be read easily without a magnifying glass.

The dimensions on a blueprint, which are given in terms of a fraction, are called *scale dimensions*. Decimals are generally used to indicate dimensions that cannot be ex-

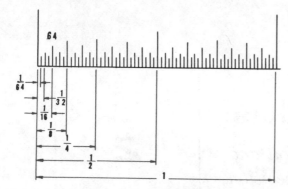

Fig. 1. American industry uses the English system of measurement. An understanding of the inch and fractional part of the inch, shown here, are a necessary part of the machinist's working knowledge.

pressed by the common fractions, such as ½, ¾, ⅞, ³⁄₃₂, ⁵⁄₆₄, etc. When a dimension on a blueprint or blackline print shows the finished diameter of a workpiece to be $5\frac{3}{64}$" (the symbol for inch or inches is written ", and for foot or feet '), the machinist has done his work accurately enough if the finished part measures $5\frac{3}{64}$ inches by measurement with the steel rule. A variation or toler-

ance of + (plus) or — (minus) $\frac{1}{64}$ inch is generally permitted on all rule or scale measurements. The trend is toward a more extensive use of decimals.

Steel Rule

The standard steel rule, Fig. 2, is graduated in inches and fractions of an inch. It is used in taking linear measurements. The recommended rule has four separate scales. The front side, Fig. 2, *bottom*, has two scales. One is graduated in eighths and the other in sixteenths. The back side, Fig. 2, *top*, has two scales graduated in thirty-seconds and sixty-fourths. For convenience in reading the rule, the fractional divisions of the inch are indicated by division marks of different lengths. For example, the division marks representing the halves are longer than those representing the quarters. The quarter divisions are longer than the eighths, the eighths longer than

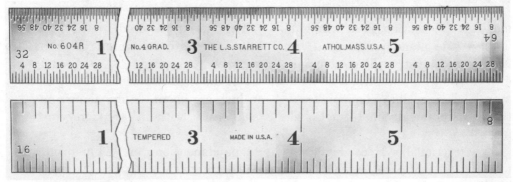

Fig. 2. The steel rule is a basic measuring tool. One side of the steel rule is calibrated in 32nds and 64ths (*top*); the other side is divided into 8ths and 16ths (*bottom*). (The L. S. Starrett Co.)

the sixteenths, and so on.

To maintain their accuracy, measuring tools must be given proper care. They should be kept in orderly arrangement, carefully used on the job, and gently set down to prevent scratching or marring. Tools with moving parts that require more lubrication than picked up on the job should be oiled from time to time. Tools should be periodically checked for accuracy and, if necessary, repaired. Tools which no longer measure accurately and which are beyond repair should be discarded.

Using a Steel Rule

When working within close limits, align the center of the graduation mark with the edge of the workpiece, as shown in Fig. 3. Otherwise the

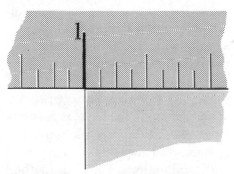

Fig. 3. For accurate measurements, center the graduation line on one edge of the part to be measured.

thickness of the graduation mark itself may make for a slight inaccuracy in the reading.

Where possible, measure from the end of the rule. Where the end of the

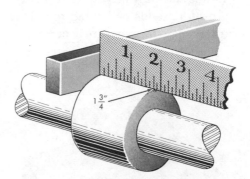

Fig. 4. Butting the end of the rule against a block flush with the end of the workpiece limits scale reading error.

rule can be butted against a shoulder, or both part and rule butted against a smooth flat surface as shown in Fig. 4, the chances of misreading the rule are limited to the mid-scale reading only. When the center of the rule is used to take a measurement, there is a greater chance of reading error.

When a length from a shoulder is measured, butt the rule against the

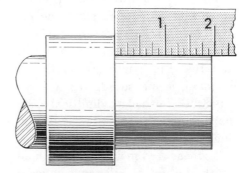

Fig. 5. Measurements from a square shoulder are easy since the rule can be butted against the shoulder.

shoulder as shown in Fig. 5, unless a radius interferes. Note that a rule worn on its ends or having rounded corners may not measure accurately. A worn rule makes for errors and should be discarded.

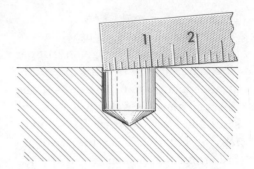

Fig. 7. When measuring across small openings, dip the rule slightly so that its end catches on the opposite wall.

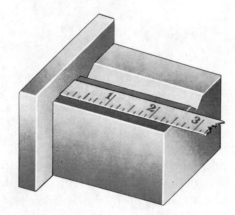

Fig. 6. Butting the entire end of the rule against a square shoulder automatically parallels the rule with the workpiece.

When measuring rectangular parts, place the rule parallel to the length being measured, not at an angle. Frequently, as in Fig. 6, there will be a shoulder from which to take the measurement. Butting the entire end of the rule against the shoulder automatically aligns the rule so that the true length, rather than a diagonal, is measured. When measuring the length of a cylindrical piece, place the rule parallel to the centerline of the part.

To measure the diameter of a hole, width of a slot, the distance between two shoulders, and similar openings approximately 1 inch wide, dip the

rule slightly so that the end of the rule butts against the far wall of the hole, slot, etc., as shown in Fig. 7. While the measurement may be increased a few ten-thousandths of an inch because of the slight angle of the rule, the chance of an error in reading is limited to the upper end only. Be sure to measure the full diameter, not a shorter chord. Again, a rule with rounded corners will measure inaccurately when used for this purpose.

When measuring across openings smaller than 1 inch wide, it is sometimes easier to lay the rule across the opening and measure it along mid-scale graduations. The same method can be used to measure the outside diameter (OD) of a rod, tube, etc. However, a more accurate method is the block-and-rule method illustrated in Fig. 8.

The depth of a blind hole is generally considered to be the full diameter depth. Accordingly, when

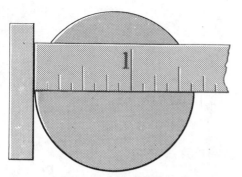

Fig. 8. Outside diameter measured by the block-and-rule method.

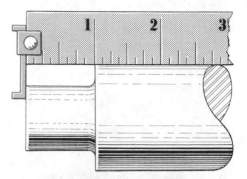

Fig. 10. The hook rule is ideal for measuring from ends with chamfers, fillets, etc. The sliding hook version is more versatile.

measuring the depth of a hole with a rule, take care to measure only the full diameter depth, not the cone-shaped void formed by the drill point or, possibly, a smaller guide hole. This is done by holding the rule firmly to the side of the hole when taking the measurement. A second precaution is to make sure there are no chips in the hole.

Types of Rules

Narrow Rule. The narrow rule, as its name implies, is used to measure depth of narrow slots, small diameter holes, etc., where the standard rule is too wide to be used.

Hook Rule. The hook rule, Fig. 9, is an especially useful tool, for

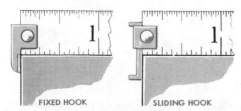

Fig. 9. The hook rule automatically aligns the end of the rule with the workpiece.

the hook at its end automatically aligns the end of the rule with an end of the workpiece. Its use precludes the errors possible when measuring a length with the mid-section of a rule, and the hook rule is faster and more convenient than the combination of block or parallel and common rule.

The hook rule is conveniently used to measure from an end having a fillet, chamfer or radius, Fig. 10. The sliding hook version is more versatile than the fixed hook rule since its hook can be flipped to either side and extends further to bridge fillets, radii, etc. too large for the fixed hook rule. The hook rule is particularly useful in situations where a plain rule cannot be read or accurately positioned along the length to be measured.

Rule Slide. The rule slide is a U-shaped, slotted holder which, when combined with a plain rule, makes a depth gage. Used with a

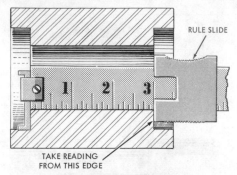

Fig. 11. Rule slide used with a hook rule. The forward edge of the slide provides an index line from which to read the rule.

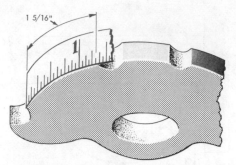

Fig. 13. The flexible rule bends to the contour of a rounded surface.

hook rule, the rule slide permits measurements difficult to obtain with a plain rule—for example, the internal measurement shown in Fig. 11.

Fillet Rule. The fillet rule, with its diagonally-cut end, permits ac-

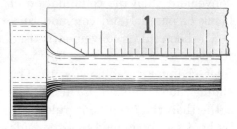

Fig. 12. The fillet rule is used to measure from a shoulder with a fillet at its foot.

curate measurements from a shoulder having a fillet at its foot, Fig. 12.

Flexible Rule. This rule, made from light-gage spring steel, can be bent to the contour of arcs and curved lengths, permitting measurements impossible to obtain with a rigid rule. See Fig. 13. Be careful not

to bend the rule too far, as it may become permanently bent.

Short Rule with Holder. This measuring tool consists of a set of short graduated rules ranging in length, and a holder, Fig. 14. The rules are made of tempered steel and graduated to read in 32nds of an inch on one side and 64ths on the other. The short rule is used to measure grooves, recesses, keyways, short lengths from shoulders, etc. The rule is held in a split chuck by a knurled

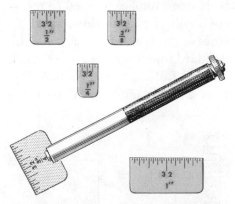

Fig. 14. Rules of varying length make the short rule and holder convenient for measuring hard-to-reach work surfaces.

nut at the top of the holder and can be set at various angles.

Rule with End Graduations. Some plain rules are graduated along one or both ends. Thus, the ends of the

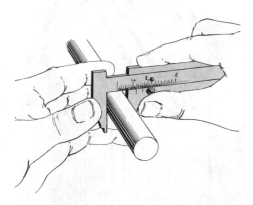

Fig. 15. The slide caliper rule permits inside and outside measurements which cannot be made accurately with an ordinary rule.

rule may be used for measuring in narrow, hard-to-get-at areas.

Slide Caliper Rule. Most of the measuring instruments discussed so far are used to sight the position of the edge or point to be measured in relation to the graduations on the scale. The rule can usually be placed on or next to the workpiece, and the measurement read without difficulty. Sometimes, however, it is necessary to measure between points in such a position that the plain rule is awkward to use. An example is measuring the diameter of a round piece of work, as in Fig. 15. In cases of this kind, the slide caliper rule is a convenient instrument to use. Its jaws are so shaped that it is possible to measure inside and outside dimensions.

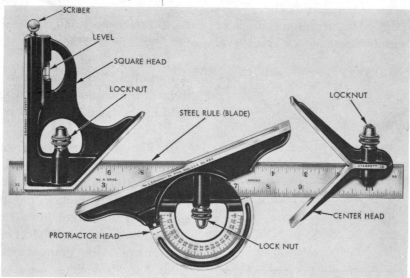

Fig. 16. A combination set consists of a steel rule, a square stock with level and scriber, a protractor head, and a center head. (The L. S. Starrett Co.)

Combination Set. The combination set shown in Fig. 16 is one of the most useful and convenient measuring tools in the machine shop. As you will note from a study of the illustration, the set consists of the following parts: a steel rule; a square stock incorporating a level, scriber and 45-degree angle; a protractor head; and a center head. The combination set is used chiefly for layout work. It tests for squareness and for angles, and can be used as a depth gage.

The steel rule of the combination set can be used as a plain rule or as a straight-edge. Combination sets are available in a range of sizes to suit the needs of the individual machinist. Figs. 17 and 18 illustrate a few of the uses of the combination set.

In the head or square stock of most combination sets, a spirit level

Fig. 18. The center head can be used to locate centers on cylindrical pieces.

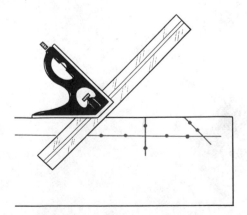

Fig. 17. A 45° dovetail slot can be laid out using the combination square and rule.

is mounted. This is useful in leveling work. For layout work, a scriber, held in the head by a friction bushing, is included.

In Fig. 16, the protractor head is shown clamped to the steel rule. So secured, the rule can be turned in an arc or half circle. The scale is graduated in degrees from 0 to 180 to read both right and left. On some instruments, the graduations read from 0 to 90 degrees in both directions. This is a very handy tool to use in checking angles.

Depth Gages

Rule Depth Gage. The rule depth gage, Fig. 19, is designed to measure depths of holes, recesses, slots, etc. Preferred to the plain rule for depth measurement, the depth gage prevents reading errors from placing the rule out of parallel with the hole.

The depth gage consists of a slotted base made of heat-treated steel and a narrow measuring rule usually six inches long and made of tempered steel. The rule slides in the base and is locked in the desired position by means of a knurled screw.

To use the depth gage, place it on the smooth surface of the workpiece, so that the rule bears against the wall of the hole as shown. Push the rule to the bottom of the hole, then lock it in place with the thumb screw. The depth of the hole is indicated by the graduation even with the bottom edge of the base.

When work cannot be moved to a convenient position for measuring, it often becomes awkward to read a plain rule inserted in the work. On the other hand, the depth gage can be removed from the work for reading, permitting internal measurements in situations where a rule could be read only with great difficulty, if at all. There is also less chance of reading error with the depth gage than with the plain rule

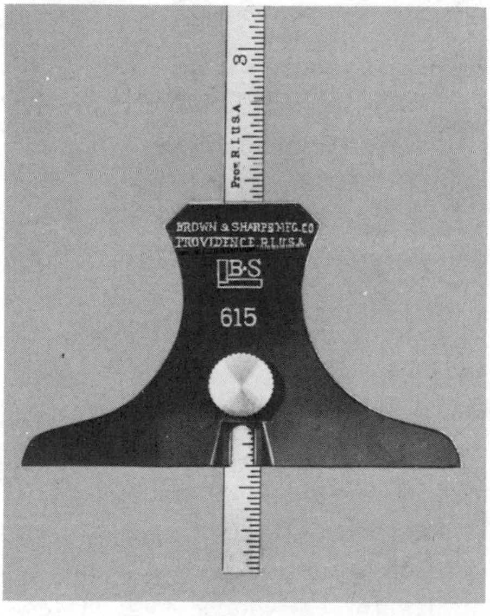

Fig. 19. The rule depth gage provides accurate depth measurement since the reading is "locked-in" when the gage is removed from the hole.

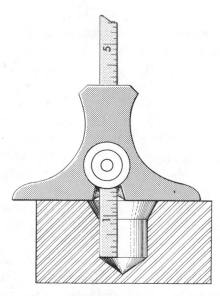

Fig. 20. The rule depth gage measures total depth of a chamfered hole.

due to the sharp reading line at the flat, lower edge of the base.

When measuring the depth of chamfered holes, Fig. 20, the depth gage again provides more accurate measurement than possible with the more common practice of sighting along the top surface of the workpiece to obtain a depth reading on a plain rule inserted in the hole.

Plain Rod Depth Gage. This commonly used gage consists of a heat-treated steel base or blade, a sliding rod with a knurled ball at its top, and a rod locking screw. The measuring rod slides in either of two holes drilled in the base at right angles to its bottom edge.

The plain rod depth gage is used like the rule depth gage described above. However, the depth reading —the distance the rod extends from the base—must be measured with a rule since the rod is not calibrated.

Spring Rod Depth Gage. The spring rod depth gage is an improved version of the plain rod depth gage. The rod is automatically forced downward by a spring in the barrel. This gage, shown in Fig. 21, is particularly useful for taking quick measurements. Its capacity is approximately three inches. Both the base and the contact end of the rod are hardened and ground.

To use this gage, place it on the workpiece, aligned over the opening to be measured, and release the lock screw located at the top of the

Fig. 21. The spring rod depth gage speeds hole depth measurement since the rod automatically plunges to the bottom of the hole.

gage. The rod will automatically plunge to the bottom of the hole or recess. Reset the lock screw to secure the rod in the extended position. As with the plain rod depth gage, the depth of the opening is determined by measuring the length of the rod extending from the base.

Protractor Depth Gage. The protractor depth gage is a combination protractor and depth gage. It consists of a graduated rule mounted on a protractor. Convenient for checking chamfers, bevels, etc., the rule may be clamped to the protractor head at any angle or depth by means of a knurled nut which also serves as the fulcrum of the protractor.

Care of Line-Graduated Measuring Instruments

The instruments so far discussed were designed to control the accuracy of measurements to the nearest graduation on the scale. All were built to withstand a considerable amount of wear and hard use, and they will give long service if properly cared for.

A trainee should observe how the experienced machinist uses and cares for his tools. The machine shop man realizes that the quality of his work depends as much upon the condition of the tools he uses as upon his personal skill. Measuring tools should not be allowed to become rusty. When not in use, they should be coated with a protecting film of oil and stored in a safe place.

Measuring by Contact

Flat surfaces are measured with common tools like the steel rule and the depth gage discussed in the preceding pages. Round work is usually measured by "feel" with non-precision tools like the spring caliper or the firm-joint caliper, which have contact points or surfaces.

The accuracy of measurements made with tools like outside and inside calipers depends on the sense of touch. With contact measuring tools, the skilled mechanic, through his keen sense of touch, can readily feel changes in dimensions.

In the human hand, this sense of touch is keenest in the fingertips. For this reason, the contact tool should be held lightly in the finger tips. If the tool is grasped too tightly, the sense of touch will be less acute. As you study the tools presented in the following pages, notice how they are held.

Calipers (Spring, Firm-Joint)

Calipers are tools used to measure work by contact rather than by scale readings. Contact measurements are made in one of two ways. By the first method, the non-precision contact tool—say, the outside

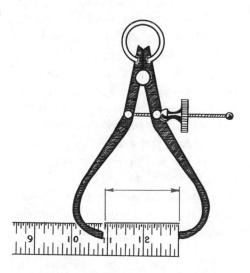

Fig. 22. Dimensions can be set with the outside caliper on the steel rule.

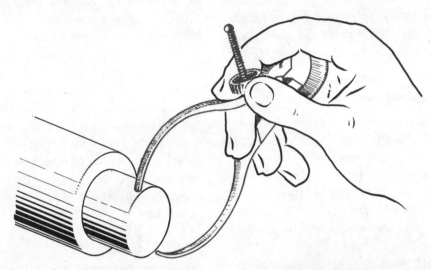

Fig. 23. The correct dimension set on the outside calipers can be transferred to the workpiece. Of course, the machinist may wish to set the caliper on the workpiece first, and then measure the actual dimension on the steel rule.

caliper—is set by the steel rule, Fig. 22. The "set" dimension of the tool is then the correct measurement for checking the size of the round workpiece, Fig. 23. By the second method, the procedure is reversed. The contact points of the caliper are set to the diameter of the round piece. Then the distance between the points is measured on the steel rule to find the size of the piece.

Outside Calipers. Outside calipers, as the name implies, are used to measure across the outside surfaces of either round or flat work. The legs of the instrument curve outward in order that the caliper may pass over large cylinders as well as small ones.

Fig. 23 shows the spring-joint,

round-leg outside caliper. The spring-joint caliper can be set more accurately and more easily than the firm-joint instrument because it has a fine adjusting lock-screw nut for changing the caliper leg setting. Therefore, it is the instrument most generally used.

Inside Calipers. Inside calipers are used to measure the diameter of holes or the width of slots, keyways, and other openings. These instruments also are of two types, firm-joint and spring-joint.

To set the inside caliper to the size of a given hole, place one leg firmly against a side wall of the hole. Bring the other leg against the opposite side wall in such a way that the leg is given a light drag at

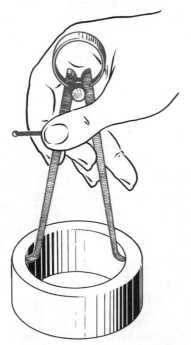

Fig. 24. Inside calipers are used to measure the diameter of round holes, the width of slots, and other openings.

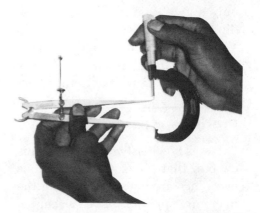

Fig. 26. Measurements on the inside caliper can also be transferred to the micrometer. (Chicago Public Schools, Manpower Training Div.)

a point directly opposite the first leg. See Fig. 24. The setting of the caliper legs (that is, the inside diameter of the hole) can then be transferred to a rule, as shown in Fig. 25. Read the graduation even

with the edge of the leg. If desired, the setting can be transferred to the micrometer (Fig. 26) by adjusting the instrument to the caliper leg points to obtain the same drag or "feel" used with the caliper in the hole. With both inside and outside calipers, it is important that the machinest develop a keen sense of touch.

Care of Calipers. Since caliper legs are slender, they are easily

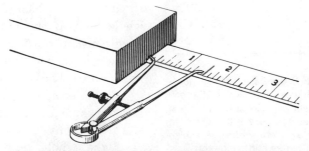

Fig. 25. Measurements taken with inside calipers are tested on the steel rule as shown here.

sprung. You must be careful to use only a light contact. Calipers of ordinary size will spring as much as $\frac{1}{64}$ of an inch unless a light touch is used in handling them. Calipers must never be forced onto the work.

When not in use, they should be hung in the tool cabinet. Do not leave them on the bench where heavy tools could be laid on them. A light protecting coat of oil will prevent them from rusting.

Precision Measuring Tools

It was previously explained in this chapter that the fractional divisions of an inch are found by dividing the inch into equal parts such as halves, quarters, eighths, sixteenths, thirty-seconds, and sixty-fourths. When smaller units of measurements are required, the decimal system is used. Measuring tools employing the decimal system are more accurate.

In machine shop work, the specifications and blueprints frequently give dimensions in decimal parts of an inch. Work of this type must be machined to very close dimensions, often to a thousandth of an inch or less of the specified size.

In the decimal system, the inch is divided into ten equal spaces. Each of these is divided into ten parts, and so on. In this way the inch is divided into spaces called tenths, hundredths, thousandths, and ten-thousandths of an inch, Fig. 27.

The Meaning of a Decimal

The decimal point is very important in the study and use of decimals. Figures to the left of the decimal point are whole numbers. The figures to the right of the decimal point represent the amount of the dimension which is less than one whole inch. If the blueprint shows a dimension of 5.025″, it means five whole inches and, in addition, twenty-five thousandths of an inch.

The dimension is always read beginning at the left and reading toward the right. This enables us to read the whole number or num-

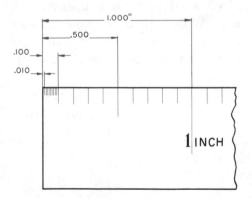

Fig. 27. The inch is the basic unit of measurement in both semi-precision and precision measurement. In precision machine shop work, dimensions are often given in decimal parts of an inch. An inch is shown here divided into tenths and hundredths.

THE MICROMETER CALIPER

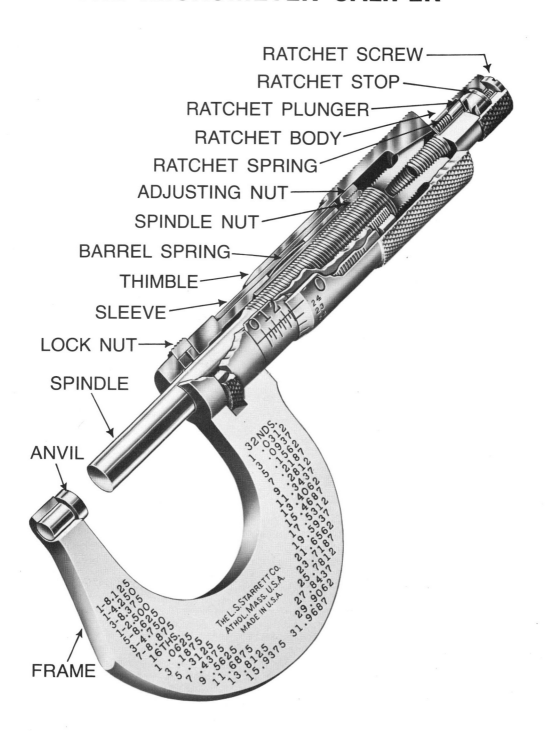

RATCHET SCREW

RATCHET STOP

RATCHET PLUNGER

RATCHET BODY

RATCHET SPRING

ADJUSTING NUT

SPINDLE NUT

BARREL SPRING

THIMBLE

SLEEVE

LOCK NUT

SPINDLE

ANVIL

FRAME

How to read the MICROMETER

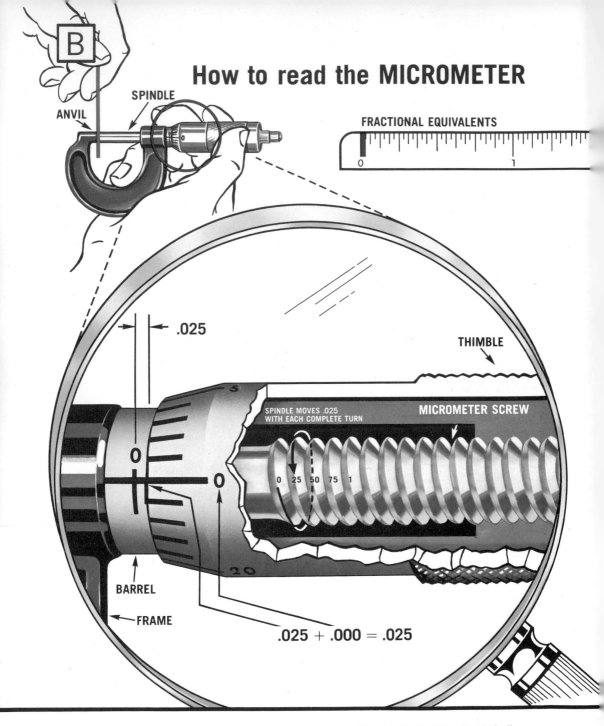

B

ANVIL SPINDLE

FRACTIONAL EQUIVALENTS

0 1

.025

THIMBLE

SPINDLE MOVES .025
WITH EACH COMPLETE TURN

MICROMETER SCREW

5

0

0

20

0 25 50 75 1

BARREL

FRAME

.025 + .000 = .025

STEP NO. 1 — After cleaning the contact faces of the Anvil and Spindle, bring the Spindle in contact with the Anvil by adjusting the Thimble so that the zero marks line up.

STEP NO. 2 — To set the micrometer to read .025, open the Thimble until the zero marks are aligned at the first graduation on the Barrel. The Thimble moves .025 with each complete revolution.

THE MICROMETER CALIPER

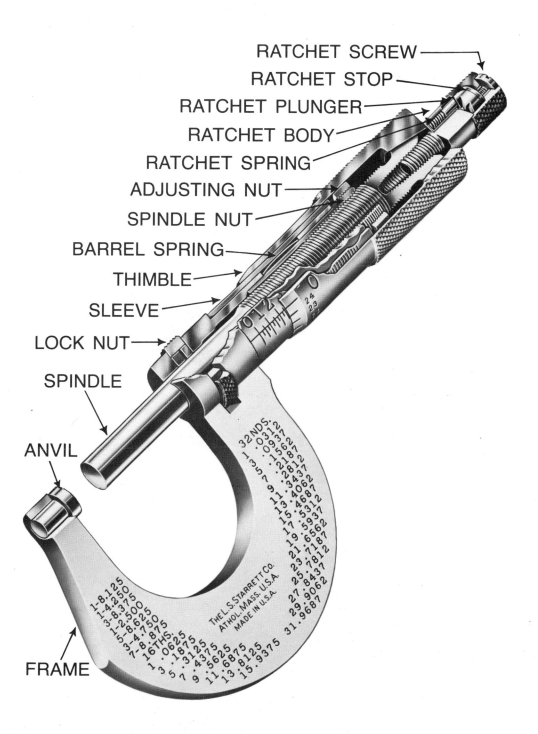

RATCHET SCREW
RATCHET STOP
RATCHET PLUNGER
RATCHET BODY
RATCHET SPRING
ADJUSTING NUT
SPINDLE NUT
BARREL SPRING
THIMBLE
SLEEVE
LOCK NUT
SPINDLE
ANVIL
FRAME

The L.S. Starrett Co.
ATHOL, MASS. U.S.A.
MADE IN U.S.A.

32NDS. 27.27
.03137 27
.09567 27
.1287 2
.2437
.3437
11. .4062
13. .4687
15. .5312
17. .5937
19. .6562
21. .7187
23. .7812
25. .8437
27. .9062
29. .9687
31. .9687

1-8.125
1-4.250
3-8.375
1-2.500
5-8.625
3-4.750
7-8.875
16THS
.0625
.1875
.3125
.4375
.5625
.6875
.8125
.9375

How to read the MICROMETER

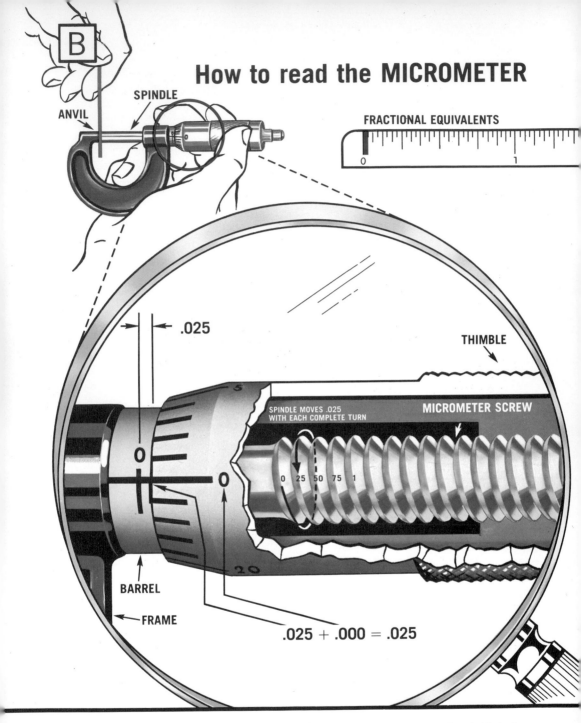

FRACTIONAL EQUIVALENTS

STEP NO. 1 — After cleaning the contact faces of the Anvil and Spindle, bring the Spindle in contact with the Anvil by adjusting the Thimble so that the zero marks line up.

STEP NO. 2 — To set the micrometer to read .025, open the Thimble until the zero marks are aligned at the first graduation on the Barrel. The Thimble moves .025 with each complete revolution.

Exercises in Reading the MICROMETER

These readings are provided to help the student learn to quickly and accurately read the micrometer. The student's interest will best be served if he tries to determine each reading **before** referring to the correct answers provided at the bottom of each page.

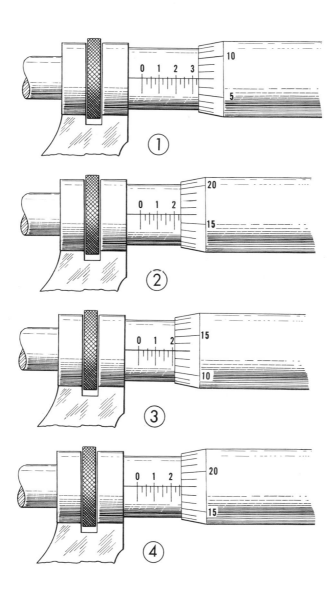

1. .300 + .025 + .007 = .332
2. .200 + .025 + .016 = .241
3. .200 + .000 + .013 = .213
4. .200 + .050 + .018 = .268

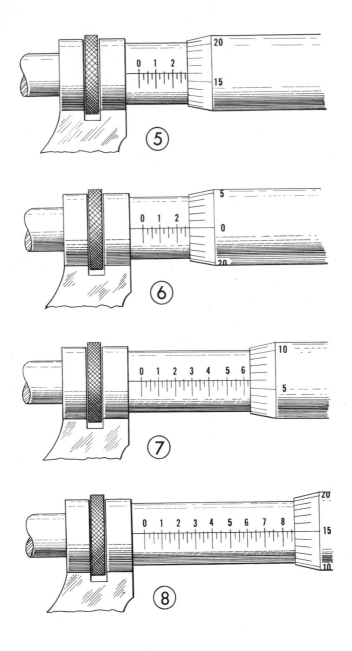

F

5. .200 + .075 + .016 = .291
6. .200 + .075 + .000 = .275
7. .600 + .025 + .006 = .631
8. .800 + .050 + .015 = .865

bers first. In reading the dimension 7.125″, we begin by saying *seven* and follow this with *and one hundred twenty-five thousandths* inches. When we come to the decimal point, we say *and* rather than *decimal point*. To illustrate further: for the dimension 12.250″, the correct reading would be *twelve and two hundred fifty thousandths inches*.

By means of the decimal system, an inch can be divided into as small a unit as one millionth of an inch, as shown below.

.1   means 1/10
(one-tenth)
.01   means 1/100
(one one-hundredth)
.001   means 1/1000
(one one-thousandth)
.0001   means 1/10,000
(one ten-thousandth)
.00001   means 1/100,000
(one one-hundred-thousandth)
.000001 means 1/1,000,000
(one millionth)

The figures at the right of the decimal have definite values. As you can see in the examples, the more places to the right of the decimal, the finer the measurement. There was a time when three places were considered sufficient, but today four places are commonly used. Sometimes in grinding, five places are needed.

Reading Decimals. In machine shop work, we generally read deci-

mals in thousandths of an inch. Sometimes one cipher must be added to a dimension. Thus, for the dimension .12″ (twelve hundredths) we would use .120 (one hundred twenty thousandths). In a dimension showing one figure to the right of the decimal, it would be necessary to add two ciphers. Sometimes the ciphers are not written but are imagined to be there. The dimension .5″ would be read by the machinist as five hundred thousandths of an inch (.500). The addition of zeros to the right of .5 does not change its value. In this example, two ciphers would be added to make three places to the right of the decimal. Study the examples given below.

.550 means five hundred
fifty thousandths
.555 means five hundred
fifty-five thousandths
.055 means fifty-five thousandths
.005 means five thousandths
.001 means one thousandth
.010 means ten thousandths
(sometimes written .01)
.100 means one hundred
thousandths (sometimes
written .1)

Where more than three decimal places are given, the machinist reads the number of thousandths first and then the part of the thousandth which remains. Thus, .4375 would be read as *four hundred thirty-seven and one-half thousandths*. The digit

5 in the fourth decimal place to the right means $\frac{5}{10}$ or one-half of the value of the third decimal place. The decimal .005 would be read *five thousandths*, but .0005 would be read as *one-half of .001 or 5 ten-thousandths*.

Quite frequently we see dimensions and specifications which have five places to the right of the decimal, as for example, .00001″. This would be read as *one one-hundred-thousandth of an inch*, 1/100,000″.

Dimension Limits. Working drawings often give two values for the same dimension, as shown at (*A*) in Fig. 28. This means that the dimension of the finished part could be as small as 1.999″ or as large as 2.001″.

Another common way of dimensioning the drawing is to use a plus sign and a minus sign, as shown at (*B*) in Fig. 28. This means that if the size produced is under 1.999″ or over 2.001″, the part will be useless. Sometimes the plus and minus signs are joined to make one sign, as at (*C*) in Fig. 28.

Tolerance. Tolerance means the difference between the high and low limits of a dimension. In the example given at (*B*) in Fig. 28, the difference between 2.001″ and 1.999″

is .002″, which gives us a tolerance of .002″. Example (*D*) of the same figure shows that the part could be made as large as 2.001″, but no smaller than 2.000″. The tolerance here is .001″.

Decimal Equivalents. In the machine shop, we frequently must convert a fraction to a decimal. In arithmetic when we write 5 over 8, as $\frac{5}{8}$, we mean that we are dividing 5 by 8. In a common fraction, such as $\frac{5}{8}$, this division is merely indicated. When we actually go about the process of division, the answer we get is called the decimal equivalent of the common fraction $\frac{5}{8}$. The fraction written as $\frac{5}{8}$ has the same value as the decimal equivalent .625, as you can see from the example which follows.

Example: Find the decimal equivalent of the fraction $\frac{5}{8}$.

Solution: Divide 5 by 8:

$$\begin{array}{r} .625 \ answer \\ 8\overline{)5.000} \\ \underline{4\ 8} \\ 20 \\ \underline{16} \\ 40 \\ \underline{40} \end{array}$$

The answer, .625, is the decimal equivalent obtained by doing the indicated division, $\frac{5}{8}$ equals .625.

(A)   (B)   (C)   (D)

Fig. 28. Four methods of indicating tolerance on a blueprint are shown here. Note that in *D* the small part may be one one-thousandth larger than two inches, but no smaller.

TABLE I. DECIMAL EQUIVALENTS OF FRACTIONS OF AN INCH

COMMON FRACTIONS	DECIMAL EQUIVA- LENTS	COMMON FRACTIONS	DECIMAL EQUIVA- LENTS	COMMON FRACTIONS	DECIMAL EQUIVA- LENTS	COMMON FRACTIONS	DECIMAL EQUIVA- LENTS
1/64	.015625	17/64	.265625	33/64	.515625	49/64	.765625
1/32	.03125	9/32	.28125	17/32	.53125	25/32	.78125
3/64	.046875	19/64	.296875	35/64	.546875	51/64	.796875
1/16	.0625	5/16	.3125	9/16	.5625	13/16	.8125
5/64	.078125	21/64	.328125	37/64	.578125	53/64	.828125
3/32	.09375	11/32	.34375	19/32	.59375	27/32	.84375
7/64	.109375	23/64	.359375	39/64	.609375	55/64	.859375
1/8	.125	3/8	.375	5/8	.625	7/8	.875
9/64	.140625	25/64	.390625	41/64	.640625	57/64	.890625
5/32	.15625	13/32	.40625	21/32	.65625	29/32	.90625
11/64	.171875	27/64	.421875	43/64	.671875	59/64	.921875
3/16	.1875	7/16	.4375	11/16	.6875	15/16	.9375
13/64	.203125	29/64	.453125	45/64	.703125	61/64	.953125
7/32	.21875	15/32	.46875	23/32	.71875	31/32	.96875
15/64	.234375	31/64	.484375	47/64	.734375	63/64	.984375
1/4	.25	1/2	.50	3/4	.75		

Many times the machinist is given a dimension such as 3⅝". To use a micrometer, he must convert the fraction to a decimal equivalent. He can divide out the fraction by this method. His answer is 3.625". With the decimal equivalent, he can measure his work with a micrometer.

Table of Decimal Equivalents. Tables of decimal equivalents (Table I) may be found in practically all machine handbooks and manuals. Charts which offer convenient means of making rapid calculations also are available as wall charts or pocket cards. Machine shop workers usually memorize the more common decimal equivalents and, as they use these tables, they find the decimal

equivalents for the more common fractions easily called to mind.

Metric Measure

The expanding technology of the metalworking industry makes it necessary for the machinist to be acquainted with the metric system of measurement.

The metric system is based on a length called the meter. All of the other units of length are either decimal divisions or decimal multiples of the meter. For example, the centimeter, as its name implies, is one one-hundredth of a meter. The kilometer is one thousand meters. The following table (Table II) gives the relationships between the different units.

A comparison of the scales of the metric system and an understanding of the prefixes will be helpful because the prefixes are used throughout the metric system.

milli—means one one-thousandth of a unit or .001

centi—means one one-hundredth of a unit or 0.01

deci—means one tenth of a unit or 0.10

meter —is the basic unit

Deka—means ten times the unit

Hecto—means one hundred times the unit

Kilo—means one thousand times the unit

Myria—means ten thousand times the unit

In Fig. 29 the centimeter scale shows 10 cm. Each of the small divisions is a millimeter: 10 mm. equals 1 cm.

In the inch scale the small units are sixteenths of an inch, and 64 sixteenths = 4 inches. You can see that 10 cm. is nearly four inches.

To facilitate inch and millimeter conversions, remember that 1 inch = 25.4001 millimeters, and 1 millimeter = 0.03937 inch. For example, a self-aligning ball-thrust bearing has an inside diameter of 55 mm. and an outside diameter of 88 mm. The machinist is to turn a shaft to fit the 55 mm. inside diameter. Since American lathes are calibrated in inches, the machinist must convert to the English system. The conversion can be made by multiplying 0.03937 inches/mm. by 55 mm. to get 2.1654 inches. Similarly, the bearing housing will be bored to a diameter of 88 mm. times 0.03937 inches/mm. or 3.4646 inches, plus or minus the necessary tolerance.

Micrometer Calipers

Micrometer calipers are measuring instruments designed to use the decimal divisions of the inch. They

TABLE II   ENGLISH AND METRIC EQUIVALENT TABLE

ENGLISH SYSTEM WITH ABBREVIATIONS	
12 inches (in.)	= 1 foot (ft.)
3 ft. or 36 in.	= 1 yard (yd.)
5 1/2 yds. or 16 1/2 ft.	= 1 rod (rd.)
320 rds.	= 1 mile (mi.)
220 yds. or 1/8 mi.	= 1 furlong
1760 yds.	= 1 mile (mi.)
5280 ft.	= 1 mile (mi.)

METRIC SYSTEM WITH ABBREVIATIONS	
10 millimeters (mm.)	= 1 centimeter (cm.)
10 cm. or 100 mm.	= 1 decimeter (dm.)
10 dm. or 100 cm.	= 1 meter (m.)
10 m. or 1000 cm.	= 1 decameter (Dm.)
10 Dm. or 100 m.	= 1 hectometer (Hm.)
10 Hm. or 1000 m.	= 1 kilometer (Km.)
10 Km. or 10,000 m.	= 1 myriameter (Mm.)

EQUIVALENT TABLE	
1 mm.	= .03937 in.
1 cm.	= .3937 in.
1 m.	= 39.37 in.
1 Km.	= 0.621 mi.
1 in.	= 2.54 cm.
1 ft.	= 30.48 cm.
1 yd.	= .9144 m.
1 mi.	= 1.6 Km.

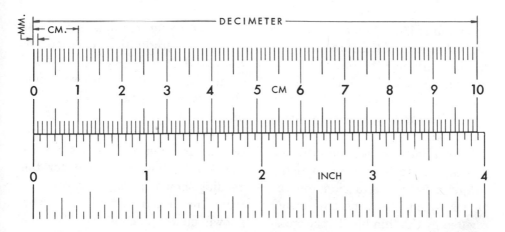

Fig. 29. The centimeter scale is compared to the inch scale.

are used to take very precise measurements. Some micrometers are graduated to read a thousandth part of an inch. Others have verniers by which measurements of one ten-thousandth of an inch can be made. Micrometers are made in different styles and sizes.

Parts of a Micrometer and Their Importance. The illustration on Page A of the preceding Micrometer Trans-Vision shows a cut-away view of a micrometer caliper. Notice the U-shaped frame to which are fastened the anvil and the barrel.

The heart of the micrometer is the precision-ground micrometer screw, also called the spindle screw, which is housed in the barrel. The spindle extends out of the front end of the barrel, and is the part of the micrometer which closes on the piece to be measured. Turning the thimble clockwise will turn the spindle screw in the fixed spindle nut, so

that the spindle advances toward the anvil. Turning the thimble counter-clockwise moves the spindle away from the anvil. The pitch of the spindle screw is exactly $\frac{1}{40}$ of an inch, or 40 threads per inch. One complete revolution of the thimble will therefore move the spindle $\frac{1}{40}$ of an inch (.025").

The index line running the length of the barrel is graduated with cross-lines. Each cross-line represents a spindle movement of .025". Every fourth cross-line is numbered and represents a spindle movement of one hundred thousandths of an inch (.100"). The beveled edge of the thimble is graduated for readings of one thousandth of an inch (.001").

Reading the Micrometer Caliper. Measuring with a micrometer means simply noting three separate readings. See Page B. The first two readings are taken from the barrel; the last reading is made on the thimble itself. The sum of the three separate readings will give the total distance, to a thousandth of an inch, that the spindle is separated from the anvil, or the thickness of the piece being measured.

In reading the micrometer, the first step is to read (left to right) the number of graduations on the barrel which are exposed, or not covered by the thimble. Every cross-line on the index line on the barrel which is uncovered by the thimble represents a spindle movement of $1/40$ of an inch (.025″). Every fourth cross-line on the barrel is numbered, from zero to nine, and represents $1/10$ of an inch (.100″).

One complete revolution of the thimble (counter-clockwise) from the closed position will uncover the first cross-line on the barrel, indicating that the spindle has moved $1/40$ of an inch (.025″). Four complete revolutions of the thimble will move the spindle $1/10$ of an inch (.100″), exposing the first numbered graduation (1) on the barrel.

The micrometer provides the person taking the measurement with a three place decimal reading, or a reading to thousandths of an inch. The first digit of the three place micrometer reading can be obtained directly from the barrel. This is the first reading to be noted. This first digit is referred to as the 'tenths' digit. Put simply, the first digit of

the micrometer reading is the last numbered graduation exposed on the barrel.

In the Trans-Vision we show three separate and unrelated micrometer readings on Pages B, C, and D. For Page B, the first digit of the three place micrometer reading is zero. The first digit for the reading shown on Page C is two (2), while the first digit for the reading shown on Page D is six (6). The first digit should be written, or considered mentally, as the first of a three place decimal:

Page B   .000
Page C   .200
Page D   .600

The second reading is also made on the barrel and is simply the decimal equivalent for the number of graduations exposed *between* the last major (numbered) graduation and the beveled edge of the thimble. On Page B only one graduation (.025″) is exposed. In C two graduations are seen. In D, again only two graduations are seen. One graduation exposed would mean that .025″ must be added to the first reading. Two graduations would mean that .050″ must be added. Three exposed would mean that .075″ must be added. Adding the second readings to the first readings, we now have:

Page B   .000 + .025
Page C   .200 + .050
Page D   .600 + .050

After the first two readings are

taken from the barrel, the third and last reading must be added. Notice that the beveled edge at the front of the thimble in Page B is divided into twenty-five equal parts. Since one complete revolution of the thimble moves the spindle twenty-five thousandths of an inch (.025″), each of the twenty-five marks or graduations on the beveled edge of the thimble represents one thousandth of an inch (.001″) of spindle movement. Every fifth graduation on the beveled edge is numbered for easy reading: 0, 5, 10, 15, 20.

Each of the twenty-five graduations represents an additional thousandth of an inch to be added to the first two readings to give the final three place micrometer reading. After having observed the first two readings, the last reading is determined by observing which of the thousandths graduations on the beveled edge lines up with the index line on the barrel. If the zero graduation aligns with the index line on the barrel, as on Page B, the final reading will be:

.000 + .025 + .000 = .025″

If the five thousandths graduation aligns with the index line, as on Page C, the final reading is:

.200 + .050 + .005 = .255″

Notice that had the fifteen thousandths graduation aligned with the index line on Page C, the reading would have affected the second digit, and would have been:

.200 + .050 + .015 = .265″

On Page D, the mark on the beveled edge of the thimble does not align evenly. The final reading appears to be midway between the six and the seven thousandths graduations. The third reading would be written as six and one-half thousandths (.006½ or .0065″). The final micrometer reading would be:

.600 + .050 + .0065 = .6565″.

Additional micrometer settings are shown on the pages following Page D. These are presented to provide practice in reading the instrument and to give additional instruction on how readings are made. Correct answers are listed at the bottom of the pages.

Holding the Micrometer. Fig. 30 illustrates the correct method of holding the micrometer caliper when measuring small parts that can be held in the left hand. The micrometer should first be opened to a size slightly greater than the diameter or thickness of the piece to be measured. Hold the part in the left hand, the micrometer in the right, as shown. Slip the micrometer over the piece. Using the thumb and forefinger, turn the thimble to close the micrometer until a slight pressure is exerted against the piece. The proper pressure of the spindle against the small stock measured in Fig. 30 is secured by turning the knurled knob with the thumb and forefinger. The ratchet built into the thimble of the

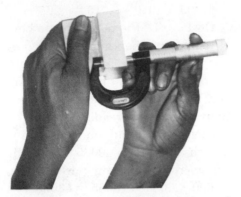

Fig. 30. Measuring small stock with a micrometer. Note how the micrometer is held.

better micrometers allows only so much pressure to be exerted. Now, read the dimension on the micrometer barrel and thimble as previously directed. For large pieces of stock, or where the part is fastened or secured in place, the micrometer should be used as shown in Fig. 31.

Fig. 31. Measuring large stock with a micrometer. (Lufkin Rule Co.)

Checking the Micrometer for Accuracy. The micrometer should frequently be checked for accuracy. To check a one-inch micrometer, advance the spindle to the anvil. If the zero line of the thimble lines up with the horizontal index line of the barrel, the micrometer is in adjustment. If the zero line fails to coincide with the barrel index line, the micrometer is out of adjustment. *Adjustment of the micrometer should be done only by a person experienced in this work.*

To check a micrometer for accuracy, a gage block (Johansson, Hoke, or other precision standard blocks or disks) of known unit size is inserted between spindle and anvil. The micrometer and the gage block must be free from dirt and grit. With the block in place, the micrometer is accurate if the reading is the same as the known size of the gage block. A new micrometer of good quality should be accurate within .0002 of an inch.

Proper Care of the Micrometer. The accuracy of a micrometer depends upon the care it receives. A good operator keeps a micrometer in a protected place where emery dust or chips cannot come in contact with it. He will not keep the micrometer in a drawer or box with heavier tools. The machinist soon learns that his micrometers measure accurately only when they are given proper care.

Micrometers should be oiled occasionally with a light grade of oil to prevent rust and corrosion. They should be checked for accuracy, and any necessary adjustments made.

To insure a correct reading when using a micrometer, the machinist should make certain that the piece being measured is at room temperature. If the workpiece is heated, it should be allowed to cool to room temperature.

When the micrometer is not in use, the spindle and anvil should not be in contact. Completely closing a micrometer and allowing it to remain closed may cause rust on the ends of the spindle and anvil.

Inside Micrometer Caliper. The caliper shown in Fig. 32, measures in thousandths of an inch. It is designed for gaging small internal dimensions, such as small holes and narrow slots. Notice in Fig. 33 that the scale on the barrel reads from right to left. This is necessary be-

Fig. 32. The inside micrometer caliper is designed for gaging small internal dimensions.

cause, as the thimble is turned over the barrel clockwise, the distance across the jaws of the micrometer becomes greater. This reversal of the scale must be remembered when using the inside micrometer.

In Fig. 33 the inside micrometer shows a reading of .500″. The numeral 5 is partly concealed by the thimble. There are four spaces between the numeral *6* on the barrel and the beveled edge of the thimble. Thus far we have a reading of .500″. Now note that the zero on the thimble is even with the index mark on the barrel. This gives us a final reading of .500″.

The micrometer jaws are opened by turning the thimble to the right. The minimum dimension measurable is determined by the width of the tips of the micrometer jaws, usually .250 inch. These tips are ground with a small radius to insure a one-point contact in a hole and to prevent cramping.

To measure a small round hole, adjust the width of the points to a dimension slightly less than the diameter to be measured. Holding the fixed jaw in one hand and the barrel in the other hand, insert the points vertically into the hole at its greatest width (diameter). Using the ratchet on the thimble, adjust the width of the jaws so that the points touch the sides of the hole with only slight pressure. Read the diameter of the hole.

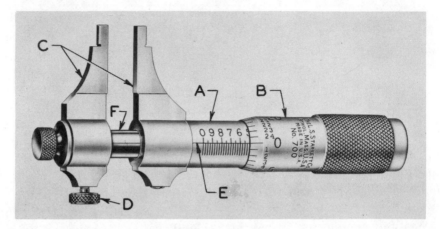

Fig. 33. Parts of the inside micrometer caliper: *A* Barrel, *B* Thimble, *C* Jaws, *D* Clamp Screw, *E* Index Line and *F* Spindle. Notice that the barrel and thimble graduations read right-to-left rather than left-to-right as on the standard micrometer caliper. (The L. S. Starrett Co.)

The four readings given in Fig. 34 show how to read the inside micrometer caliper. Study them carefully.

Inside Micrometer. The inside micrometer is inserted into and across the hole being measured. This micrometer (not to be confused with the similarly named *inside micrometer caliper*) is equipped with an extension rod. See Fig. 35. In using a micrometer of this type, select an extension rod of the proper length and insert the rod in the head of

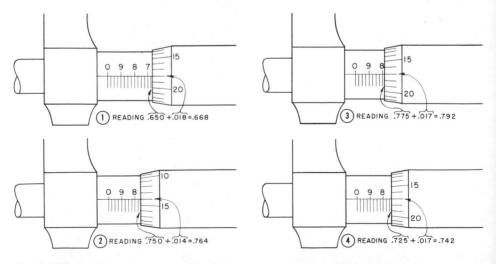

Fig. 34. Reading the inside micrometer caliper is similar to reading the standard micrometer caliper. Study these readings. Learn to read either micrometer caliper easily.

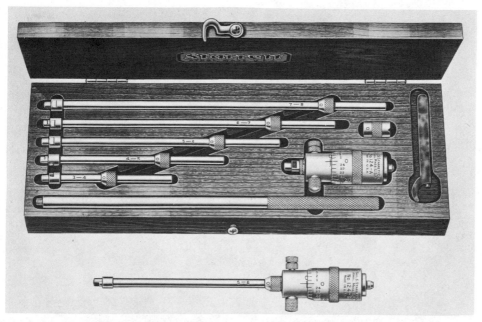

Fig. 35. The inside micrometer is used to measure inside dimensions of round holes and other openings. This micrometer is equipped with a set of extension rods. (The L. S. Starrett Co.)

the micrometer. Lock the rod in the head of the micrometer by turning the small knurled locknut. Now set the tool slightly less than the diameter to be measured and insert it in the hole. See Fig. 36. Hold the tool accurately on the center line (diameter) of the hole. Adjust the thimble of the micrometer until you feel a slight pressure. Now remove it and read the dimension on the barrel and thimble of the micrometer. To this add the length of the extension rod.

The average inside micrometer has a capacity of from 1½ to 10 inches, but with long extension rods, 100 inches or more can be accurately measured.

Micrometer Depth Gage. The micrometer depth gage, Fig. 37, is an instrument for measuring the depth of blind holes and slots. The gage illustrated has a screw movement of 1 inch and a range of from 0 to 3 inches in thousandths of an inch. Greater dimensions are obtained by use of an extension rod. (Three rods and the head comprise the set.) Gages having a range up to 9 inches are obtainable. The rods are generally .100 of an inch in diameter. They are easily inserted in the head by unscrewing the cap.

Fig. 36. The inside micrometer is shown in position in the hole. The type shown features a convenient handle which is used when measuring holes not easily accessible. (The L. S. Starrett Co.)

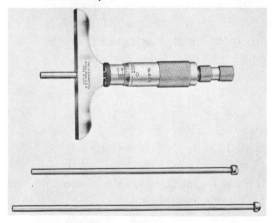

Fig. 37. The micrometer depth gage measures the depth of blind holes and slots. (The L. S. Starrett Co.)

The graduations on the barrel of this instrument read from right to left, as on the inside micrometer caliper, Fig. 32. The rod advances from the head as the thimble is turned to the right.

Follow these steps in using a micrometer depth gage: First, insert the proper extension rod in the micrometer head. Second, place the gage over the hole. Third, while you hold the base of the gage firmly on the workpiece with your left hand, screw the thimble down with your

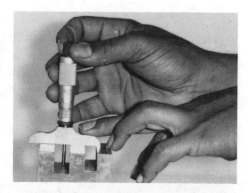

Fig. 39 shows two different readings for measurements taken with the micrometer depth gage. Reading *A* was taken with a two inch extension rod. You should add two inches to the reading. Reading *B* was taken with a five inch extension rod. Add five inches to the reading. Study the correct readings for *A* and *B*.

Fig. 38. When using the micrometer depth gage, hold the base of the gage with your left hand and screw the thimble down with your right hand. (Chicago Public Schools, Manpower Training Div.)

The Vernier Micrometer

right hand. Fourth, when the rod touches the bottom of the hole, get the micrometer reading. Then add the length of the rod. Fig. 38 shows the micrometer depth gage being used without an extension rod to measure the depth of a slot machined into a workpiece.

The standard micrometer discussed up to this point is used to take measurements to the nearest one-thousandth part of an inch. This is a very small unit of measurement and is usually fine enough. To work to even finer measurements, a vernier micrometer, graduated in "tenths" (.0001), should be used, Fig. 40.

On the ordinary micrometer, the

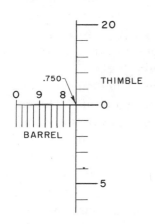

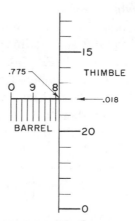

$$2.000 \left(\begin{smallmatrix} \text{EXTENSION} \\ \text{ROD} \end{smallmatrix} \right) + .750 \left(\begin{smallmatrix} \text{BARREL} \\ \text{READING} \end{smallmatrix} \right) = 2.750$$

(A)

$$5.000 \left(\begin{smallmatrix} \text{EXTENSION} \\ \text{ROD} \end{smallmatrix} \right) + .775 \left(\begin{smallmatrix} \text{BARREL} \\ \text{READING} \end{smallmatrix} \right) + .018 \left(\begin{smallmatrix} \text{THIMBLE} \\ \text{READING} \end{smallmatrix} \right) = 5.793$$

(B)

Fig. 39. Two readings of measurements taken with the micrometer depth gage are demonstrated. (The L. S. Starrett Co.)

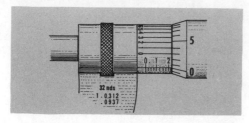

Fig. 40. The vernier scale on the barrel of a vernier micrometer makes it possible to take measurements to the nearest ten-thousandth of an inch.

twenty-five divisions marking the thimble serve to measure a part of a division on the barrel. On a micrometer with vernier scale, the additional marks on the barrel measure a part of a division on the thimble. In other words, a thousandth of an inch is itself divided by means of the vernier scale so that it is possible to measure tenths of a thousandth of an inch.

The venier proper consists of ten divisions which equal, in over-all space, nine divisions on the thimble. These ten equally spaced lines are etched on the barrel, Fig. 40. The first nine lines are numbered from 0 to 9, the tenth division being marked with another 0. One division is equal to $\frac{1}{10} \times .009$ inch, or .0009 inch. The difference between the thimble and the vernier divisions equals $.0010'' - .0009'' = .0001''$.

Reading the Vernier Micrometer. In reading the micrometer with a vernier, the first two steps are the same as in reading the ordinary micrometer. Read the divisions on the

barrel; then the divisions on the thimble. *Now take a third reading.* Look for a line on the thimble that coincides with a line on the vernier scale of the barrel, Fig. 41.

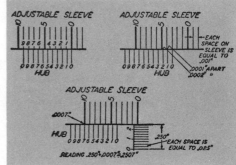

Fig. 41. To read to ten-thousandths on the vernier micrometer, an additional reading must be made using the vernier etched on the barrel.

The Metric Micrometer

Micrometers are also available which provide readings in the metric system. The metric micrometer provides readings to hundredths of a millimeter (.01 mm. = .00039 in.). See Fig. 42.

Each cross-line on the index line on the barrel indicates 0.5 mm., and the thimble moves the spindle toward or away from the anvil 0.5 mm. per complete revolution. To advance the spindle 1 mm. requires two complete revolutions of the thimble. The graduations, or cross-lines, are numbered every 5 millimeters (every tenth cross-line) from 0 to 25 mm.

The beveled edge of the thimble has fifty graduations, with every

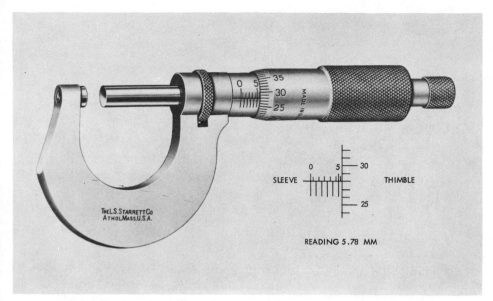

Fig. 42. The metric micrometer provides measurements to hundredths of a millimeter. (The L. S. Starrett Co.)

fifth graduation numbered for easy reading. Since one complete revolution of the thimble moves the spindle 0.5 mm., each of the fifty graduations on the beveled edge represents one hundredth of one millimeter (0.5 mm./50), or .01 mm.

Reading the Metric Micrometer. The procedures for reading the met-

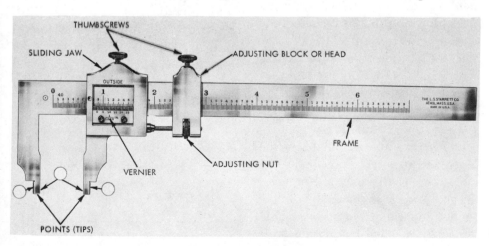

Fig. 43. The vernier caliper uses a vernier scale to make measurements to thousandths of an inch. Principal parts are identified. (The L. S. Starrett Co.)

ric and the English micrometers are similar. The metric micrometer, however, will give a reading (in millimeters) to two decimal places (hundredths). To read the metric micrometer, simply add the number of millimeters read on the barrel to the reading obtained from the beveled edge when one of the fifty graduations aligns (or nearly aligns) with the index line on the barrel. In the example of a reading shown in Fig. 42, *inset,* the reading taken from the barrel is 5.00 mm. + 0.5 mm. Finding that the 28th graduation on the beveled edge of the thimble aligns with the index line on the barrel, the total reading is then:

5.00 + 0.5 + 0.28 = 5.78 mm.

The Vernier Caliper

The vernier principle of measurement is used on various precision measuring instruments including the vernier caliper, vernier height gage, and gear tooth vernier caliper. These instruments are widely used for taking measurements as close as .001″ accuracy.

The vernier caliper is a beam caliper with jaws at right angles to the frame and is used in practically the same way as the micrometer, with the exception that the reading to the nearest thousandth of an inch is obtained by means of a vernier scale instead of a rotating thimble. Principal parts of the vernier caliper are

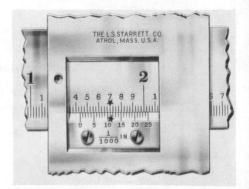

Fig. 44. Close-up view of the frame and vernier scales. (The L. S. Starrett Co.)

identified in Fig. 43. The frame, sometimes called the "beam" of the vernier caliper, is graduated like the barrel of a micrometer. See Fig. 44. Each line is divided into tenths (.100″), and each tenth has four sub-divisions equal to 25 thousandths of an inch (.025″).

The vernier scale, Fig. 44, like the thimble of the micrometer, is divided into 25 graduations, with every fifth graduation numbered for convenience. Note in Fig. 45 that the 25 graduations of the vernier are equal in length to 24 graduations of the beam scale (24/40″ or .600″). Each interval of the vernier scale is there-

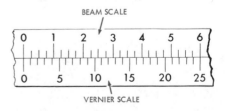

Fig. 45. Each beam scale division is .001″ wider than a vernier scale division.

fore equal to 1/25 of .600", or .024". Therefore the difference between the width of a beam scale division and a vernier division is the difference between .025" and .024", or .001".

In the diagram of the two scales, Fig. 45, the zero graduations on each coincide. When the beam scale is moved so that the first scale graduation coincides with the first vernier graduation, the scale will have moved .001", the difference between a scale division and a vernier division. Likewise, if the beam scale is moved so that the 6th lines of the vernier and beam scales coincide the scale will have moved .006", or, the vernier caliper jaws will have opened .006". Thus, when reading the vernier scale, that vernier graduation which coincides with a scale graduation indicates the number of thousandths of an inch to add to the scale reading.

Reading the Vernier Caliper. The vernier caliper may be used for both outside and inside measurements. Ordinarily, outside dimensions are

read from that side of the instrument with graduations reading from left to right. Inside dimensions are read on the side of the instrument with graduations reading from right to left. The sides of the vernier are marked *Outside* and *Inside* to prevent confusion.

The vernier caliper reading is determined by the position of the vernier scale "zero" graduation relative to the beam scale. The measurement to the nearest 25 thousandths of an inch is read directly from the beam scale. Then the vernier reading, to the nearest thousandth of an inch, is added to the beam scale reading to obtain the total measurement.

For example, in Fig. 46, an *outside* measurement is read from left to right as follows:

Last inch graduation passed
by vernier 0 is 2 ....2.000"
Last tenth graduation passed
by vernier 0 is 3 (x .100) .... .300
Intermediate graduations
passed by vernier 0 is 2
(x .025) .... .050
Vernier graduation that
coincides with beam scale
graduation is 18 (x .001) .... .018
Total external reading ....2.368"

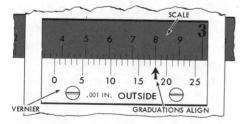

Fig. 46. Outside vernier caliper reading. Vernier "zero" points to correct beam scale reading; vernier graduation that aligns with beam scale graduation determines final thousandths reading.

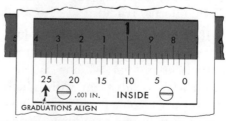

Fig. 47. Inside vernier caliper reading. Scales read right to left.

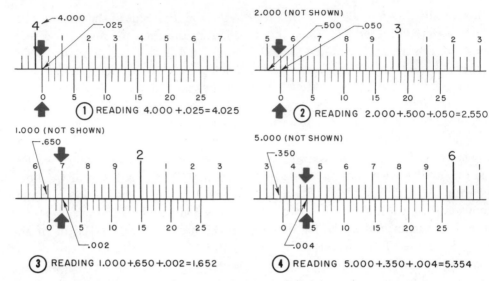

Fig. 48. Examples of vernier caliper readings are provided to help the student become familiar with the tool.

The *inside* measurement is made using the reverse side of the scale and is read from right to left as demonstrated in Fig. 47:

No inch graduations passed
by vernier 0 ....0.000″
Last tenth graduation passed
by vernier 0 is 7 (x .100) .... .700
Intermediate graduations
passed by vernier 0 is 1
(x .025) .... .025
Vernier graduation that
coincides with scale
graduation is 25 (x .001) ...... .025
Total internal reading ....0.750″

If the vernier 0 and 25 graduations both coincide with scale graduations, the final reading is that beam scale graduation with which the vernier 0 coincides. Fig. 48 illustrates four outside readings. Study each of the readings carefully.

Some vernier calipers are constructed with both vernier scales on one side so that both internal and external measurements can be read from the same side of the instrument. The reading method, however, remains the same.

The 50-Division Vernier Caliper

Many vernier tools now utilize a 50-division vernier, corresponding in length to 49 beam scale divisions. See Fig. 49. The beam scale is graduated in inches and tenths of an inch, but has only one intermediate graduation—which makes two .050″ intervals—instead of the .025″ intermediate graduations of the standard vernier caliper. This makes reading the 50-division vernier easier because graduations are more widely

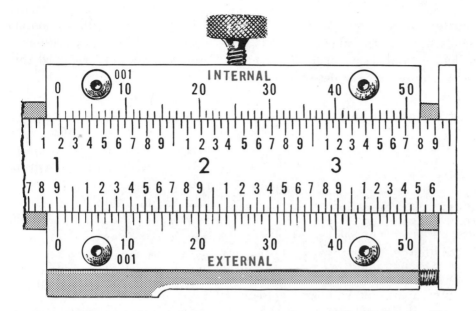

Fig. 49. The 50-division vernier caliper permits internal and external measurements from the same side. All scales read left to right. (Brown & Sharpe Mfg. Co.)

spaced than on the 25-division vernier.

Reading the 50-Division Vernier Caliper. The 50-division vernier caliper permits external and internal measurements from the same side of the instrument. Both internal and external scale readings are left to right. External measurements are

made using the *lower* scale as demonstrated in Fig. 50:

No inch graduation passed by
vernier 0 ....0.000″
Last tenth graduation passed
by vernier 0 is 8 (x .100) .... .800
Plus intermediate graduation
passed by vernier 0 .... .050
Plus vernier graduation which
coincides with beam
graduation: 20 (x .001) .... .020
Total external reading ....0.870″

Fig. 50. External reading using the 50-division vernier caliper. (Brown & Sharpe Mfg. Co.)

Fig. 51. Internal reading using the 50-division vernier caliper. (Brown & Sharpe Mfg. Co.)

Internal measurements with the 50-division vernier caliper are made using the *upper* scale, again reading from left to right, as demonstrated in Fig. 51:

Last inch graduation passed
by vernier 0 is 1 ....1.000″
Last tenth graduation passed
by vernier 0 is 2 (x .100) .... .200
No intermediate graduation
passed by vernier 0 .... .000
Plus vernier graduation which
coincides with beam
graduation: 30 (x .001) .... .030
Total internal reading ....1.230″

Note: If vernier 0 and 50 graduations *both* coincide with scale graduations, the final reading is that beam scale graduation with which the vernier 0 coincides.

Working with the Vernier Caliper. Loosen the clamp screw on the adjusting head (refer to Fig. 43), then move the sliding jaw to the approximate measurement. Tighten the clamp screw on the adjusting head, then turn the adjusting screw to bring the jaws into light contact with the work. Take the reading.

When positioning the vernier caliper, avoid manipulative errors. Keep the instrument scale in proper alignment with the work. Set the jaws exactly on the line of measurement, and do not bend or twist the tool. Adjust carefully with proper "feel." Bring the jaws into contact with the workpiece but do not set up undue pressure. Sensitive contact assures reliable readings and avoids undue wear.

The Universal Vernier Caliper

The universal vernier caliper, shown in Fig. 52, is a 3-way measuring instrument: it measures hole depth as well as inside and outside part dimensions in thousandths of an inch. The universal caliper has

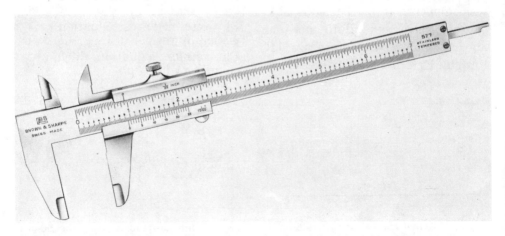

Fig. 52. The universal vernier caliper permits internal and external measurements, and can also be used as a depth gage. (Brown & Sharpe Mfg. Co.)

two integral pairs of jaws, one each for inside and outside measurements, and both readings are made from the same side of the instrument.

The Vernier Height Gage

The vernier height gage is used for accurately laying out work that is to be machined, and for taking outside measurements. The gage

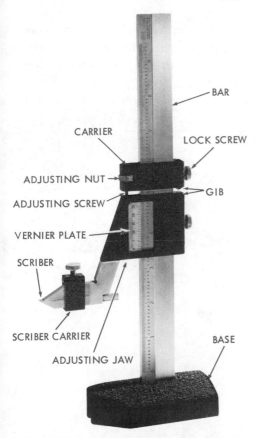

CARRIER

ADJUSTING NUT

ADJUSTING SCREW

VERNIER PLATE

SCRIBER

SCRIBER CARRIER

ADJUSTING JAW

BAR

LOCK SCREW

GIB

BASE

Fig. 53. The vernier height gage is used for layout work and external measurements. Major parts are identified. (Brown & Sharpe Mfg. Co.)

consists of a base, a graduated beam, a sliding jaw and scriber point, a movable vernier scale, and thumb screw adjustments and locks for the vernier scale, Fig. 53. When laying out work where great accuracy is required, the gage is usually used in combination with a surface plate.

The beam of the vernier height gage is graduated on both sides, like the standard vernier caliper. On the outside scale, the zero reading is 1″ from the bottom of the base, making it impossible to read or layout work less than 1″ high unless a parallel or gage block is used to raise the work. In this case the thickness of the parallel or gage block is added to the reading of the gage. The height of the gage block or parallel must total 1″ or more.

On the inside scale, when the sliding beam and the tip of the base are run together, the zero readings of the beam and vernier scales will coincide. This means that if the gage is set to any dimension, such as 1.500″, the distance from the top of the base to the bottom edge of the sliding jaw is 1.500″. The inside scale, it should be pointed out, can also be used for measuring outside part dimensions.

Metric Vernier Calipers

Some vernier calipers have both English and metric scales, permitting measurements in both inches and millimeters (mm). The method for reading the scale in either system is identical. The metric scale is usually graduated in increments of .02 mm. Metric micrometers and metric

vernier calipers are widely used in shops making parts for export to countries on the metric system.

Angular or Circular Measurement

All of the previous measuring tools explained in this chapter are used in taking linear measurements, or the measurement of parts in a straight line. Before discussing the next measuring tool, a brief explanation of circular measurement will be given.

The system of circular measurement is, as the name implies, based on the circle. A true circle is first divided into 360 degrees. The *degree*, then, is the largest unit of measurement in this system. The *minute* is the second largest unit of measurement. It is obtained by dividing one degree into sixty equal parts. One minute in other words is $\frac{1}{60}$ of one degree. The smallest unit of circular measurement is the *second*. A second is obtained by dividing one minute into 60 equal parts. In other words, the second is $\frac{1}{60}$ of one minute. For convenience, symbols are used to write these three units of measurement. For example: an angle of twelve degrees, 35 minutes, 15 seconds would be written 12° 35′ 15″.

Vernier Bevel Protractor

The vernier bevel protractor, Fig.

54, is a precision measuring tool to check angles with an accuracy of 5 minutes. The instrument shown consists of the base or stock, slotted to receive the blade, and the dial. The face of the dial has a scale starting at zero and going to 90 degrees both to the right and to the left of the zero. This scale, often called the *true scale*, is divided into

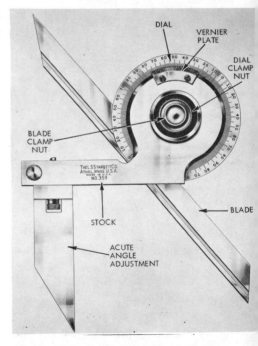

Fig. 54. The vernier bevel protractor is a precision measuring tool used in measuring angles. (The L. S. Starrett Co.)

graduations of 1 degree, with each 10-degree mark indicated. The *vernier plate* or *scale*, which is attached to the true scale, is divided into twelve equal parts and can be read to the left or to the right of the zero. Each small graduation on the vernier scale represents $\frac{1}{12}$ of 60 minutes or 5 minutes. Each third graduation is marked, beginning with zero and continuing through equally spaced intervals marked 15, 30, 45, and 60. The vernier scale is read in the same direction as the true scale.

Reading the Vernier Bevel Protractor. In the preceding material, it has been made clear that the divisions on both the true scale and the vernier scale are numbered to the right and also to the left of the zero. When reading the vernier bevel protractor, the reading is taken to the right of the zero on both scales if

the zero on the vernier scale has been moved to the right of the zero on the true scale. The reading is taken to the left of the zero on both scales if the zero on the vernier scale has been moved to the left of the zero on the true scale.

Fig. 55 shows three readings of the bevel protractor. When the zero on the vernier scale exactly coincides with a graduation on the true scale, as in reading *No. 1*, the reading will be in degrees only. The reading shown is 20 degrees. When the zero graduation on the vernier scale does not coincide with a degree graduation on the true scale, as in reading *No. 2*, we must check the vernier scale to find which line does coincide with a line on the true scale. Then we add the proper number of minutes. The reading is 20 degrees and 15 minutes in reading *No. 2*.

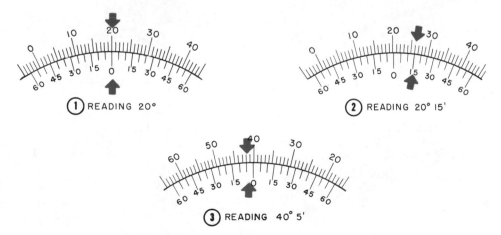

Fig. 55. Study these vernier bevel protractor readings carefully. Try to determine each reading before referring to the answers provided.

So far all of our readings have been from left to right. In reading *No. 3* of Fig. 55, we will use the other part of our scale which reads from right to left. In this example, we have 40 degrees and a fraction. Checking for a line on the vernier scale which coincides with a line on the true scale, we see that this fraction over 40 degrees is 5 minutes. Our reading is 40 degrees and 5 minutes.

Dial Indicator

The dial indicator, Fig. 56, is a precision measuring instrument often used in the machine shop to check any size variation of a machined part from the desired di-

mension. It is also used to determine the accuracy of alignment of work and machine in machine setups. Dial indicators range in size from about one inch to four and one-half inches in diameter. Each important part has been named and indicated in the illustration. Dial indicators are usually classified as either one thousandth-type or one ten-thousandth-type indicators. This is marked on the dial face, as shown in Fig. 56. The dial may be of the balanced type, Fig. 56, in which the figures read both to the left and right of zero, or it may be of the continuous reading type, in which the figures read from the zero only in the clockwise direction. The

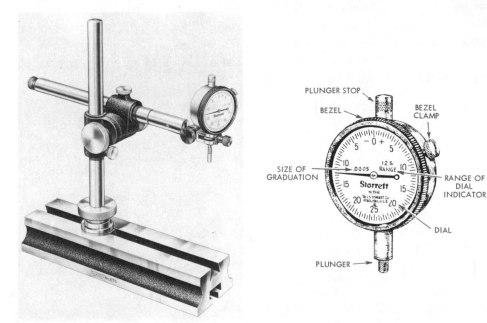

Fig. 56. The dial indicator is used to measure alignment between machine and work as well as provide a final check on the dimensions of finished parts (The L. S. Starrett Co.)

dial is graduated, and each graduation represents a definite movement
of the plunger. The plunger is designed so that it will slide both in
and out. On *one-thousandth-type*
indicators, each graduation may represent a plunger movement of one
thousandth, one-half thousandth, or
one-quarter thousandth of an inch.
On one *ten-thousandth-type* indicators, each graduation may represent a plunger movement of one
ten-thousandth, one-half of one ten-
thousandth, or as small as one-quarter of one ten-thousandth of an inch.

How to Read the Dial Indicator.

The value represented by each
graduation on an indicator dial is
stated on the dial face as a decimal.
Using Fig. 56 as an example, we
first look at the figure on the dial
face (which appears to the left of,
and slightly above, the center of the
dial face). The figure is .0005, or
one-half of one thousandth of an
inch per graduation. This indicator
is a one thousandth-type indicator;
every second graduation represents
one thousandth of an inch. The first
numbered graduation from zero is
5, indicating five thousandths of an
inch of plunger movement (.005″).
The range of the indicator is also
stated on the face of the indicator,
and is twenty-five thousandths of
an inch (.025″). The reading shown
in Fig. 56 is .0125″, or one hundred
and twenty-five ten-thousandths of
an inch.

Since it is obvious that such small
movements of the plunger could not
be seen, the plunger is linked to the
indicator hand or pointer through a
train of small gears (or through a
set of linking levers) which multiply
any small movement of the plunger
into a larger, more easily seen movement of the indicator hand. For this
reason, it can be understood that
the dial indicator must be mounted
rigidly to some support, such as
shown in Fig. 56, so that only the
plunger can move. Any movement
of the indicator body, in relation to
the plunger, would cause an error
in the reading.

When not in use, the plunger
spring will naturally push the
plunger outward, away from the indicator body, to the limit of plunger
travel. On most standard indicators,
the pointer will come to rest in the
"zero reading" position. From this
position, however, the hand can only
travel to the right or clockwise. To
use the indicator properly, so that
the pointer could move either to the
right or left of the zero (to indicate
plus or minus), the indicator must
be set up for use so that the plunger
has been pushed inward far enough
so that the pointer has equal travel
to the right or to the left of the
zero. When using the indicator, remember that as the plunger moves
in toward the indicator body, the
pointer moves clockwise, and as the
plunger moves out away from the

indicator body, the pointer moves counterclockwise. Handle dial indicators carefully when using them, and store them in their proper cases when you have finished using them.

Use of Dial Indicator. The dial indicator can be used for detecting differences in the size of various parts of a workpiece. For example, assume we have a 6-inch long shaft of ¾-inch diameter which we wish to check for concentricity or roundness, and also for taper. To check or indicate for concentricity, you might use a setup *similar* to Fig. 57. V-block clamps (not shown in the figure) would be loosened to permit you to rotate the end of the shaft by hand as it rests in the blocks. The spindle lockscrew of the pedestal is loosened, so that you can position the indicator at the proper

height. This can be determined by watching the pointer of the indicator. The button of the indicator is brought to bear on the workpiece between the V-blocks until the pointer moves clockwise .015, at which point you lock the indicator to the spindle of the pedestal by turning the spindle lockscrew. Now loosen the bezel clamp and turn the bezel until the zero on the face of the dial lines up directly under the pointer of the indicator; after this, the dial is clamped in place by turning the bezel clamp screw. To check the shaft for concentricity, simply turn the end of the shaft by hand and notice the amount of variation detected by deflection of the pointer as the workpiece revolves.

To check this same piece for taper, first remove the V-block clamps,

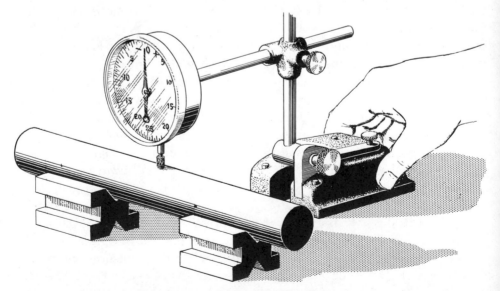

Fig. 57. A dial indicator is used to test the roundness of a workpiece.

as it is only necessary to *lay* the piece in the blocks. Then follow the procedure outlined above. Adjust the indicator to the proper height on the pedestal spindle and bring the button of the indicator to bear on one end of the workpiece. Adjust the dial so that the zero on the dial lines up with the pointer of the indicator. Now move the pedestal base on which the indicator is mounted so that the button of the indicator is brought to bear on the opposite end of the workpiece. By detecting the deflection of the needle from zero, you can readily see the amount of taper in the shaft.

NOTE: Please see review questions at end of book.

Bench Tools:

Including Layout Tools

Many operations performed in the machine shop require the use of hand tools. The machinist should learn what these tools are, and how they should be used. Tools generally classed as hand tools are the hammer, file, chisel, wrench, and similar simple instruments. To select and use the proper tool for any job which may arise requires skill. This is particularly true of filing and scraping operations as performed in bench work. Manual skill in the use of hand tools can only be developed through constant use and practice. The material in this section, however, will serve as a guide in the selection and use of the various hand tools.

Hammers

Hammers used in the machine shop are of two types, the hard and the soft. The hard hammer has a

head made of tool steel. It is used in various bench work and layout operations. The ball-peen hammer, Fig. 1, is an example. It is named from the small, ball-shaped head, called the peen. The flat end of the hammer head, opposite the peen, is called the face of the hammer. The hammer face is used in striking

Fig. 1. The ball-peen hammer is used in layout work, chiseling operations, and riveting.

the end of prick punches in laying out work and in chiseling operations. The peen of the hammer is used in riveting, where it is necessary to peen or stretch the end of the rivet to prevent it from working loose.

The ball-peen hammer is available in sizes ranging from 4 ounces to 2½ pounds. The smaller sizes from 4 to 10 ounces are used in layout work. The heavier hammers are used in general bench work.

Soft-headed hammers are used in assembly work where it is necessary to preserve the surface or finish of machined parts. They are also used in setup work to set a workpiece in a vise. The soft hammer head will not mar the surface of a workpiece because it is softer than the metal on which it is used. The hammer head is made of brass, babbitt, lead, rawhide, or celluloid.

When using the hammer, always grasp the handle at the extreme end. This grip provides greatest leverage and helps to bring the face of the hammer down flat upon the object being struck. The face of the hammer should be kept free of nicks and bumps, and the hammer head should fit tightly on the handle.

Screw Drivers

Because of the wide use of various metal fasteners in the form of flat-head, round-head, and other types of screws, various kinds of screw drivers are required in the driving of these parts, and in their removal as well.

Screw drivers consist of three parts: the *handle*, the steel portion extending from the handle called the *shank*, and the end which fits into the head of the screw called the *blade*.

Screw drivers are made with blades of various widths and in lengths suited to special requirements, such as working in close

Fig. 2. The standard screw driver consists of three parts: the handle, the shank, and the blade.

Fig. 3. The offset screw driver is made so that screws may be inserted or removed in difficult-to-reach places. (General Motors Corp.)

Fig. 4. The blade of the screw driver should fit snugly into the screw slot. (General Motors Corp.)

quarters or in deep holes. Fig. 2 shows a regular-type screw driver.

The larger size screw drivers generally have square shanks. With a shank of this kind, a wrench can be used to obtain added leverage when it is desired to turn a screw to its limit.

The offset screw driver, Fig. 3, is made so that screws can be inserted or removed in places that are difficult or impossible to reach with the ordinary straight-shank screw driver. The offset screw driver has two blades which are positioned at right angles to each other, one being at right angles to the handle and the other parallel thereto.

Since the blade of an ordinary screw driver frequently becomes damaged or misshaped in use, it is important that the student of machine shop learn to repair the damage and restore the shape by grinding. The flat sides should be ground parallel at the tip and not tapered or wedge-shaped, as a point so shaped tends to force the blade out of the screw slot when a turning motion is applied, thus raising a burr on the slot of the screw. The blade thickness at the tip should fit the screw slot to which it is applied, Fig. 4. Never apply a screw driver with a small tip to a screw having a large slot. Also, do not use a screw driver as a substitute for a chisel or pinch bar, as the tool is not intended for such use.

Bench Vise

When the machinist or the tool and die maker performs bench work operations such as filing, sawing, and chipping, he employs a holding device to grip the work securely.

The bench vise, Fig. 5, is commonly used to hold the workpiece. Vises are of many varieties and sizes. All consist essentially of a fixed jaw, a movable jaw, a screw, a nut fastened in the fixed jaw, and a handle by which the screw is turned to bring the movable jaw into the desired position.

In the machinist's vise, both jaws are made of cast iron having removable faces of hardened tool steel. The faces are usually serrated to provide a firm grip for heavy work, but may be smooth to prevent marring of the surface of certain workpieces. When holding soft metal or finished steel surfaces, even the

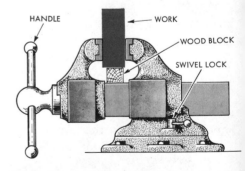

Fig. 5. The machinist's bench vise is used to hold work securely. Note that the vise pictured here has a swivel base so that the work can be turned to a convenient working position.

smooth steel jaws will mar the work. In such cases, it is customary to use false jaws of brass, copper, or lead, Fig. 6, or to place leather or paper directly over the steel jaws as a means of protecting the work-

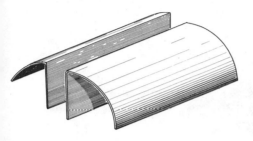

Fig. 6. When holding soft metal or finished surfaces, false or lining jaws are used on the vise to prevent marring the workpiece.

piece. The false jaws are merely placed over the steel jaws. They are not fastened in any manner.

Some vises have a swivel base, permitting the machinist to position the work to suit his convenience.

Files

Files differ in length, shape, and the cut of the teeth. The length of a file is the measurement taken from heel to point, Fig. 7. This measure-ment excludes the tang, the part which fits into the handle.

Machinists' hand files are classi-fied according to tooth form as *sin-gle, double, rasp,* and *curved.* Fig. 8, *top,* shows the two types most frequently used—the single cut and double cut files. A single-cut file has rows of parallel teeth extending at an angle across the face. A double-cut file has parallel rows of teeth which cross each other. The first row, known as the *over-cut,* is usu-ally coarser and deeper than the sec-ond row, known as the *under-cut.* The teeth are shaped to make a cut-ting edge like that of a tool bit, and they have both rake and clearance angles.

According to the coarseness of the teeth, they are *rough, coarse, bas-tard, second-cut, smooth,* and *dead smooth.* Fig. 8 shows the types of files most often used in machine shop work. A magnified view of each illustrates the differences in form and coarseness. The different file *shapes* are shown in Fig. 9.

File Handles. Files should not be used without a handle. The round

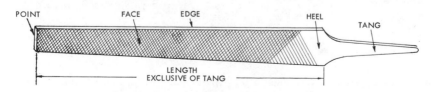

Fig. 7. Files are used in cutting, grinding, or smoothing.

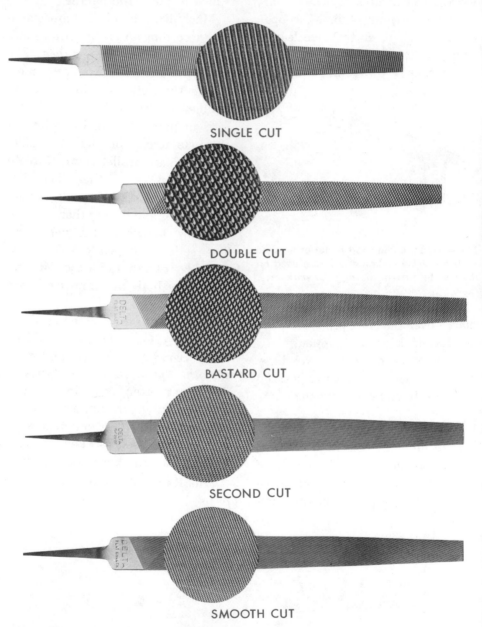

SINGLE CUT

DOUBLE CUT

BASTARD CUT

SECOND CUT

SMOOTH CUT

Fig. 8. Files are classified according to the type of tooth form, and the degree of coarseness of the teeth. (Delta File Works, Inc.)

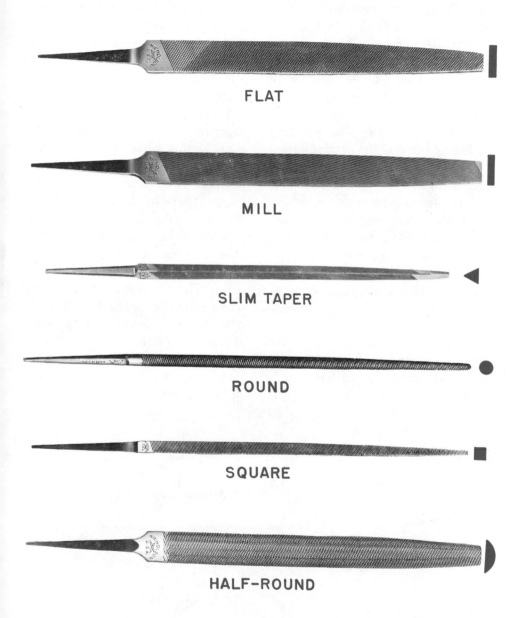

FLAT

MILL

SLIM TAPER

ROUND

SQUARE

HALF-ROUND

Fig. 9. The different file shapes used in machine shop work are shown here. (Nicholson File Co.)

wooden handles are a safety measure. They prevent accidents which may injure the hand.

Work Determines Choice of File. The kind of metal to be filed is the important factor in deciding what file cut to use. Cast iron, especially if the scale has not been removed, is hard on a new file. The glasslike scale tends to dull the cutting edges. New files should not be used on such a surface.

Two general rules for selecting the proper coarseness of file cut are:

1. When filing any hard metal such as steel, use a second-cut file.

2. When filing any of the soft metals, such as brass, bronze, or copper, use a coarse or rough-cut file.

Nearly all files used in the machine shop are double-cut. The 10-inch or 12-inch bastard file is generally used for rough filing in bench work. The second-cut file gives a fairly smooth finish. A finer file will give a smoother finish.

Suggestions on How To File. First, select the proper file for the job. Make sure that it is fitted with a good handle.

Stand in a comfortable position, especially if the job will require considerable time. It is impossible to file accurately in an awkward position. The workman (if right-handed) should stand with his feet slightly apart, with the left foot forward. The weight of the body should be balanced so that the arms may move easily forward and backward. The file should be regarded as an extension of the arm. That is, file and arm, from elbow to wrist, should be in line.

Grasp the file handle in the right hand, thumb on top. For light work, hold the point (tip) of the file with

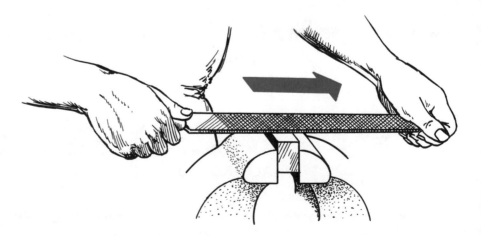

Fig. 10. The correct method of holding a file is shown.

the thumb and first two fingers of the left hand, the thumb on top and the fingers below. For heavy cuts, the heel of the left hand may be placed on top and the fingers closed on the underside of the file, Fig. 10. *Do not drag a file back over the work or attempt to cut on the return stroke. This will dull the file.*

Never rub your hand or fingers over the filed surface of cast iron or brass. Oil from the hand may be deposited on the metal. This will cause the file to slide over the work instead of cutting it.

Rough Filing. When rough fil-

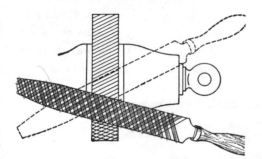

Fig. 11. A crossing stroke should be used in rough filing.

ing, cross the stroke at short intervals, Fig. 11. This will help to keep the surfaces flat and straight while you are learning to file. *Bear down only on the forward stroke.*

Testing the Work. Inspect the work occasionally with a scale, Fig. 12, to determine whether the filed surface is flat and straight. Use a steel square or a combination square to check squareness.

Rounding a Corner. To round a corner, rough file across the workpiece. Reduce the corner by filing a series of angles until the proper radius is secured. See Fig. 13. Now finish the corner by following along the rounded corner with a fine-cut file.

Draw Filing. Draw filing, Fig. 14, consists of grasping the file firmly at both ends and alternately pushing and pulling the file sideways across the work. The file is held at right angles to the line of stroke. The pressure should be the same for both forward and return strokes.

Very little stock is removed in

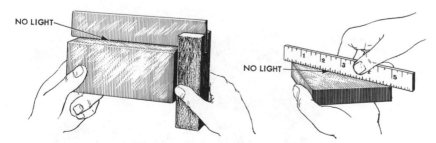

Fig. 12. The filed surface is flat if no light is seen between the steel square or combination square when the surface is tested as shown.

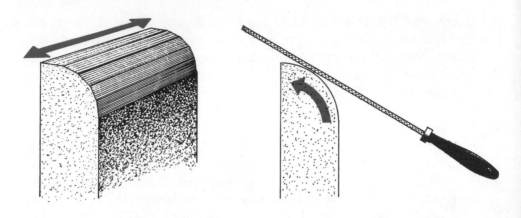

Fig. 13. Rounding a corner with a file is usually done in two steps.

draw filing. The purpose of this operation is to produce a straight or square surface. Draw filing gives a smoother finish on edges and narrow surfaces than "straight" filing. Single-cut files are preferred to double-cut files when draw filing, as a better finish is produced.

Cleaning the File. The particles of metal removed by the file frequently lodge in the teeth and lessen the cutting power and quality of finish. With hard metals, these particles, or *pins*, may scratch the work. It is important, therefore, to clean files regularly. This is best accomplished by using a stiff brush or file card, Fig. 15. When filing steel, the

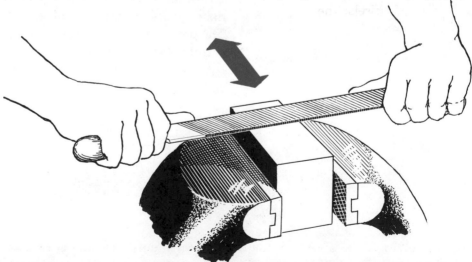

Fig. 14. Draw filing is used to give smooth finishes to edges and narrow surfaces.

Fig. 15. A file should always be cleaned with a brush—never with the hand alone. (Nicholson File Co.)

file may be rubbed with chalk to prevent pinning.

Chisels and Chipping

Types of Chisels. One of the most common processes in the machine shop is cutting away excess metal with a chisel. There are several types of chisels; the machinist must know how to select the proper chisel for the job at hand. Chisels are classified by the width and shape of the cutting edge.

The *flat chisel*, Fig. 16, can be used for cutting off heads of rivets, for cutting thin metals, and for split-

Fig. 16. The flat cold chisel is used for cutting away excess metal.

ting nuts that have rusted. In bench work, it is used for chipping. The cutting edge has a surface slightly wider than the stock from which the chisel was forged.

The *cape chisel*, Fig. 17, is narrow. It is used for cutting keyways in shafts and for chipping narrow grooves and channels in metal.

The *round-nose* or *grooving chisel*, Fig. 17, is used for cutting oil grooves in bearings and for chipping inside corners having a radius or fillet. It is used also for making grooves with round bottoms, and for cutting in curved surfaces.

The *diamond-point chisel*, Fig. 17, is made square at the end and ground on an angle. This produces a diamond-shaped cutting face. Chisels of this type are used for cutting V-grooves and square corners.

How to Chip or Cut with a Chisel. Work to be chipped should be held in a vise or otherwise secured. When mounting finished work or soft material in the vise, be careful to use soft metal guards to prevent damage to the surface. It is good practice to put a block under the work held by the vice. This prevents the work from slipping down.

Hold the chisel firmly enough to guide it but, at the same time, lightly enough to ease the shock of the hammer blows. Grip the hammer handle near the end and strike with a force suited to the metal being chipped.

When chipping, hold the chisel on a slant. By raising or lowering the head of the chisel, the cut can be made deeper or shallower. Avoid taking too deep a cut. Pull the chisel away from the work after every second or third blow. CAUTION: Wear safety glasses when chipping.

Cutting Edge. A surface which produces a cutting edge on a chisel is called a *facet*. Flat and cape chisels have two facets, while the diamond-shaped and round-nose chisels have but one facet. The facets on flat and cape chisels should be ground alike to form the same angle with the axis of the tool.

For chipping, flat and cape chisels should be ground so that the facets form an angle of 70 degrees for cast iron, 60 degrees for steel, 50 degrees for brass, and about 40 degrees for babbitt, copper, and other soft metals. When there are two bevels, they should be alike in width

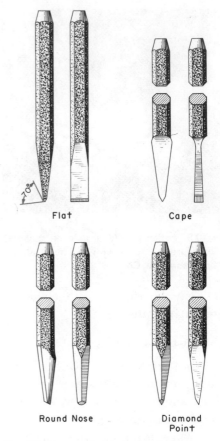

Flat   Cape

Round Nose   Diamond Point

Fig. 17. Types of chisels in common use.

CONVEX

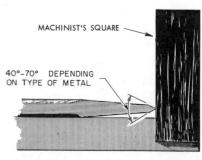

MACHINIST'S SQUARE

40°-70° DEPENDING ON TYPE OF METAL

Fig. 18. When grinding a chisel, hold it securely against the wheel and move it back and forth across the wheel, keeping the correct grinding angle. The angle can be checked using a machinist's try square.

and should form equal angles with the center line of the chisel. Small round-nose chisels, and some of the slotting chisels, are ground single-sided; that is, with but one bevel or facet.

Grinding. To grind a flat chisel, Fig. 18, hold it securely. Press the facet against the face of the revolving wheel at the proper angle. The cutting edge should be ground con-

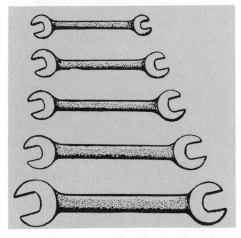

Fig. 19. A set of open-end wrenches.

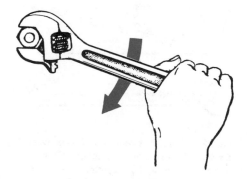

Fig. 20. Always apply force on an adjustable wrench so that the force is applied in the direction of the adjustable jaw.

vex. As the grinding proceeds, cool the chisel by dipping it in water frequently.

When grinding a chisel, do not hold it in one place on the wheel. Move it back and forth across the wheel face, keeping the correct angle. If the chisel is held in one place on the wheel, a groove will be worn in the wheel face. Never use the side of the grinding wheel. Such procedure is unsafe and inefficient. CAUTION: Always wear safety glasses when grinding.

Wrenches

Many different kinds of wrenches are made for turning nuts, bolts, pipes, and so on. Wrenches are named according to their shape, use, or construction.

Open-End Wrenches. Solid wrenches with an opening in both ends are called *open-end wrenches*, Fig. 19. A set of 5 open-end wrenches will have openings ranging from 5/16 of an inch to 1 inch in width.

Open-end wrenches have the head end opening turned at an angle. Most open-end wrenches are turned 15 degrees, and some are turned 22½ degrees. If you have ever used a wrench of this type in close quarters, you probably noticed the advantage the angles afford.

It is important that the wrench be of a size and type suited to the nut. If it is not, the corners of the nut may be rounded or damaged. Also, if you use an unsuitable wrench, it may slip off the nut and cause an accident. If possible, pull rather than push on the wrench. If pushing cannot be avoided, use the base of the palm and hold the hand open.

To tighten or loosen a bolt or nut, a quick jerk or a sharp blow with the ball of the hand will be more effective than a steady pull or push. To loosen a bolt or nut that proves stubborn, try dropping a little oil on the thread. Avoid using a hammer on a wrench.

Adjustable Wrenches. Adjustable wrenches have one stationary and one adjustable jaw. Fig. 20 illustrates the direction in which force is applied when using this wrench. The angle of the opening to the body of the wrench is 22½ degrees.

Adjustable wrenches are useful for various purposes. They are not, however, intended to take the place of the standard open-end wrench, box wrench, or socket wrench. They are used mainly for nuts and bolts which do not fit a standard wrench.

Box Wrenches. Box wrenches are popular because they can be used in very close quarters. They are called *box wrenches* because they enclose the bolt head or nut. Instead of a six-sided opening, they have twelve notches arranged in a circle. This type of wrench is known as a 12-point wrench, Fig. 21, *top.*

It can be placed on a nut or on the head of a bolt in any of twelve holds to provide continuous tightening or loosening action in several positions. The sides of the box wrench are very thin. Thus, the tool is convenient in inaccessible places.

In addition to the straight-handle box wrench, there are box wrenches with heads set at an angle of 15 degrees to the handle. See Fig. 21, *center.* This tips the handle of the tool upward and provides clearance for the hand of the mechanic.

A *combination wrench* has both box and open ends, Fig. 21, *bottom.* It is sometimes decribed as a half-and-half wrench.

Socket Wrenches. Greater improvement has been made in socket wrenches than in most commonly used hand tools. The socket-wrench

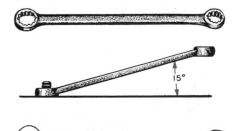

Fig. 21. *Top:* A twelve-point box wrench. *Center:* Some box wrenches have their heads set at a 15° angle; this provides clearance for the mechanic's hand. *Bottom:* A combination box and open-end wrench.

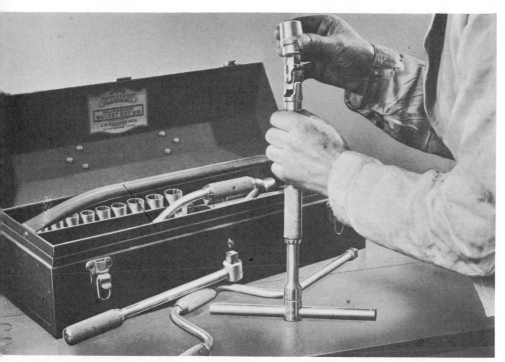

Fig. 22. Socket wrenches permit interchangeability of various socket sizes on the handle for different sizes of nuts. (J. H. Williams & Co.)

set contains useful accessories in addition to the various sockets, such as the ratchet and speed handles. See Fig. 22.

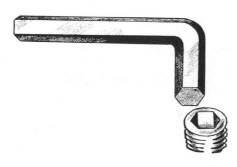

Fig. 23. When using a setscrew (Allen) wrench be sure the screw socket is clean.

Set-Screw Wrenches. Set-screw wrenches are L-shaped bars of tool steel, Fig. 23. The type most commonly used is hexagonal because the set screws have hexagonal sockets. This wrench is also called an *Allen wrench,* and it comes in sizes matching the socket sizes of the set screws. When using a set-screw wrench, be sure the dirt and grease are picked out of the screw socket.

Spanner Wrenches. Spanner wrenches, Fig. 24, are used with notched adjustable nuts. These wrenches must be of a size and type

63

Fig. 24. The adjustable spanner wrench has a knuckle which flexes to fit nuts of varying sizes. (J. H. Williams & Co.)

proper to fit the part to be adjusted. Spanner wrenches come in hook, face, and pin styles, and they may have either pins or lugs fitting the notches of the nut. Adjustable spanners are made to fit nuts of various diameters.

Hand Taps and Tapping

A tap is a tool used to cut internal threads, Fig. 25. The threads on a tap are not continuous. Three or

four flutes are cut lengthwise across the threads, forming the cutting edges. The shank is square at one end so the tap can be turned with a wrench.

Various standards and styles of taps are available. The discussion which follows relates to the Unified form of screw thread. The Unified thread form is the present American industrial standard. The Unified form, however, is essentially the same as the previous standard, the American National thread form, and the two are interchangeable for most pitch-diameter combinations. The two standard *series* of Unified screw threads most widely used today are the Unified National Coarse (UNC) and the Unified National Fine (UNF) Series—more familiarly the NC and NF Series respectively. Tables I and II of the Appendix list the basic thread dimensions of the Unified Coarse, Fine, and Extra Fine Series, with the required tap drill sizes given in Table III.

Tap and Screw Thread Sizes.

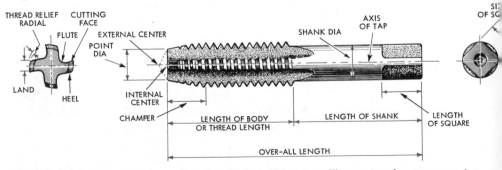

Fig. 25. Internal threads are cut with a tool called a tap. The parts of a common tap are shown here.

Taps are made with two, three, or four flutes and are available in standard machine screw sizes.

The term *machine screw size* refers to a system of numbering used to identify machine screws smaller than ¼ of an inch. The number system refers to a series of individual screw sizes within both the Unified National Coarse (UNC) and the Unified National Fine (UNF) Series. For example, the smallest screw size listed in Table II of the Appendix is identified by the number 0-80. The largest screw in the machine screw series is identified by the number 12-28. The first part of the number specifies the gage size

of the body of the screw. The second part of the number specifies the number of threads per inch. For the machine screw 0-80, 0 specifies the size (diameter) of the screw, and 80 specifies the number of threads per inch. In this case, the thread is a Unified National Fine thread. In the number series, diameters range from .060 to .216 inches.

There is a variation in the number of threads per inch for a screw of a given size. A No. 12 screw in the NC series is made with 24 threads per inch; a No. 12 screw in the NF series is made with 28 threads per inch. These screws would be designated as 12-24 and

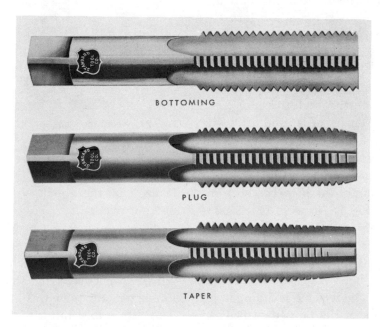

Fig. 26. The standard tap set consists of bottoming, plug, and taper taps. (Standard Tool Co.)

12-28 respectively. This system of numbering is used to identify small threads, or machine screws, up to ¼ of an inch in diameter within both the NC and NF series. More examples are: 6-32 (NC) and 6-40 (NF), and 10-24 (NC) and 10-32 (NF).

Hand Tap Sets. Standard hand tap sets are composed of three units: a *taper* or *starting tap*, a *plug tap*, and a *bottoming tap.* These three taps are used in the order indicated when tapping a blind hole (a hole which does not go entirely through the workpiece).

The taper tap, Fig. 26, is used to start the thread. It is tapered or chamfered back from the end approximately six threads before the full diameter of the tap is reached. This taper gives the tap an easy start and forms threads gradually.

The plug tap, Fig. 26, is tapered back from the end three or four threads. This tool is used after the taper tap has cut the threads as far as possible.

When it is necessary to thread to the bottom of the hole, the bottoming tap, Fig. 26, is used. It is chamfered only one thread. It is used last to cut the thread to the bottom of the hole.

A left-hand tap cuts left-hand threads. The letter *L* is stamped on the shank of the tap. Identification by letter is not essential, however. Careful study of the cutting edge

of the flute will show whether the tool is a right-hand or left-hand tap.

Tap Wrenches. The adjustable T-handle tap wrench is shown in Fig. 27. The jaws are made to hold the square end of the tap by clamping onto it. The jaws should be kept free of dirt and grease, and the

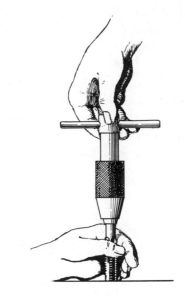

Fig. 27. Keep the jaws of the adjustable T-handle tap wrench free of dirt and grease.

threads in the chuck shell should be kept oiled.

When working on a surface which provides ample room, a double-ended or T-handle wrench is best. Grasp it between the thumb and forefinger and turn the tap slowly. The adjustable tap wrench shown in Fig. 28A is used for larger taps. By turning the handle at one end,

Fig. 28A. The adjustable tap wrench is used for larger taps.

the movable jaw is forced tight against the square end of the tap.

Selecting Tap Drills. Before a tap can be used, a hole must be drilled. For this purpose, it is very important to choose a drill of the correct size. As the size of the tap is the same as the outside diameter of its threads, the hole drilled for tapping must be smaller than the tap diameter. If the drill hole is too small, the tap will bind in the hole and probably break. If it is too large, the threads in the hole will not be deep enough.

Table III of the Appendix illustrates the system of numbering adopted by the National Screw Thread Commission. The table should be consulted whenever a tap drill is to be used. It is not good practice to trust to memory. If a drill of the size indicated in the table cannot be obtained, consult a table of drill sizes for the next larger drill. Remember you must use a drill that is a little smaller than the outside diameter of the tap thread.

Lubrication of Tap. When tapping ferrous, non-ferrous, or non-metallic materials it is important

that you pay careful attention to the selection of the proper lubricant for the material you expect to use. Manufacturers of lubricants will be glad to give you specifications, prices, and their recommendations regarding their products.

How To Cut a Thread with a Tap. To begin, clamp the work in a bench vise so the drilled hole is in an upright position. Select a tap of the proper taper and clamp its square

Fig. 28B. When starting the tap make certain that it is straight. Turn the tap with the palm, applying a steady downward pressure.

end in the tap wrench. The T-handle tap wrench is used for small taps, the adjustable for larger taps.

Grasp the tap wrench with the right hand directly over the tap and place the end of the tap in the hole.

Position the tap square with the top surface of the worpiece. See Fig. 28B.

Give the tap one full turn, using a steady downward pressure. Then sight it to see whether the tap is entering the hole straight. If the tap is straight in the hole, give it another turn, holding it steady and straight. To check even more carefully in seeing that the tap is

Fig. 28C. Regularly reverse the tap to clear the chips from the flutes of the tap.

straight, place a square against the tap at various places. If the tap is not square, back it out of the hole, straighten it, and with careful pressure, screw it back straight in the hole.

Use the lubricant previously recommended. Continue to turn the tap with one hand until the thread has made a good start. It is not necessary to continue downward

pressure, as the tap will pull itself in. Back up the tap now and then to break off the chips. It is a good idea to back up one step or part of a turn, then go forward two steps or a full turn. See Fig. 28C.

Removing a Broken Tap. Because the tap is a somewhat brittle and fragile tool, it is easily broken. A broken tap can be removed by using a tap extractor, a small prick punch, or a cape chisel. Sometimes it must be drilled out.

To remove a broken tap with a tap extractor, first clean out any chipped or broken particles of the tap remaining in the hole. This can be done with a magnetic scriber. If some chips have become lodged in the hole it may be necessary to loosen them first with a prick punch. Then they can be taken out with a magnetized scriber.

After the chips have been removed from the hole, the tap extractor is put in place. The movable fingers of the extractor are inserted in the flutes of the broken tap as deeply as possible. Then the collar of the extractor is brought against the surface of the work. The broken tap is backed out by turning the extractor with a tap wrench. To loosen the tap in the hole, twist the extractor back and forth gently a few times.

If the method just described does not prove effective, the tap will have to be annealed (softened by heating and cooling) and drilled out. In view of the time and effort involved in removing a broken tap, handle a tap carefully.

CAUTION: In removing a broken tap with a small prick punch or a chisel, remember that both tap and punch (or chisel) are very hard. When the tools are brought together in sharp impact, small pieces of steel may break off and fly into the air at great speed. *To avoid injury, wear safety glasses when working with punch or chisel.*

Sharpening a Tap. Dull taps break easily. The student or machinist should inspect a tap closely when he receives it from the tool crib. If the cutting edges are not sharp, he should refuse the tool and call for another, or have it sharpened at once.

The tap is sharpened by grinding the face of the cutting edge. Only enough metal should be ground away to produce a sharp cutting edge, and all edges should be of uniform height. If all edges are not even in height, too great a load will be thrown on a single cutting edge, and the tap may break.

Threading Dies

A threading die is a tool with an internal thread like the thread in a nut. It is used to cut external threads on bolts and similar round stock. To cut the thread, the die is turned or screwed on the round workpiece. As the die advances, the thread is cut by the teeth of the die.

Solid Dies. The solid threading die shown in Fig. 29 is a square block of hardened tool steel with a threaded hole. The cutting edges of the die take the form of flutes. The general rules for the use of taps apply also to the use of dies.

Dies for small work are usually solid and of fixed size. They cannot be sharpened, but offer compensating advantage in that they center on the work readily. As the full thread is cut in one passage of the die, con-

Fig. 29. The solid threading die is a block of hardened tool steel with a threaded hole used to cut external threads. (Pratt & Whitney Div., Niles-Bement-Pond Co.)

siderable power is needed to operate solid dies of large size. For this reason, hand-operated solid dies exceeding ½-inch diameter are seldom used.

Split Dies. The split type of die, generally known as the *adjustable die*, Fig. 30, can be sharpened easily, has limited adjustment for size, and cuts the thread by easy stages. The holder for dies of this type is called

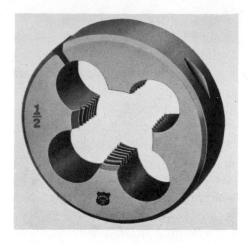

Fig. 30. The split die can be adjusted by turning a small setscrew. (Standard Tool Co.)

a *die stock* or *die holder*. On some holders, a guide or ring under the die fits around the bar or bolt to be threaded and guides the die so it will go on the work squarely.

The round, adjustable, split threading die is split on one side. In cutting, it can be closed gradually to the required size of thread. Most dies are set to cut the thread slightly over or slightly under the designated size. They can be adjusted by turning a small set screw. Two-piece threading dies are also available.

Left-Hand Dies. Occasions arise when a die is needed to cut a left-hand thread. Left-hand dies are stamped with the letter *L* as the means of identification.

Sizes of Threading Dies. The size of the threading die and the number of threads per inch which the die will cut are stamped on the die. The same system of numbering is used in designating the thread as that used for threads of taps.

Cutting a Thread with a Die. The procedure for cutting threads by hand is as follows: Fasten the workpiece securely in a vise. Select the proper die for the job, as indicated on the blueprint or specification sheet. Place the die firmly in the die holder.

After placing the die in the die holder, drop some cutting oil on the chamfered end. Place the chamfered side of the die squarely on the end of the workpiece. Then start turning the die slowly but firmly until the thread takes hold. After several turns, stop and see if the die is square with the work.

If necessary, adjust the die to cut the thread to the proper depth. Turn the die back frequently to clear away the chips. During the cutting, use plenty of cutting oil, *except on cast iron.* Oil is not used on cast iron.

Some die stocks or die holders are provided with a guide line. This line must correspond with the axis of the bolt being threaded. Some die stocks do not have a guide, and unless great care and skill are used, the thread on the bolt will be cut too deep on one side and not deep enough on the opposite side. A thread having this fault is known as a "drunken thread."

Cutting Pipe Threads. If you examine a piece of threaded pipe, you will see that the thread is rounded slightly at top and bottom. The threaded part of the pipe is tapered three-quarters of an inch per foot. The smaller dies are set in a holder, and the die stocks are provided with a ring that fits over the pipe and holds the die square. This prevents the threads from being cut at an angle to the axis of the pipe.

Threading Lubricants. When threading steel with a die, use lard oil, but use only enough to keep the work moist. Oil that runs out of the die is wasted. A mixture of white lead and cutting oil also makes a good lubricant.

Hand Reamers

The hand reamer, Fig. 31, is a finishing tool for drilled holes. The reamer is ground straight for nearly the full length of the teeth. For a distance about equal to its diameter, it is tapered to enable it to enter the hole before starting to cut, and to prevent chattering. The hand reamer is used when not more than 0.003 to 0.005 of an inch is to be removed in the drilled or bored hole.

The hand reamer should never be operated by mechanical power. The end of the shank is machined square to fit a tap wrench. It is easy to tell a hand reamer from a machine reamer, as the machine reamer has a tapered shank with a tang or a straight shank.

Hand reamers have straight or spiral flutes. The advantage of the spiral-fluted reamer is that the tool will not chatter. A reamer should always be checked for size with a micrometer.

Hand Hack Saws

The sawing of metal is one of the most common operations performed in the shop. Because the hack saw is used extensively by both skilled and unskilled mechanics, it is more or less taken for granted. It does, however, require proper care and

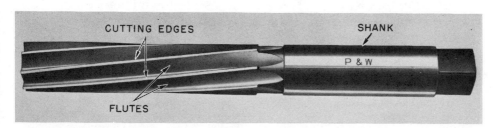

Fig. 31. The hand reamer is used to finish drilled holes. (Pratt & Whitney Div., Niles-Bement-Pond Co.)

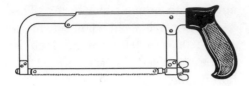

Fig. 32. The hand hack saw has an adjustable steel frame and pistol grip.

use. Since different kinds of blades can be mounted in the frame quickly and easily, the hand hack saw cuts different kinds and shapes of material. The saw shown in Fig. 32 consists of an adjustable steel frame with pistol grip. Hack saws having solid frames are also available. In both types of frame, the blade is secured by pins or hooks and is drawn taut by means of wing nuts

or by turning the handle. Eight-, ten-, or twelve-inch blades can be mounted in the frame.

Hack Saw Blades. Hack saw blades are made from high-grade tool steel that has been hardened or tempered. Since the blades are hard, they are also brittle.

The hack saw blade in general use is ½ inch wide, .025 inch thick, and 8 to 12 inches long. The length of a blade is the distance between the centers of the holes in the ends of the blade. Hand blades are made in four different tooth spacings for different jobs.

The number of teeth per inch is called the *pitch*. Standard pitches are 14, 18, 24, and 32 teeth per linear inch. *Blades should be chosen*

TABLE I   PITCH FOR VARIOUS MATERIALS

STOCK TO BE CUT	PITCH OF BLADE (TEETH PER INCH)	EXPLANATION
MACHINE STEEL COLD ROLLED STEEL STRUCTURAL STEEL	14......	THE COARSE PITCH MAKES SAW FREE AND FAST CUTTING.
ALUMINUM BABBITT TOOL STEEL HIGH SPEED STEEL CAST IRON	18......	RECOMMENDED FOR GENERAL USE.
TUBING TIN BRASS COPPER CHANNEL IRON SHEET METAL (OVER 18 GAGE)	24......	THIN STOCK WILL TEAR AND STRIP TEETH ON A BLADE OF COARSER PITCH.
SMALL TUBING CONDUIT SHEET METAL (LESS 18 GAGE)	32......	

according to the metal to be sawed. See Table I. A course blade, such as a 14-pitch blade, will cut fast and free, but will leave a rough, jagged finish.

An 18-pitch blade (18 teeth per linear inch) is common around the shop. It can be used for general work, such as cutting tool steels, iron pipe, and light angle iron. A blade having 24 teeth per inch is used to cut thin sheet metal (over 18 gage) and to cut tubing and drill rod.

A 32-tooth per inch blade is used on small tubing, conduit, and sheet metal under 18 gage. Sometimes for very thin metal, two blades can be used together, with teeth pointed in opposite directions.

Hack Saw Blade Sets. All hack saw blades have the teeth set to give a saw kerf slightly wider than the thickness of the blade. This prevents the blade from pinching in the cut. On a blade with coarse teeth, you can see that one tooth is bent over to the right and the next bent to the left. This setting is referred to as *regular-alternate*. On blades having fine teeth (24 and 32 teeth per inch), the teeth are set over in pairs, first to the right and then to the left. This setting is specified as *double-alternate* setting. On some blades, groups of teeth rather than individual teeth are set to prevent pinching in the cut. Teeth set in this manner are *wave set* teeth.

Wave set teeth are especially adapted to cutting thin materials, as they do not bind.

Mounting the Blade. After the proper blade is selected, it is mounted in the hand hack saw frame by hooking the blade on the pins at the ends of the frame. The teeth should point away from the handle. Thus, the saw cuts only on the forward stroke. The blade is then stretched and tightened by turning the wing nuts or the handle.

Do not draw the blade too tight. Under severe tension the frame might spring or the pins which secure the blade might break off. On the other hand, a loose blade may buckle or break. After one or two cuts have been taken with a new blade (particularly when a flexible-back blade is used), tighten the blade slightly.

Using a New Blade. Sometimes it is necessary to mount a new saw blade part way through a job. Take care not to force the blade into the cut, as the set of the new blade is greater than that of the old. Use very light pressure until the saw reaches the bottom of the cut.

Holding the Work. The work to be sawed should be held securely in the bench vise. The piece should not vibrate or shift while being sawed.

A wooden support block is sometimes placed under the work to keep it from moving down in the vise under pressure of the saw. When sawing hollow tubing, do not tighten the vise too much.

Starting the Cut. Apply the blade to the work to engage as many teeth as possible at one time, Fig. 33, and use a short stroke in the beginning. When sawing work in a vise,

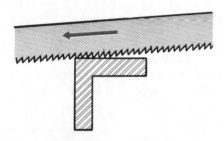

Fig. 33. When starting the cut apply the blade to the work to engage as many teeth as possible.

the cut should be made about ¼ of an inch from the jaws of the vise to prevent the work from springing. *Best results are obtained with a steady, forward stroke.*

Suggestions on Use of the Hand Hack Saw. Select a blade suited to the material to be cut and mount it in the saw frame. Then mount the work in the vise. Now grasp the handle of the saw in one hand and hold the front end of the saw frame in the other hand to guide the saw. Keep sufficient pressure on the blade on the forward stroke to produce a cut, and let up on the back stroke.

Take long, steady strokes. Avoid short strokes which wear a part of the blade. Start the cut easily, par-

ticularly if you must begin on a sharp corner. Do not let the blade slide over the work. Teeth that scrape, but do not cut, soon become dull. Keep the cut straight.

Do not twist the blade in the work after the cut has been started, for the blade may break. Four common causes of blade breakage are: (1) using too coarse a blade on thin

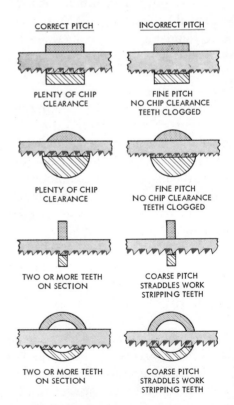

Fig. 34. Some examples of correct and incorrect pitch are shown.

stock; (2) trying to straighten a cut that has angled; (3) exerting too much pressure; and (4) insecure clamping of the work.

Fine-tooth saws should not be used on soft materials. Soft materials make large chips which clog fine teeth and cause them to bind. See Fig. 34 for examples of correct and incorrect pitch.

Layout Tools and Layouts

In machine shop work, *layout* means the marking or scribing of lines on a piece of metal to indicate in full-scale the area to be machined. Before drilling with the drill press, for example, it is necessary to mark the place where the hole will be drilled unless, of course, a drill jig is used.

Before mounting pieces in a lathe between centers, the center holes need to be located by the layout process. Before taking a cut with hand tools or machine tools, it is often necessary to determine carefully and accurately the areas to be cut away.

Use of Blueprints

The source from which the machinist gets his information before scribing lines on the metal is generally a black-and-white print or a blueprint. The layout dimensions for the workpiece are taken from the blueprint and from specifications covering the job. The dimensions given on the blueprint represent the actual size of the finished workpiece.

Use of Coating Materials

To make the lines scribed on the metal stand out clearly, a colored coating is applied to the surface of the workpiece. Thus, the form outlined by the scribed lines stands in sharp contrast to the background. The solution used will depend on the kind of metal the workpiece is. The coating material should not wash or rub off.

A coating commonly employed is called *layout blue* or *layout dope*. This material is inexpensive, dries fast, and produces clear-cut layout lines. Several different commercially prepared solutions of this kind are available.

For laying out large-size castings or a large number of small pieces, an inexpensive coating can be made by mixing alabastine or whiting (powders) with water or alcohol.

A copper sulphate solution ($CuSO_4$), called *blue vitriol*, has for many years been used as a coating for iron and steel. The solution is applied like paint to the surface to be scribed and produces a thin copper plating. Before applying the solution, be sure that the surface is clean.

Metals such as aluminum and copper can be easily coated with vermilion and shellac thinned with alcohol. This treatment produces a surface upon which the layout lines can be scribed without danger of scratching the underlying body of metal. To clean the coating substance from the metal, use alcohol as a wash.

Layout or Surface Plates

To insure accuracy of layout work, plates on which the work can be laid out are used. Layout plates or tables, Fig. 35, are generally made of cast iron or granite. Black granite is not affected by temperature change and remains perfectly flat.

Surface plates are used frequently in making layouts and measurements, and are essential inspection tools. Some surface plates are scraped and accurate to millionths of an inch. These plates must be checked frequently for flatness where *extreme* accuracy is important.

Surface plates made of stress-relieved cast iron have a serious limitation—they are easily nicked, and

Fig. 35. Surface plates are used in laying out precision work. (Collins Microflat Co.)

burrs are raised that damage bases of height gages. Every precaution should therefore be taken when laying out workpieces to prevent dropping them on the surface of the plate.

Granite surface plates, Fig. 35, are gradually replacing cast-iron plates. The main advantages are that granite plates have less shrinkage than cast-iron, and they will not burr when workpieces are accidentally dropped.

The work is positioned directly on the surface of the plate or it may be supported by parallels or V-blocks, Fig. 36. Small jacks or shims which permit fine adjustments are used to support some parts of a casting. In some cases, it is convenient to clamp the work to knee or angle plates. The holding device is then placed upon the layout plate and adjusted to obtain proper positioning of the workpiece.

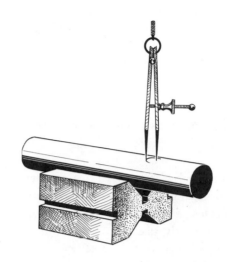

Fig. 36. Work can be supported on V-blocks.

Layout Tools and Their Uses

The layout tools most commonly used are the steel rule, scriber, dividers, prick punch, center punch, square, combination set, surface gage, and surface plate.

Scriber. The scriber, Fig. 37, is

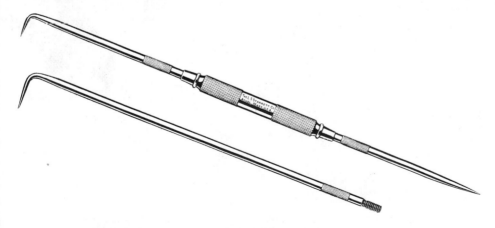

Fig. 37. Lines are drawn with a scriber in laying out work. (The L. S. Starrett Co.)

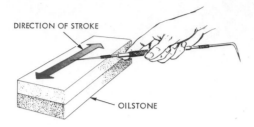

DIRECTION OF STROKE

OILSTONE

Fig. 38. An oilstone is used to sharpen the point of a scriber.

the tool with which lines are drawn in laying out work. Of the many styles available, the scriber most frequently used consists of a tool-steel rod about 8 or 10 inches long and $\frac{3}{16}$ of an inch in diameter, the ends of which form a long, hard point. One end is usually bent at right angles to the shank. In laying out work, this end is used for reaching into places not readily accessible. To keep the point of the scriber sharp, it is sharpened as shown in Fig. 38.

Scribing the Line. A line can be scribed on the metal surface by drawing the scriber along the edge of the rule at a slight angle, as you would a pencil. See Fig. 39. The point of the scriber must be kept tight against the guiding edge of the rule.

The process of laying out calls for extreme accuracy and demands intelligent and careful planning. The surface to be machined is shown by a straight line or a curved line. You can readily see that a double or indistinct line would not be helpful. More distinct lines will be made if the point of the scriber is kept sharp.

Prick Punch. The prick punch is made of tool steel and has a hard, conical point with an included angle of about 30 degrees, Fig. 40, *top*. It is used for making small indentations at intervals on a line or at the point where two layout lines will

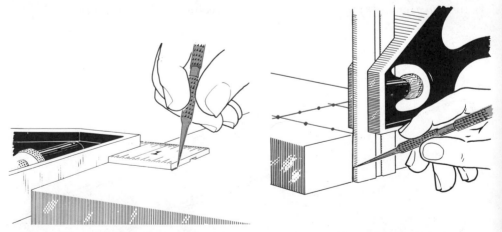

Fig. 39. When scribing lines the scriber must be kept tight against the edge. The combination square is used in scribing parallel lines (*left*) and for scribing lines at right angles (*right*).

intersect. In these indentations, the points of the measuring instrument —dividers, for example—are placed. The prick punch is not as heavy as a center punch, and its point is shaped differently. See Fig. 40, *bottom*.

Setting Over an Incorrect Prick Punch Mark. If the prick punch mark spotting the hole is made at a point slightly removed from the hole's true location, it will be necessary to correct the fault. With a center punch and a ball-peen hammer, punch the hole toward the more accurate center point.

Center Punch. When the true location of the hole is established, a deep center punch mark is made at the correct center location. If drilling is to be done, this center punch mark will help to start the drill at the true center. The following rules should be kept in mind when using the prick punch or the center punch:

1. Make sure that the point of the punch is sharp.

2. Hold the punch in one hand, with the point on the scribed line at the place where it is desired to make the indentation.

3. With the punch in a vertical position, tap it gently with a light hammer.

Automatic Center Punch. Some center punches are designed so that a hammer is not needed to make the center punch hole. A spring and movable weight inside the handle can be adjusted so that when the top of the punch is depressed, the weight trips and hits the punch proper. By varying the spring tension, it is possible to make light or deep impressions with the punch point.

Dividers. Dividers are commonly used for transferring measurements, comparing distances, and scribing circles. They usually vary in length from $2\frac{1}{2}$ inches to 10 inches. For a spread exceeding 10 inches, trammel points offer greater convenience.

In scribing an arc or a circle, grasp the top of the dividers between the thumb and the first finger of one hand. Then place the point

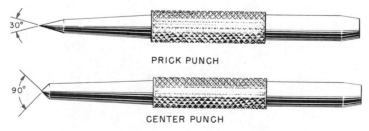

PRICK PUNCH

CENTER PUNCH

Fig. 40. The prick punch (*top*) is used for making small indentations in which measuring instruments may be placed. The center punch (*bottom*) is used to help start a drilling hole. Note the difference in point angles of the two tools.

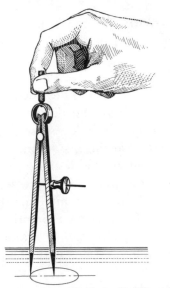

Fig. 41. The correct method of holding dividers to scribe a circle is shown.

of one leg on the prick punch mark. Scribe a circle by swinging the free leg of the divider in a circular movement. See Fig. 41.

Steel Beam Trammels. In laying out large work, the steel beam trammel, Fig. 42, will be needed. This instrument permits the scribing of large circles.

Trammels are not difficult to use. The points are mounted on a round steel shaft (or beam) having a flat side for alignment of the pointers along the beam. A steel rule is used to set the points to the desired measurement. Trammels come equipped with interchangeable points. They are used not only to scribe large circles, as in Fig. 43,

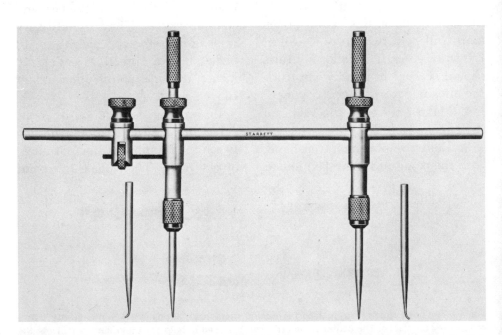

Fig. 42. Steel beam trammels are used to scribe large arcs and circles. (The L. S. Starrett Co.)

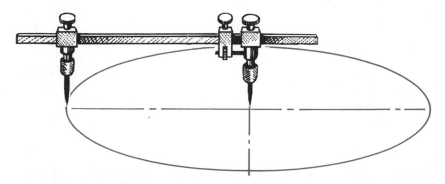

Fig. 43. Steel beam trammels can be used to scribe large circles as shown here, but can also be used to locate points along a line.

but also for locating different dimensions in a straight line. Moreover, they can be used as either inside or outside calipers, simply by removing the standard trammel points and inserting curved caliper legs in their place. See Fig. 42.

Keyseat Rule. For drawing lines and laying off distances parallel to the axes of round shafts, a combination of two straightedges, or a straightedge and a rule, may be used. Fig. 44 shows this tool being used. This instrument is called a keyseat rule because its chief use is to lay out keyways on round shafts.

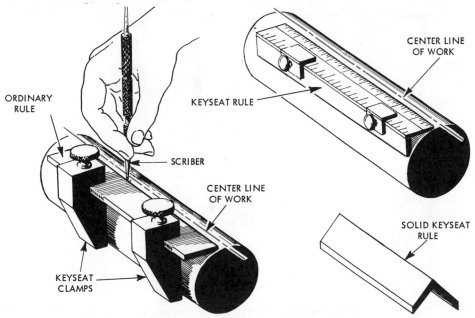

Fig. 44. Three types of keyseat rules are shown.

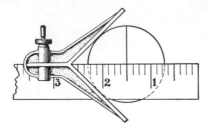

Fig. 45. The center head and rule of the combination set can be used to locate the center of a cylindrical workpiece.

The machinist should know how to use this rule.

Center Head and Steel Rule. The center head and steel rule of a combination square set is another important layout tool. See Fig. 45. It is useful in locating the center of the cylindrical piece of work. To do this, hold the head firmly against the work. Then by drawing a scriber along the blade of the square, scribe a line across the end of the stock. In the same way, scribe a second line at right angles (90 degrees) to the first. Turn the piece and scribe two more lines to give a total of four. This last step is especially important if the stock is out of round. The intersection of the four scribed lines will be the center of the work. Mark the center with a prick punch. The center head method can also be applied to square and octagonal stock.

Hermaphrodite Caliper. The hermaphrodite caliper is a caliper

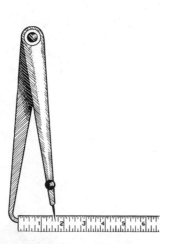

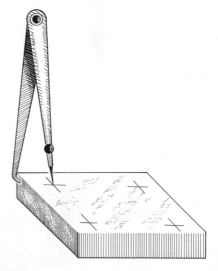

Fig. 46. The hermaphrodite caliper can be set to a certain dimension (*left*) which is then transferred to the workpiece to locate centers of corner holes (*right*).

having one curved leg and one straight leg, Fig. 46. The straight leg of the caliper has a point so it can be used as a scribe. Calipers of this kind are used to locate centers of cylindrical work, to scribe lines on a block parallel to its side (Fig. 46, *right*), and other applications.

To set the tool, the hooked leg is set on the end of the scale and the leg containing the scriber point is adjusted to the desired graduation on the scale. See Fig. 46, *left*.

To locate the center of a shaft or other cylindrical stock, the caliper is spread a distance slightly less than the radius of the stock, Fig. 47.

With the end of the stock already coated with chalk or layout ink, hold the hooked leg against the round of the stock very close to the coated end and scribe an arc on the chalked surface across the end. Now move the hooked end around the stock a quarter of the circumference and scribe another arc. This proce-

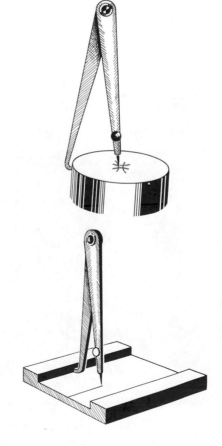

Fig. 47. The hermaphrodite caliper can be used to locate the center of a cylindrical workpiece as shown.

Fig. 48. The hermaphrodite caliper can be used to scribe a line parallel to an inside shoulder.

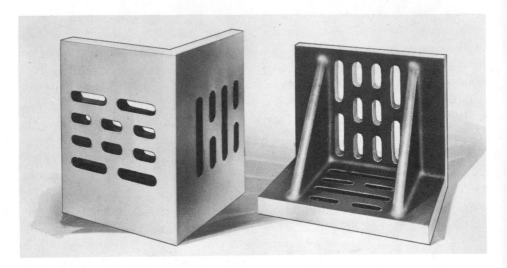

Fig. 49. Angle plates are used to hold work for layout, machining or inspection. (Taft-Peirce Mfg. Co.)

Fig. 50. A surface gage is being used here to scribe a line.

dure is repeated twice more to obtain a total of four arcs. Inside the arcs is the true center. This method is accurate enough for centering lathe work. Fig. 48 shows a hermaphrodite caliper being used to scribe a line parallel to an inside shoulder.

Angle Plate. The angle plate, Fig. 49, is a cast-iron or forged piece. Both surfaces are positioned to form an angle, usually a right angle. The work is clamped to the angle plate for laying out and for inspection. Some angle plates are slotted for bolts.

Surface Gage. The surface gage,

Fig. 50, is made to scribe lines in layout work. It is used on the flat surface of a surface plate. Fig. 51 shows a surface gage in use. The base of the gage has a V-slot for use on cylindrical work. See Fig. 51. By means of pins in the base, the surface gage can be used against one edge of a work surface or in a slot, as on a milling machine table slot.

Fine adjustment is made by turning the thumbscrew at one end of the rocker after the spindle has been clamped in its approximate position. This screw works against a stiff spring at the other end of the rocker

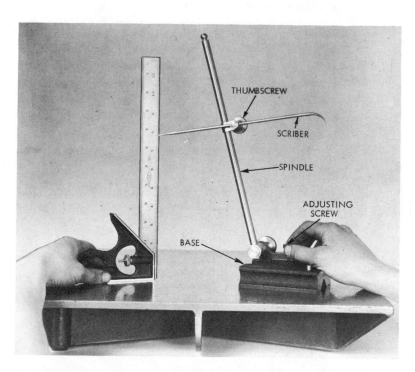

Fig. 51. The surface gage is used in scribing lines and in inspection. The height of the scriber is shown being checked with a combination square. (Brown & Sharpe Mfg. Co.)

and permits an accurate adjustment.

The scriber point with which the line of the work is scribed is usually set to the correct position with a combination square, Fig. 51. The setting is considered to be correct only within scale tolerance.

Location of Holes To Be Drilled

Careful layout is required in locating holes to be drilled without a drill jig. The hole must be drilled in the proper place if parts are to fit together as intended. Different means are used for locating the drill hole, depending upon the location of the hole in relation to other holes or to certain surfaces. It will be as-

sumed in this discussion that the surface has been properly prepared and coated.

Center lines on a blueprint are frequently dimensioned and located from two adjacent edges, and the same method is followed in layout, Fig. 52. It is important that these edges be square and smooth or machined accurately first.

Using edges A and B, Fig. 52, two lines can be drawn that will intersect at right angles and represent the two lines on the blueprint which cross at the center of the hole. For this, the square head and steel rule should be used. The hole will be located at the point where the two lines intersect.

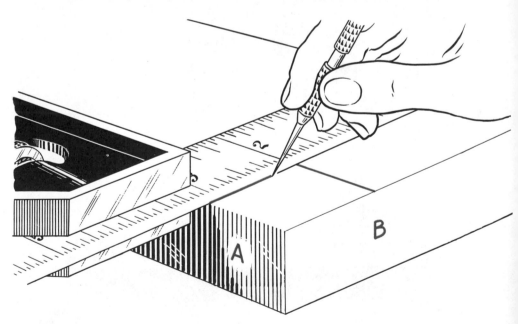

Fig. 52. The center of a hole can be located by using the combination square to develop two intersecting lines.

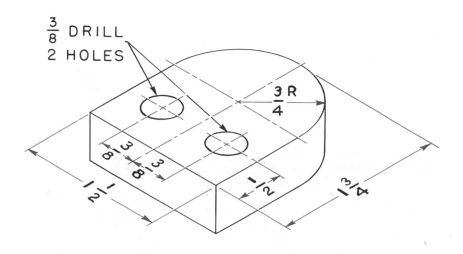

Fig. 53. Center lines are often used to locate hole centers.

Next, a small indentation is made at the intersection by using a prick punch and striking it lightly with a hammer. The prick punch mark should be checked to see if it is at the exact point of intersection. If the mark is accurate, a center punch is used to deepen the indentation to seat the drill at the start.

Frequently two holes must be laid out accurately in relation to one another. Having located one hole by the method just described, the other hole can be located in relation to the first by measuring carefully from the center line and layout lines of the first hole. This measuring must be accurate. For this work, the spring dividers can be set with close accuracy and the measurements transferred from the dividers to the workpiece.

On some jobs, once a center line is established for the part, holes will be laid out in both directions from this line. See Fig. 53. Again, there are times when it becomes necessary to work from one finished side or one finished shoulder.

Diagonal-Line Method of Locating a Center. When locating the center of a piece of square or rectangular stock, scribe diagonal lines from corner to corner. The intersection of the two lines gives the center. See Fig. 54. The center can be marked by a light prick punch mark at the point of intersection.

Dividers Method of Locating a Center. The dividers are set for slightly more than one-half the diameter of the cylindrical piece. The workpiece is laid on its side on the layout plate. One leg of the dividers

87

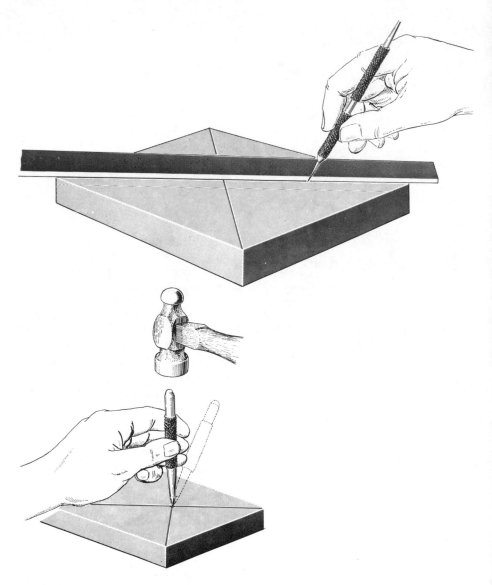

Fig. 54. *Top:* Lines are scribed between opposite corners. *Bottom:* A prick punch mark at the intersection of the two lines accurately spots the center of the workpiece.

rests on the layout plate and the other scribes a line across the end of the workpiece. The workpiece is revolved a quarter of a turn, and another line is scribed. This is re-peated until four lines have been drawn, making a small square on the end of the cylinder. The true center is situated in the square and is marked with a prick punch.

Safety Precautions—Hand Tools

Even though one might think that no injury would come from using small hand tools, a great many injuries have occurred from improper operation of them. The many safety rules needed to cover the great variety of tools presented in this chapter would take considerable space. The procedure discussed in the preceding pages is the recommended procedure and if carefully followed, no injury will occur.

Because of the great danger to the workman's eyes in performing almost any operation, goggles should be worn when doing any work in the shop. The following precautions are also given when using tools.

1. All files should be fitted with a suitable handle on the tang.

2. When using the chisel for chipping, keep flying chips from hitting people working in front of you.

3. When using a wrench, be sure it fits the nut closely. A loose-fitting wrench will round the corners of the bolt or nut and slip, resulting in hand injuries.

Courtesy of the National Safety Council

NOTE: Please see review questions at end of book.

<table>
<tr><td>

Chapter

4

</td><td>

Power Saws:

Power Hack Saws, Band Machines

</td></tr>
</table>

For a great many years the only way to saw off metal was to use a hand saw. The development of power-driven machines for driving metal-cutting saw blades has made the task considerably easier. Power saws can do the work more rapidly and with greater accuracy. Today, the power saw is an indispensable piece of production equipment, employed in a wide range of high-production cutting-off and cut-to-length operations. Recent improvements in efficiency and productivity for one saw in particular, the vertical band machine, have prompted many to speak of it as the eighth basic machine tool.

The Power Hack Saw

The power hack saw, Fig. 1, is used for cutting a wide range of ferrous and nonferrous bar stock. The blade is secured in a frame which is attached to a drive wheel. The blade undergoes a reciprocating motion, cutting on the back or return part of the stroke. Power hack saws are made in many styles and sizes and, with suitable stock feeding and clamping mechanisms, can be

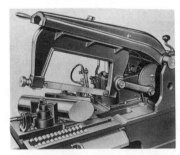

Fig. 1. Power hack saws cut metal more accurately and efficiently than hand saws. (Racine Hydraulics & Machinery Co.)

adapted to semi- and fully-automatic operation.

Power Hack Saw Capacities. The capacity of power hack saws ranges from 3″ x 3″ square or round stock for a small machine, to 16″ x 16″ square or round stock for a large machine. A machine of 3″ x 3″ capacity will saw stock as small as ⅛″ square or round, and as large as 3″ square or round.

Band Machines

The band machine is one of the most versatile machine tools. Essentially a flexible saw blade driven in a continuous loop around two wheels —like a fanbelt in an automobile engine—the band machine provides a continuous cutting action that slices through all materials from Asbestos to Zinc — even the difficult-to-cut metals.

Types of Band Machines

Band machines are available in horizontal models, Fig. 2, and vertical models, Fig. 3. Shaping, slotting and cut-off operations can be performed on both types. Both horizontal and vertical saws are available in a wide range of sizes, with fixed or power-driven work tables. Saws with power tables are particularly useful when producing repetitive parts and are commonplace in high-production operations.

The Vertical Band Machine

While the horizontal band machine is used mainly for cut-off and cut-to-length operations, the vertical band machine, frequently referred to as the *contour saw*, is used for a much wider range of sawing operations, including cutting metal to any desired contour. Holes of any size and shape can also be cut from the center of the workpiece.

Many three-dimensional shapes can be more easily produced on a contour saw than on other types of machine tools, Fig. 4. On some machine tools, for example, shapes with radii and comparable angles must be produced by setting up a series of steps on the surface of the work. By

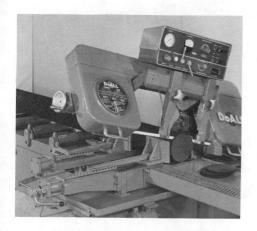

Fig. 2. The horizontal metal-cutting band saw cuts in a clockwise motion; cutting action is visible from all angles. (The DoALL Co.)

contrast, irregular surfaces and angles can be cut directly on the contour machine.

One of the main advantages of the contour saw is that the workpiece can be cut directly to shape. An economical machining process, "waste" material is removed in whole, reusable sections rather than as chips. The band machine is being used increasingly for shaping, slotting and other metal cutting operations that were formerly performed on the shaper.

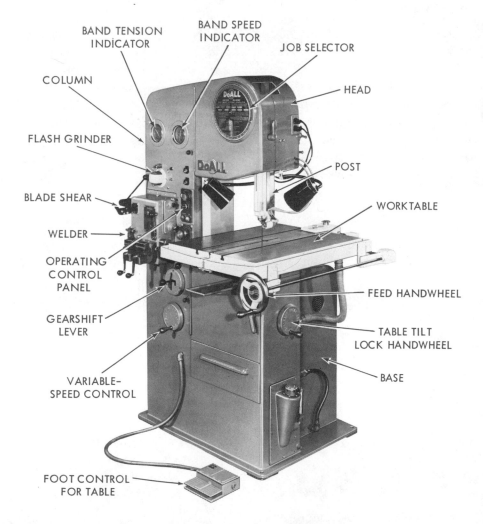

Fig. 3. The vertical band saw, referred to as the contour saw, performs many metal removal jobs formerly done on other machine tools. Principal parts are identified. (The DoALL Co.)

Fig. 4. Many three-dimensional parts can be cut directly on the vertical band machine. (The DoALL Co.)

And by replacing the saw band with a filing or abrasive band, the contour saw can also be used for filing, polishing and abrasive cutting of even difficult-to-work materials.

Principal Elements of the Vertical Band Machine

The vertical band machine is available in a wide range of sizes and modifications. Size of the band machine is measured by its throat width —the clearance between the column and the saw blade — which in turn determines the size of the workpiece the machine can handle. While they may vary with respect to power options and controls, contour machines have the following basic elements:

Base. The base of the contour machine, Fig. 3, supports the column and houses the drive for the lower band wheel, the band speed adjusting control, gearshift, and main power elements. On machines with power-fed work tables and circulating coolant systems, the base will also house the hydraulic actuating system.

Column. The column, Fig. 3, encloses the return route for the saw band and serves as a mounting point for the operating controls and dials and the band welder, shear and grinder.

Head. The head of the contour machine houses the upper band wheel which can be adjusted for saw band alignment. The post, which extends from the head to guide and support the saw band, is adjustable for various workpiece heights.

Band Guides and Supports. Driven by the lower carrier wheel, the saw band rotates clockwise from the upper carrier wheel through the post blade guard, upper blade guides, work table saw kerf, and the lower guides. This provides a straight-line direction for the saw band.

The saw band guides, which are discussed in greater detail in a later section, guide the saw band, resist the pressure applied when feeding the work into the band, and prevent the band from deflecting.

Worktable. The worktable, Fig. 3, is mounted above the power transmission with the table saw kerf (slot) in line with the saw band. Primary function of the table is to provide a convenient work space for feeding the workpiece into the saw band while following the layout for the shape to be cut.

On fixed-table machines, the workpiece is fed into the saw band by hand; on machines equipped with hydraulic or pneumatic table feeds, the workpiece is power fed into the saw band. T-slots machined into the worktable surface provide for work guides or clamps. Power-fed tables are also equipped with table stops for pre-setting the desired length of cut.

The worktable of the contour machine shown in Fig. 3 can be tilted up or down for angled cuts. Its power feed is activated by foot control. The handwheel mounted at the lower right of the worktable rotates the workpiece on the table by means of a chain and sprocket linkage. The workpiece can be "steered" through contour cuts as it is power fed into the saw band.

Butt Welder

In day-to-day operation, the saw band must frequently be cut, then rejoined by welding. The butt welding system built into the column of the machine is therefore essential for rejoining blades for internal sawing. The elements of the butt welding system are shown in Fig. 5.

Band Shear. The band shear, Fig. 5, is used to cut the saw band when it must be threaded through a workpiece for internal sawing operations. Although the shear provides a square cut, the ends of the band are usually dressed on the grinder before they are rejoined by welding.

Welding Unit. A pair of welder jaws supports the blade ends as they are rejoined by welding, then annealed. The stationary jaw gives the blade support; the movable jaw

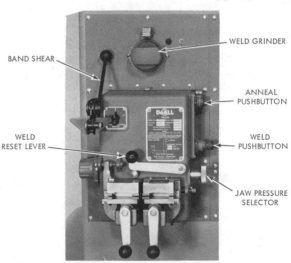

Fig. 5. The butt welding assembly used to join the saw band. Principal parts are identified. (The DoALL Co.)

moves approximately .040″ during the welding operation, pressing the molten blade ends together. On some machines a welder operating lever is thrown which both activates the welder and the movable jaw. For the machine shown in Fig. 5, a pushbutton is used to activate the welder and movable jaw.

Jaw Pressure Selector. The jaw pressure selector, Fig. 5, regulates the tension of the movable jaw according to the width of the saw band being welded. This dial is mechanical and enables the operator to control the pressure with which the movable jaw advances toward the stationary jaw during the band welding operation. Wide bands require greater tension than narrow bands. Too much tension on narrow bands will cause climbing or lapping.

Line Voltage Regulator. The line voltage regulator controls the amount of heat generated at the weld. This switch regulates the low-voltage, high-amperage current through the blade clamping jaws from 2½ volts at the "less" position to 3¼ volts at the "more" position, and compensates for variations in the regular 220 line voltage. The regulator setting is adjusted by screwdriver but, for shops with a fairly constant line voltage, the regulator setting need not be changed once it's properly set.

Annealing Switch. After the blade is welded, it must be annealed to soften the otherwise brittle joint. The annealing switch is a pushbutton control which is pushed into a hard stop and held in during annealing, or until the weld reaches a dull cherry color.

Weld Grinder. Above the welder jaws is the grinder, Fig. 5, which is used to grind the ends of the saw band before welding, and to grind off the excess metal after the weld. Above the grinding wheel guard is a gage to test the saw thickness after grinding. When properly ground the area of the weld will pass freely through the gage.

The Saw Band

The saw band is literally an endless band of single-point cutting tools, and is installed as a continuous loop 125″ long or longer depending on the size of the machine. The saw band cuts continuously and fast, and wear is spread over all of the teeth to extend tool life.

Saw blades are available in a variety of widths, alloys and tool geometries for the purpose of improving the machinability of the material being cut. Considerable research has gone into improving the materials and cutting tool geometry of the saw band, extending the life of the saw band and making possible cleaner cuts and better surface finishes. Cutting band features and characteristics are defined and illustrated in Table I.

TABLE 1. COMMON BAND TOOL TERMS AND DEFINITIONS

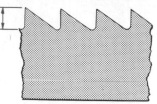

TEETH: Protuberances on one edge of the band which do actual cutting.

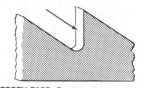

TOOTH FACE: Tooth surface on which the chip impinges as it's cut away from work.

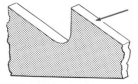

TOOTH BACK: Surface opposite the tooth face.

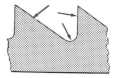

TOOTH GULLET: Area between teeth; acts to remove chips from cut.

SET: Bend given teeth to create side clearance for back of band.

WAVE SET PATTERN: One group of teeth to right, next group to left, etc.

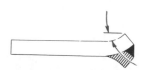

STRAIGHT SET PATTERN: Teeth set alternately one to right, one to left, etc.

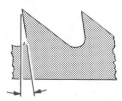

TOOTH SIDE CLEARANCE ANGLE: Degree of bend of each set tooth. Angle depends on saw band pitch, kerf size desired.

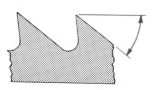

TOOTH RAKE ANGLE: Angle of tooth face measured from line perpendicular to back edge of saw band.

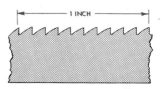

TOOTH BACK CLEARANCE ANGLE: Angle of tooth back measured from line parallel to back edge of saw band.

PITCH: No. teeth per inch. Generally the greater the work thickness, the fewer the no. of teeth per inch needed.

GAGE: Thickness of band back, usually measured in thousandths of an inch.

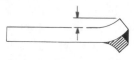

SIDE CLEARANCE: Difference in dimension between tooth set and side of band tool; provides space on sides of band back to enable maneuvering work, minimizes transfer of heat to work, and prevents leading-off when making straight cuts.

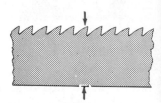

WIDTH: Is measured from tooth tip to back edge of band.

SWAGED TOOTH: Type of set common to wide bands and circular saws to create side clearance.

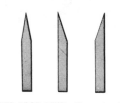

KNIFE EDGE BEVEL: Type of cutting edge on knife-edge bands.

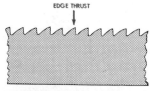

BEAM STRENGTH: Resistance to band back deflection due to edge thrust or feeding pressure.

Determining Saw Band Length. To determine the proper saw band length for installation on any two-wheel band machine, measure the distance between wheel centers, multiply by two, then add the circumference of one wheel to that figure. This gives the total band length. If the band length measurement is calculated when the wheels are at their greatest distance apart, deduct a fraction of an inch to allow for stretching when the saw band is placed under tension.

Joining the Saw Band. Cut the band tool to the proper length for the machine (allow ¼″ for stretching). Use the band shear to obtain ends that are square and flat.

The squarely trimmed ends are then wiped clean and clamped in the welder jaws. Before clamping the jaws, make sure the jaw contact surfaces are in alignment and clean and free from pits. Also make sure that the front and back edges of the band ends are in proper alignment once secured in the welder jaws. The weld is then made by pressing the "Weld" pushbutton. Note in Fig. 5 that the weld reset lever must be moved to the down position for welding.

After welding, the saw band must be annealed at the weld, otherwise the brittle joint will break when the band is flexed. To anneal the weld, release the clamped saw blade sufficiently to center the welded area between the jaws. Then reclamp the band and press and hold in the "Anneal" pushbutton.

Time and temperature for annealing must be carefully controlled. Hold in the button only long enough for the weld to reach a dull cherry red color. Then jog the button, gradually increasing the time between jogs, to allow the temperature to drop off slowly and gradually over a period of approximately 5-10 seconds. Exact time will depend on the metallurgical characteristics of the saw band and its width. The technique is easily acquired after a few practice welds.

After annealing, remove the saw band from the welder jaws, then grind the bead or flash from the weld. Care should be taken not to undercut the welded area during grinding. The welded area after grinding should be the same thickness as the rest of the band. This is checked by passing the welded area through the band thickness gage near the grinder. Make certain that the back edge of the band is burr-free and in proper alignment.

The welded area will be shiny after grinding. To eliminate hardening stresses which may have developed during grinding and to blue that area, place the band back in the welder jaws and re-anneal. The saw band is now ready for use.

Installing the Saw Band. Make sure that the teeth are pointed downward in the direction of band travel before mounting the band on the machine. If the teeth are pointed in the wrong direction, simply turn the band loop inside out to point the teeth in the right direction for cutting.

When mounting the saw band, it should be placed on the crown of the wheels as close to the center as possible, then place a slight tension on the band and revolve the upper wheel by hand until the band moves into its natural operating position on the wheels. If the band fails to track in the right position, use the upper wheel tilt control to tilt the wheel. As soon as the band tracks properly and is in slight contact with the top and bottom back-up bearings of the band guides, band tension should be increased to the recommended setting for the machine being used.

Band Guide Adjustments. Band guides, mounted on the tool post and just below the worktable, fall into three categories: roller types, insert types, and file band guides. The roller types are generally used for anti-friction high-speed applications; insert types are used for low-speed precision work. Regardless of type, the same basic adjustments are required to guide the band properly.

For slow band speed applications, the inserts should be adjusted to support the width of the band from the gullet area to the back edge, with a total clearance between band and insert guides of no more than .001-.002". Roller guides should be set close to the band with just enough clearance so that they turn freely by hand when the band is at rest.

In all cases, the band should not exert undue pressure on the guide back-up bearings, as determined when the band is moving but not in the process of cutting. There should be only a slight contact between the back of the band and the guide back-up bearings, otherwise the resulting thrust will shorten bearing life and damage the back of the saw band. Under no circumstances should the band be permitted to operate with teeth in contact with the guide inserts as this would quickly ruin the cutting edges of the teeth and damage the inserts.

Use a feeler gage when making guide adjustments, particularly where fine-tolerance sawing is required. A strip of cellophane can be used if a feeler gage is not available. Set the left-hand inserts or rollers to barely touch the band and then tighten. Place the cellophane between the band and the right-hand inserts or rollers, adjust them firmly to the band, and tighten. Then remove the cellophane, which will give a total clearance between band and guide of approximately .001-.002" as previously specified.

For band filing operations, a file guide corresponding to the width of the file band is used in place of the roller or insert guide. File guides serve only to support the back of the file band at the point of work. For band polishing operations, a guide similar to the file guide is used.

Care of Band Guides. Proper maintenance of the guides is essential. Worn guides promote inefficiency in sawing and should be replaced as soon as detected. Have replacement sets on hand for immediate installation when the worn set must be reground or replaced.

Storing the Saw Band. When changing bands, the removed band must be coiled for storage. In order to loop the saw band into a tight coil, hold the band in front of you with the teeth pointing away. Then, with one foot holding the band to the floor, grasp the top of the band with your hands 2-3 ft apart, palms up and wrists firm, and with a forward, inward twisting downward motion the band will take a natural coiled position. The saw band is then ready for storage.

Vertical Band Machining Operations

Band sawing to a straight or gradually curved line, often referred to as two-dimensional sawing, is one of the basic operations performed on the band machine. The job selector dial mounted on the upper column, Fig. 6, indicates the correct saw and file to be used for each of 55 basic materials, along with recommended

Fig. 6. The job selector dial, located on the column of the machine, indicates the correct saw band, band speed and feed for many different materials. (The DoALL Co.)

cutting speeds. After the correct blade is installed on the machine and properly tensioned, the machine is set to the correct cutting speed for the work.

When an as-sawed finish is acceptable, the work is cut so that the kerf just splits the layout line on the waste side of the line, Fig. 7, *left*. If a filed finish is desired, approximately $1/64''$ should be allowed on the waste side of the layout line for removal during the subsequent filing operation, Fig. 7, *right*.

When straight sawing is performed, a constant pressure should be maintained on the workpiece as it is fed into the blade. Permitting the saw to ride without cutting dulls the saw teeth. On band machines with fixed worktables, a push stick should be used when applying pressure to

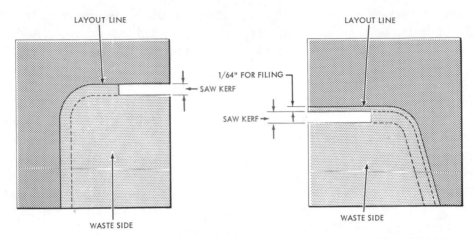

Fig. 7. *Left:* For an as-sawed finish, the saw kerf should just split the layout line. *Right:* Leave $\frac{1}{64}$" to the layout line if the cut is to be filed afterward.

the work to eliminate the danger to the operator's hands while sawing. Danger to the eyes from flying chips can be avoided by adjusting the blower to point away from the operator. Saw band breakage is rare, but should it occur stand clear of the machine and press the stop switch.

Contour Sawing

Contour sawing is sawing to a layout line of regular radius or irregular contour. The term is often used to describe any band machining operation more complicated than a simple straight-line cut.

Cutting Corners. Contour sawing frequently involves cutting away sections of the workpiece, producing one or more corners in the process. The procedure for cutting a corner with a tight radius starts with drilling a hole at the corner tangent to

the layout line, as shown in Fig. 8. The drill should approximate the radius of the layout line.

The drilled hole allows the workpiece to be turned freely at the corner to continue the cut in another direction. As a general rule, holes are

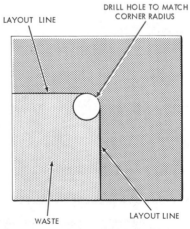

Fig. 8. When cutting corners, first drill a hole tangent to, and roughly the same radius as, the layout line.

drilled wherever sharp corners are to be navigated.

Cutting Square Corners. To cut a square corner, the corner is drilled as described in the above section. After the waste section is cut away, the rounded corner is then notched

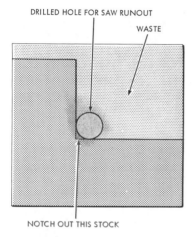

Fig. 9. For square corners, cut off the waste stock then notch the corner square.

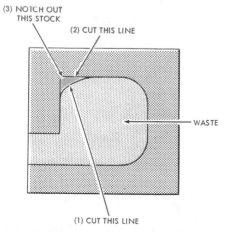

Fig. 10. A square corner can be cut without first drilling a corner hole using the method shown.

square with the saw, as shown in Fig. 9.

Square corners may also be cut *without* drilling. Rounded cuts are made at corners to first remove the waste section, then the corners are notched square as shown in Fig. 10.

The size of the radius to be cut and the thickness of the material are basic considerations in the selection of the saw band for the contour sawing operation. The smaller the radius, the narrower the blade must be to cut the curvature properly. The correct blade is the widest possible blade consistent with the contour involved and the material thickness. This information is given on the job selector dial.

When work is laid out, a coat of layout compound will make the scribed lines visible and easy to follow. Work that is to be machined to a close tolerance should be scribed only once to maintain a sharp layout line. Double lines, caused by scribing over a line, should be avoided. Often, prick-punching the layout lines will aid the operator in the event that the lines are rubbed off.

Internal Sawing

Internal sawing is a special area of contour sawing in which a hole or opening is cut in the workpiece entirely within the boundaries of the external contour. The cut begins not at the edge of the workpiece, as in

Figs. 8, 9 and 10, but at some internal point.

The internal sawing operation begins by drilling a starting hole into the workpiece tangent to the layout line. The diameter of the hole depends on the width of saw used for the sawing operation, since it must be large enough to permit the saw band to be inserted. The saw band is then cut, threaded through the starting hole, and re-welded to position it "inside" the workpiece for the internal cut.

It is customary to start the cut parallel to the layout line as the blade leaves the starting hole. When starting the cut on a curve, a section around the hole curvature is notched out to the layout line, Fig. 11, to permit starting the saw parallel to the line.

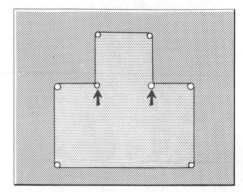

Fig. 12. In general, holes should be drilled at all corners to facilitate turning the work. Note that square corners will be produced at the points indicated simply by turning the work.

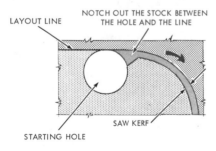

NOTCH OUT THE STOCK BETWEEN THE HOLE AND THE LINE

LAYOUT LINE

STARTING HOLE

SAW KERF

Fig. 11. To begin the internal cut parallel with the layout line, the starting hole should be notched as indicated.

Procedure then follows that for contour sawing: holes are drilled wherever corners are to be cut, Fig. 12. If a corner requires a small radius, the proper size drill to produce

that radius should be used. Corner hole diameters must be at least as large as the saw is wide, but can be larger to meet corner specifications.

Square inside corners are sawed by notching out the drilled hole after the waste slug has been removed. Note that while most of the inside corners in Fig. 12 will have to be notched out, two square corners (*arrows*) can be cut simply by turning the work 90°.

Saw selection for internal sawing is generally consistent with the curves to be cut and thickness of the material. The selector dial on the machine will indicate the proper saw to use.

When a saw band is severed to position it for internal sawing, it's cut on both sides of a previous weld so that the old weld is completely removed. After the sawing operation is completed it is necessary to cut the band again in order to remove the work from the machine.

Stack Sawing

Stack sawing consists of cutting multiple pieces of sheet stock at a time. This method of producing identical parts saves time and money, especially when just a few pieces are needed and the job does not warrant the expense of a blanking die.

Stack sawing is best suited to flat sheet stock. The number of pieces that may be stack sawed at one time

is limited only by the capacity of the machine to saw a solid of equal height. However, the material must be fastened together to prevent movement of one or more of the workpieces during sawing. For large work it is best to place the stock in an arbor press rather than rely on the weight of each piece for the close contact desired. Holes may be drilled in the waste section and the stack bolted together to hold the work. Small stacks for small parts can be placed in a vise and fastened by soldering.

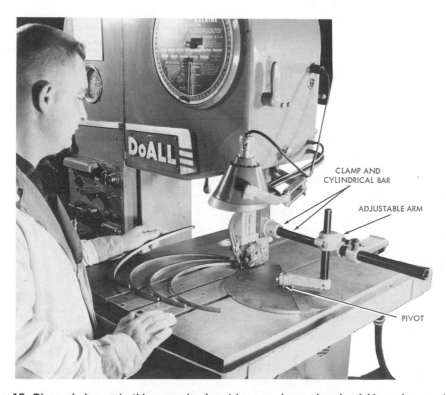

Fig. 13. Discs, circles, or in this case circular strips, can be produced quickly and accurately using the disc cutting attachment. (The DoALL Co.)

The operator must decide from the nature of the work which method is best suited for fastening the pieces together. If bolts are selected, to cite one consideration, their protruding heads may prevent the work from laying flat on the table. (The heads could, however, be countersunk flush with the bottom piece of work to eliminate the problem.)

Disc Sawing

The disc cutting attachment, Fig. 13, consist of three parts: a cylindrical bar which fastens to the saw's guide post, an adjustable arm that slides along the cylindrical bar, and a pivot or centering pin. This attachment permits the cutting of discs or circles, either internal or external. The diameter of the circle that can be cut is limited by the length of the cylindrical bar on the attachment and by the throat depth of the machine.

The discs must be laid out and a center hole drilled with a combination center drill to a depth of $\frac{1}{8}$-$\frac{3}{16}''$ to serve as a pivot point for the centering pin. The work can be fed into the saw by hand or power feed.

Safety Rules — Band Machine

1. Keep worktable clean, especially the T-slots, by removing chips with a brush.

2. After chips have been brushed away, wipe the worktable with an oily rag to prevent rust.

3. Doors housing band wheels should be closed at all times.

4. Always adjust the saw blade guard opening slightly above the metal thickness to be cut, power off on the machine.

5. Safety goggles should be worn at all times when sawing.

6. Short workpieces should be held in a workholding fixture; keep fingers at least 6-8″ away from the cutting band.

7. Use the proper saw blade, blade tension and cutting speed for the material being cut.

8. Provide sufficient working space around the machine: 4 ft at front and back, 3 ft on each side.

9. Machine should never be operated by more than one operator at a time.

10. Follow the builder's operating and maintenance manuals.

11. Replace dull saw bands. They require heavy feed pressures which result in band breakage.

12. Inspect butt welds carefully before mounting saw band on machine.

13. Use the proper saw band width

for the radius to be cut.

14. Wear gloves when handling saw bands.

15. Have cuts and bruises treated immediately.

16. Remove burrs from cut workpieces; they can cause serious cuts.

17. If the machine is equipped with a blower, make sure that it points away from the operator.

18. On machines with power-fed worktables, make sure that all work-holding fixtures are secure and properly positioned.

19. If you are uncertain about the setup or some aspect of the cutting operation, check with your instructor before proceeding.

NOTE: Please see review questions at end of book.

Drill Presses:

Types, Setups,
and Operations

The drill press is one of the most frequently used machine tools. Its principal purpose is the cutting of round holes into or through materials. This machine employs a variety of cutting tools; the twist drill is most common. Holes which do not need to be accurately sized may be drilled without finishing. But when close dimensional tolerances are required, secondary operations are necessary which also can be performed on the drill press. Drilled holes may be finished or modified by reaming, boring, countersinking, counterboring, spot-facing, and tapping.

A machinist must be able to locate holes to be drilled accurately and to use a drill correctly ground for the material he is working with. Successful operation of the drill press not only requires knowledge of various setups, selection of correct speeds and feeds, selection of the most effective coolants, or lubricants, proper drill geometry for machining a variety of work materials, but also a working knowledge of the machine itself.

Drill Press Types and Construction

Basic Types of Drill Presses

Drill presses for performing drilling and allied operations vary in design. There are four basic types: the Sensitive Drill Press, the Vertical Drill Press, the Radial Drill Press, and the Multi-Spindle Drill Press.

The Sensitive Drill Press. This is a small manually operated machine designed for drilling small holes, usually not to exceed ½ inch

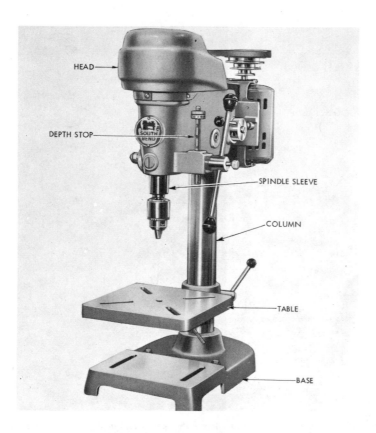

HEAD

DEPTH STOP

SPINDLE SLEEVE

COLUMN

TABLE

BASE

Fig. 1. The sensitive drill press is a small machine designed for drilling small holes; the feed of this type drill is controlled by hand. (South Bend Lathe, Inc.)

in diameter. See Fig. 1. The sensitive drill press is available as a bench or floor type, with one or more spindles. The term *sensitive* is used because the operator can "feel" the action of the drill point as it penetrates the work while he manually controls the feeding of the spindle into the work. This is due to the accurate balancing of the spindle. The precision qualities built into the machine make it possible for the operator, by controlling the

feed of the drill by hand, to reduce breakage of small diameter drills.

The Vertical or Upright Drill Press. The vertical drill press is a floor type machine designed to perform drilling and allied operations. See Fig. 2. This machine differs from the sensitive drill press in that it has power feed on the spindle, and is capable of handling drills larger than ½ inch in addition to the smaller drills.

The power feed on the spindle

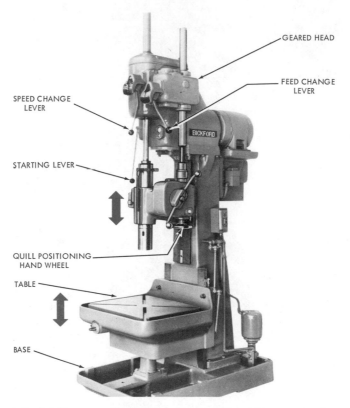

SPEED CHANGE
LEVER

GEARED HEAD

FEED CHANGE
LEVER

STARTING LEVER

QUILL POSITIONING
HAND WHEEL

TABLE

BASE

Fig. 2. The vertical drill press differs from the sensitive drill press in that it has power feed. (Giddings & Lewis Machine Tool Co.)

applies the thrust or feed force required to drill large diameter holes. Machines of this type are designed with dual feeds, manual and power, so that if twist drills of small diameter are to be used in the press, manual feed can be used.

The Radial Drill Press. The construction of the radial drill press is such that the arm carrying the spindle head can be swiveled about the vertical column, as shown in Fig. 3. In other words, the arm is the radius of a circle having the

column as its center. The outstanding feature of this machine is that the head that carries the spindle can be located at any point along the arm, while the arm itself may be rotated in the horizontal plane, raised or lowered on the column, and then clamped tightly in position.

Work performed on a radial drill press is usually large and heavy. It is more advantageous to swing the arm of the radial drill press and bring the drill into position over the

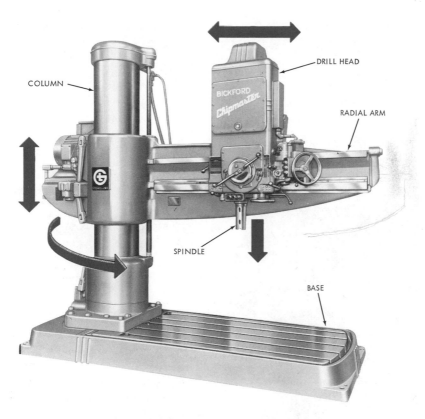

COLUMN

DRILL HEAD

RADIAL ARM

SPINDLE

BASE

Fig. 3. The radial drill press is designed so that the spindle can be located in many positions over a workpiece. (Giddings & Lewis Machine Tool Co.)

work than to move the work under the spindle, as must be done with other types of drill press. The flexibility of the radial drill press includes: manual feed for small drilling jobs, conventional power feed on the spindle, power driven arm, a head which can be positioned either manually or by power, a rapid traverse control to quickly reposition the head, and a table which clamps to the base when drilling small work, or which is removed when heavier work is fastened to the base.

The Multi-Spindle Drill Press. This is a drill press with *several* spindles mounted in a single head. See Fig. 4. Modern production requires drill presses that will supply topmost production and efficiency. The multi-spindle drill press was developed to meet this need.

A multi-spindle drill press allows several different operations to be

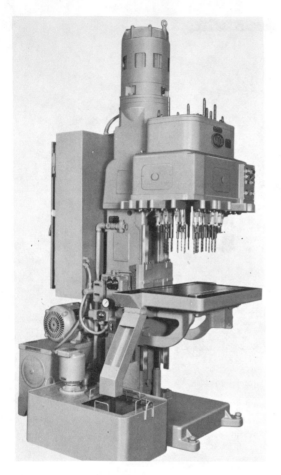

Fig. 4. The multi-spindle drill press can perform drilling and related operations simultaneously on different holes. (National Automatic Tool Co., Inc.)

Essential Elements of Drill Presses

The spindle carries the drill or other cutting tools and revolves in a fixed position in a sleeve. The drill press spindle is vertical and work is supported on a horizontal table, Fig. 1.

The sleeve does not revolve but may slide in its bearing in a direction parallel to its vertical axis, Fig. 1

The column supports the drill press head and also the table on some machines. See Fig. 1. Most drill presses have circular columns, but some feature a square bar-section construction.

The head is the part which carries the sleeve, spindle, and feed gears, Fig. 1. In many models the head is bolted to the column.

The table is supported on an arm which is adjustable on the finished lower section of the column. To accommodate different heights of work, the table is vertically adjustable or may be swung out of the way so that the work can be seated on the base. See Fig. 1.

Size Designation

The sizes of the bench and floor type sensitive drill presses are usually given in terms of the maximum diameter of holes that can be drilled with the twist drill. It may also be expressed as a maximum distance from the center of the spindle to the nearest point on the column of

performed on several different holes in a single workpiece being machined. The drill press head may hold any combination of tools so that, while some holes are being counterbored, other holes may be reamed or tapped. Note the following classifications of the drill presses discussed (Table I).

TABLE I.   DRILL PRESS CLASSIFICATIONS

TYPE	TRADE NAMES	DRILL CAPACITY	SPINDLE FEED	SET-UP TOOLING	DRILL HOLDING DEVICES
Sensitive (single and Multiple Spindle)	Atlas Press Co. Buffalo Forge Co. Cincinnati Lathe & Tool Co. Edlund Machine Co. South Bend Lathe, Inc.	#80 to 1/2"	Manual	Vise Table Drill Jigs	Drill Chucks Pin Chucks Morse Taper #2 Spindle
Vertical or Upright	Avey Drilling Machine Co. Barnes Drill Co. Giddings and Lewis Machine Tool Co. National Automatic Tool Co.	1/2" to 1 1/4"	Manual or Power	Vise Table Drill Jigs and Fixtures	Drill Chucks Pin Chucks Morse Taper #2 to #4 Quick Change Chucks
Radial	American Tool Works Giddings and Lewis Machine Tool Co. Fosdick Machine Tool Co.	1/2" to 3"	Manual or Power	Vise Table Fixture	Drill Chucks Pin Chucks Morse Taper #2 to #4 Quick Change Chucks
Production Automatics Multiple Spindle Multiple Head Transfer	Barnes Drill Co. National Automatic Tool Pratt & Whitney	1/8" to 1 1/4"	Power	Fixtures	Adapters Quick Change Chucks

the drill press. For example, a drill press with a work capacity of 16″ diameter is called a 16″ drill press.

The size of the radial drill press is designated by the length of the arm measured from the center of the spindle to the edge of the col-umn. For example, a 3 foot radial drill press will have a reach of three feet and will permit drilling a hole in the center of round work up to six feet in diameter. The sizes of radial drill presses vary from three to twelve feet.

Drills and Drilling Fundamentals

Since the drill rotates into the material being machined, drilling may be classed as a turning operation. A drill press performs as a rough turning tool. It is capable of removing more metal per minute per pound of its weight than any other cutting tool. Machining may be analyzed in terms of the following variables, which must be carefully controlled to achieve optimum machinability:

(1) The behavior of the work material in cutting,

(2) Operating characteristics of the machine tool, and

(3) The performance of the cutting tool itself.

Twist Drills

Twist drills are rotary-end cutting tools with two or more cutting lips and two or more helical or straight flutes for the removal of chips and the admission of a cutting fluid. See Fig. 5. Twist drills are made of three types of steel: carbon steel, high speed steel (with molybdenum or cobalt added for hardness), and carbon steel with cutting edges of durable tungsten-carbide. The last two combinations mentioned have a tendency to retain their hardness at high temperatures. This characteristic contributes to longer tool life, or less frequent drill sharpening.

All twist drills have three major parts, Fig. 5. These are the *point*, the *body*, and the *shank*. The term point refers to the entire cone-shaped cutting end. The body of the drill extends from the point to the shank. The shank end of the drill fits into the holding device that revolves the drill. The shank can be tapered or straight.

Taper-shank drills are mounted directly into the taper hole of the drill press spindle or into a tapered sleeve. The tang fits into a slot in the spindle or sleeve to prevent the drill from slipping in the spindle and to provide a more positive drive.

Drills with straight shanks, such as those shown in Fig. 6, are held in a key type drill chuck. This type

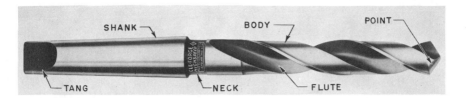

Fig. 5. Twist drills have three parts: the point, the body, and the shank. (Cleveland Twist Drill Co.)

of chuck has three jaws that move simultaneously to center and grip the drill.

Dead Center. The cone-shaped end which forms the cutting part of the drill is the point. The shape and condition of this point are very important. As the two cone-shaped surfaces of the point are ground, their surfaces intersect to form a sharp edge of short length in the exact center of the axis of the drill. This extreme top end of the drill point forming the one sharp edge is called the *dead center*. See Fig. 7.

The dead center acts as a flat drill and cuts its own hole in the workpiece. When drilling large holes 5/8″ in diameter and up, it is common practice to drill a lead hole first in the workpiece. This lead hole provides clearance for the dead center of large diameter drills; as a result,

less feeding pressure is required. It also prevents the larger drills from running off center.

Lips. The cutting edges of the drill point, extending from the dead center (chisel edge) to the periphery, are called the *lips*, see L and L' of Fig. 7A. The side view of the drill point in the figure shows the standard (helix) angle of 59 degrees, which the two cutting edges or lips make with the center line of the drill body. This gives an included angle or point angle of 118 degrees. In deep-hole drilling with a single drill, the included angle should be larger than the standard angle of 118 degrees to produce a chip that readily passes up the flutes.

Drill Clearance. It may appear at first glance that the full point of the drill is cone-shaped. The two lips or cutting edges do lie on the

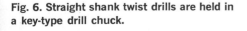

Fig. 6. Straight shank twist drills are held in a key-type drill chuck.

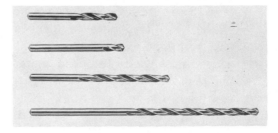

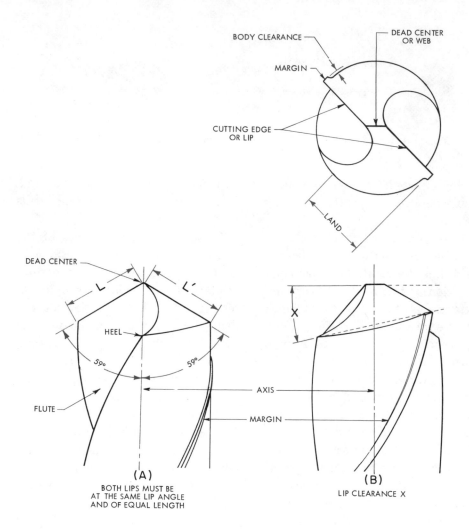

Fig. 7. Essential drill point geometry is illustrated in the side and top views of the twist drill shown.

surface of the cone, but the metal behind these cutting edges does not lie in the cone. This metal is "backed off," or ground off the surface of the cone at an angle to the dead center line to provide clearance for the cutting lips of the drill, see x in Fig. 7B. Without this clearance, the drill

would not cut, as the metal behind the lip would rub on the bottom of the newly drilled hole.

At the extreme diameter of the drill, the angle of lip clearance is generally established at 12 to 15 degrees. A drill ground in this manner will handle the bulk of shop

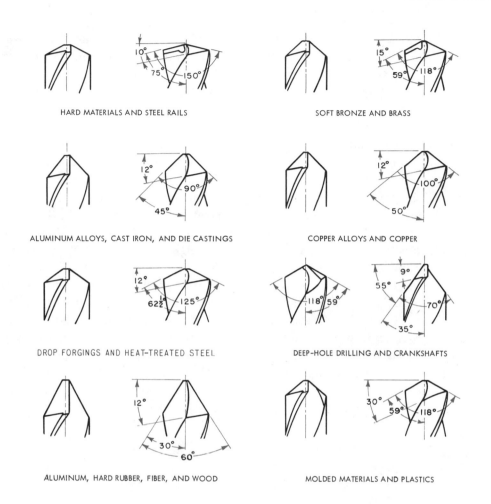

HARD MATERIALS AND STEEL RAILS

SOFT BRONZE AND BRASS

ALUMINUM ALLOYS, CAST IRON, AND DIE CASTINGS

COPPER ALLOYS AND COPPER

DROP FORGINGS AND HEAT-TREATED STEEL

DEEP-HOLE DRILLING AND CRANKSHAFTS

ALUMINUM, HARD RUBBER, FIBER, AND WOOD

MOLDED MATERIALS AND PLASTICS

Fig. 8. Different materials will be more easily drilled using recommended point angles.

work. Fig. 8 shows recommended drill points for drilling various materials. More detailed information is provided in the Appendix Table IV.

Flutes. The *flutes*, Fig. 5, are the helical grooves running along opposite sides of the drill. They are shaped to help form the proper cutting edges on the cone-shaped point and to provide a means for the escape of drill chips from a hole when drilling.

Margin, Body Clearance, and Land. The area called the *margin* lies on the full diameter of the drill, Fig. 7. When measuring a two-fluted drill with a micrometer, the measurement must be taken across both margins of the drill. When the drill is manufactured, the margin is

115

formed along the entire length of the flute. The margin assures that the hole will be the correct size.

Immediately back of each margin is the *body clearance*. This lessened diameter reduces the friction between the drill and the wall of the hole so the drill does not bind.

The term *land* means the periphery of that portion of the drill body not cut away by the flutes. In other words, the land consists of both margin and body clearance. There are two lands on a two-fluted drill.

The body of the drill is usually made slightly smaller near the shank to give a slight taper from the cutting end to the tang end. This *back taper*, as it is called, allows a slight clearance for the drill in the hole.

Web. The metal column at the center of the drill is called the *web*. This is the supporting section of the body joining the lands. The web thickness is increased toward the shank of the drill to give it needed strength. As the drill is ground more and more, the web must be ground thinner at the point so as to produce the desired chisel edge, Fig. 9.

Carbon Steel and High-Speed Steel Drills. Twist drills in common use are made of carbon steel or high-speed steel. High-speed drills are designated by the words "High Speed" or by the letters HS or HSS near the size markings. They are used almost entirely in fast production work because they may be run at twice the speed of carbon drills. If the drill shank is not stamped with these designations, meaning high-speed steel, it

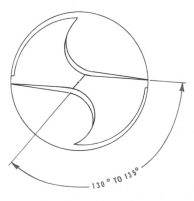

Fig. 9. As a drill is ground, the web must be ground thinner to maintain the proper chisel edge. A line across the dead center of the drill should always be at an angle of 130 to 135 degrees to the cutting edge.

can be assumed that the drill is made of carbon steel.

Drill Standards. Over a period of years, the twist drills made by many manufacturers have become more and more standardized. Drill diameters available as standard now are classified in a decimal (inch) series. Standard over-all and flute lengths have been provided for each diameter.

The *American Standard of twist*

TABLE II. NUMBER, FRACTION, AND LETTER DRILL SIZES

Drill	Diameter (Inches)	Drill	Diameter (Inches)	Drill	Diameter (Inches)	Drill	Diameter (Inches)
80	.0135	49	.0730	20	.1610	I	.2720
79	.0145	48	.0760	19	.1660	J	.2770
1/64	.0156	5/64	.0781	18	.1695	K	.2810
78	.0160	47	.0785	11/64	.1719	9/32	.2812
77	.0180	46	.0810	17	.1730	L	.2900
76	.0200	45	.0820	16	.1770	M	.2950
75	.0210	44	.0860	15	.1800	19/64	.2969
74	.0225	43	.0890	14	.1820	N	.3020
73	.0240	42	.0935	13	.1850	5/16	.3125
72	.0250	3/32	.0937	3/16	.1875	O	.3160
71	.0260	41	.0960	12	.1890	P	.3230
70	.0280	40	.0980	11	.1910	21/64	.3281
69	.0292	39	.0995	10	.1935	Q	.3320
68	.0310	38	.1015	9	.1960	R	.3390
1/32	.0312	37	.1040	8	.1990	11/32	.3437
67	.0320	36	.1065	7	.2010	S	.3480
66	.0330	7/64	.1094	13/64	.2031	T	.3580
65	.0350	35	.1100	6	.2040	23/64	.3594
64	.0360	34	.1110	5	.2055	U	.3680
63	.0370	33	.1130	4	.2090	3/8	.3750
62	.0380	32	.1160	3	.2130	V	.3770
61	.0390	31	.1200	7/32	.2187	W	.3860
60	.0400	1/8	.1250	2	.2210	25/64	.3906
59	.0410	30	.1285	1	.2280	X	.3970
58	.0420	29	.1360	A	.2340	Y	.4040
57	.0430	28	.1405	15/64	.2344	13/32	.4062
56	.0465	9/64	.1406	B	.2380	Z	.4130
3/64	.0469	27	.1440	C	.2420	27/64	.4219
55	.0520	26	.1470	D	.2460	7/16	.4375
54	.0550	25	.1495	E	.2500	29/64	.4531
53	.0595	24	.1520	1/4	.2500	15/32	.4687
1/16	.0625	23	.1540	F	.2570	31/64	.4844
52	.0635	5/32	.1562	G	.2610	1/2	.5000
51	.0670	22	.1570	17/64	.2656		
50	.0700	21	.1590	H	.2660		

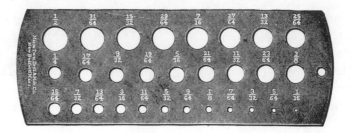

Fig. 10. A twist drill gage is used to check the sizes of twist drills. (Morse Twist Drill & Machine Co.)

drills covers fractional sizes which come between number and letter sizes. With this assortment of drills, it can be seen from Table II that the machinist has a wide range from which to choose. The size of the drill is usually stamped on the shank. Very small drills, however, are not identified by size. Therefore, they must be measured with a drill gage or a micrometer. Small drills should be kept in sets and preferably in holders.

All number, letter, and fractional drills up to ½″ diameter are available with a straight shank. Fractional drills over ½″ diameter are made with a taper shank.

Drill Gage. A twist drill gage, Fig. 10, is used to check the size of a twist drill. The gage consists of a plate having a series of numbered, lettered, or fractional size holes into which the cutting end of the drill can be placed to test for size.

A micrometer caliper can also be used to check the size of a twist drill by measuring over the margins of the drill.

Basic Cutting Principles of the Twist Drill

In order to establish optimum operating conditions in mass-producing drilled holes, one must understand the basic cutting action of the tool. Two distinct actions are observed in the process of drilling. The chisel edge of the drill point is forced against the metal to be drilled, creating forces which tend to deform the metal at the point of contact. Secondly, the cutting edges of the drill are more durable than the material being machined, and are pressed against the material with sufficient force or stress to shear off the material, pushing it aside and into the flutes of the drill for removal. The action of the cutting edges is by far the more important of these two actions.

The Chisel Edge. Note the effect of the downward motion due to feed shown in Fig. 11. Here the drill acts like a center punch indenting a piece of metal. As the drill bit moves into the metal, the drill cutting edges take a larger and

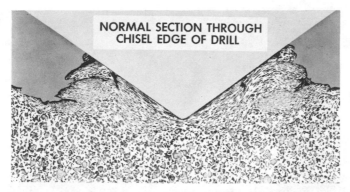

NORMAL SECTION THROUGH
CHISEL EDGE OF DRILL

Fig. 11. The combined stresses of the chisel edge and the cutting lips cause the deformation of the metal shown by this microphotograph. (National Twist Drill & Tool Co.)

larger bite, shearing off more and more material until the body of the drill has completely entered the workpiece.

The physical change in the metal is too severe and complex to treat in detail. The cutting action at the point of the drill is relatively inefficient compared with other tools due to the chisel edge. This inefficiency partly explains the high thrust forces that are developed by the drill.

Chip Formation. Segment-type chips come from the cutting tool as distinct fragments. These fragments are produced by fracture of the metal ahead of the cutting edge. This type of chip is usually produced when drilling brittle materials or when drilling ductile materials at low drill speeds.

Whether the chip is segmental, continuous, or continuous with a built-up edge, the cutting action is similar; the material just ahead of

the cutting edge is deformed by the process of shear. In the case of the continuous chip, the metal may or may not be fractured by the action of shear. A continuous chip is produced when the material is not fractured. Fracture can occur intermittently due to a build-up of workpiece particles on the cutting edge of the drill; the continuous chip is thus broken up. Chips produced by drilling always show a defined shear zone and an elongated crystal structure (indicating stress) and may also have a small built-up edge. Fig. 12 shows magnified section views of chips from two different kinds of steel.

Chip Flow. To obtain chip sections like those shown in Fig. 12, it is first necessary to determine the direction of chip flow. The discussion on chip flow observations and effective rake angle which follows is adapted from the pamphlet, *Metal*

119

 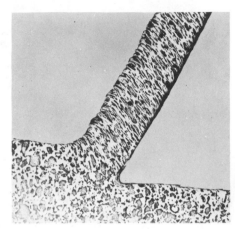

Fig. 12. Note the elongated crystal structure of these drill chip sections. The jagged effect on the shear surface indicates where the built-up edge breaks away to discharge the chip. (National Twist Drill and Tool Co.)

Fig. 13. A chip is shown leaving the cutting edge of a twist drill. Since the cutting edge is ahead of center, the chip is discharged at an obtuse and oblique angle to the cutting edge. (National Twist Drill and Tool Co.)

Cuttings, Vol. 1, No. 2, through the courtesy of the National Twist Drill and Tool Company.

Drill chip flow has long been assumed to be substantially perpendicular to the cutting edge, but Fig. 13 shows that this is distinctly not the case. Near the center of the drill, the chip flow direction is inclined approximately 60° to the perpendicular with a gradual falling off to about 15° inclination to the perpendicular at the outer corner of the cutting edge. A little reflection on drill point geometry shows that this is because the drill cutting edge is ahead of center and is thus oblique to the direction of rotational motion. Hence, the cutting edge has a shearing or slicing action. This action is more pronounced near the center because here the obliquity of the cutting edge becomes greater. Other investigations have shown that an oblique cutting edge always causes inclined chip flow.

Effective Rake Angle.   With inclined chip flow and oblique cutting edges, the effective rake angle becomes quite different from the actual tool rake angle. Effective rake angle is measured between the direction of tool motion and the direction of

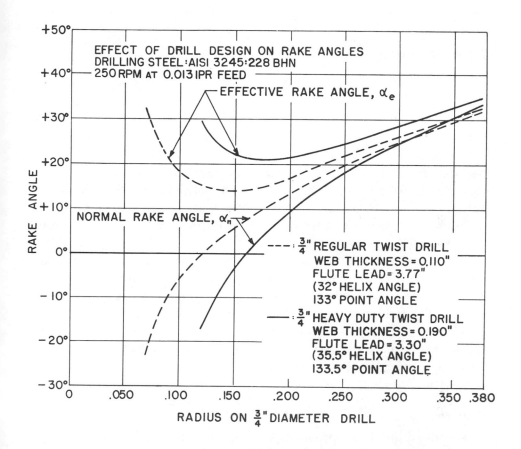

Fig. 14. The chart illustrates the change in effective rake angle as the tool enters the workpiece and the cutting edges progress into the material. (National Twist Drill and Tool Co.)

chip flow. An analysis of the complex geometry of the drill point coupled with experimentally-determined data on drill chip flow direction has made possible an evaluation of the effective rake angle of the twist drill.

Fig. 14 is a plot of effective rake angle along the cutting edge for two different drill designs. At small radii, just outside the chisel edge, the effective rake angle is relatively high because of the high shearing action of the cutting edge. This angle gradually decreases out to about the half diameter point, because of the progressive reduction of the shearing effect. Approaching the outside diameter, the influence of the drill helix angle becomes significant and the effective rake angle again increases. From the graph it will be noted that the normal rake angle (roughly, the tool rake angle), is highly negative near the center of the drill, but that the effective rake angle is quite positive. This suggests that the cutting action of a twist drill is much more efficient than has been generally supposed.

121

Speeds, Feeds and Coolants for Drilling

Speed refers to the revolutions per minute of the drill press spindle. For efficiency in specific drilling operations, the spindle should rotate at a certain number of revolutions per minute, and the drill should feed into the work a certain distance (usually in thousandths of an inch) for each revolution. On a sensitive drill press where a hand feed is used, the spindle is turned as fast as possible without undue wear to the drill. The feed is determined by feel through the hand feed lever.

When using the power feed, the speeds and feeds are set by control levers and must be determined before starting the job.

Drill Speeds

The speed at which various metals may be cut is expressed in terms of feet per minute (F.P.M.) or surface foot speed (S.F.S.). Both numbers indicate the distance that a point on the outer edge of the cutting tool will travel in one minute. A small drill must rotate much faster than a large one in order to cut the same F.P.M.

No hard and fast rule can be laid down for drilling speeds suitable for the various materials. The proper speed will usually vary with the conditions of each specific operation. Table V of the Appendix gives suggested peripheral speeds in surface feet per minute for drilling various materials with high speed drills. Speeds for ordinary carbon steel drills should be half those suggested for high speed drills.

Determining R.P.M. of a Drill. To find revolutions per minute, the drill press operator either consults a table to find the speed (in F.P.M.) at which the tool should revolve for the material to be worked, or he makes his own calculations. From the speed determined in F.P.M., he can determine the number of R.P.M. for the size drill being used.

Using the Tables to Find R.P.M. Suppose, for example, that the operator wishes to drill a 1-inch hole in cast iron, using a high-speed drill. From Table V (Appendix) he knows that the recommended cutting speed for this kind of metal is 100 feet per minute. The operator next consults Table VI of the Appendix to find out how fast the 1 inch drill should turn to attain a cutting speed of 100 feet per minute. This is done by locating the column giving 100-foot ratings for various diameters of drills and following down the column to a point in line with the 1-inch drill in the column farthest left. There the most favorable number of revolutions per minute is given as 382. When the R.P.M. has been determined, the machine is set at the nearest spindle speed obtainable. To obtain drill speeds, Table V must be used with Tables VI, VII, and VIII, which are placed in the Appendix for ease of reference.

Calculations to Find R.P.M. In addition to finding correct drill speeds from the tables, the speeds can also be calculated from a simple formula. To calculate the proper speed in revolutions per minute (R.P.M.) at which to run a drill for a particular metal, a single formula is used:

$$\text{R.P.M.} = \frac{\text{Feet per Minute (Surface Foot Speed)}}{\text{Diameter of Drill (in ")} \times 0.2618}$$

where the number 0.2618 is a constant.

For example: To determine the speed in R.P.M. at which a ½″ high speed steel drill should be turned to drill medium carbon steel (annealed), multiply the recommended surface foot speed of 80 F.P.M. by 4 (which is roughly equivalent to dividing by the constant 0.2618) and divide by the drill diameter:

$$\text{R.P.M.} = \frac{80 \text{ F.P.M.} \times 4}{0.5}$$
$$= \frac{320}{0.5}$$
$$= 640$$

When the R.P.M. has been determined, the machine is then set for the nearest spindle speed obtainable.

Feeds for Twist Drills

Feed is the rate at which the drill is advanced vertically into the work per revolution of the drill. It is usually expressed in thousandths of an inch. Feed is an important factor in drilling. Since the feed partially determines the rate of production and is also a factor in tool life, it should be chosen carefully for each job.

The feed for twist drills is governed by the size of the drill, the material being drilled, and the R.P.M. of the drill. The feed should increase as the drill diameter increases. The feeds shown in Table IX (Appendix) are generally used.

Feeds for alloys and hard steels should be less than given. Cast iron, brass, and aluminum may be drilled with a heavier feed. Use a good coolant to maintain speeds and feeds. A coolant is not used with cast iron, brass, or aluminum.

Feeds can also be determined from formulas. Table X (Appendix) lists several formulas which can be used to find feeds when tables are not available.

Cutting Fluids for Drill Press Work

As the drill cuts into the metal, the cutting edge becomes very hot. Heat may soften the cutting edge and ruin the drill. A cutting fluid should be used to cool the drill and, consequently, make the drill cut easier and more smoothly. The drill will stay sharp longer if a good cutting fluid is used. The fluid tends to wash away the chips, prevents undue friction, and permits faster cutting speeds. Recommended cutting fluids for drilling, tapping, and reaming are given in Table XI of the Appendix.

Drill Grinding

Suggestions on How To Sharpen Drills

A properly sharpened drill is very important. A drill that is not properly ground requires considerable effort to force it into the metal, and it may produce an oversize hole with a rough wall. See Fig. 15. An improperly ground drill cuts slowly, does poor work, becomes overheated, and may break. Drills that are often used require frequent grinding to keep them in shape.

Drills can be machine or hand ground. Both methods produce sat-

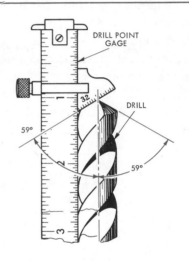

Fig. 16. The drill point gage is used to check the angle and length of drill lips.

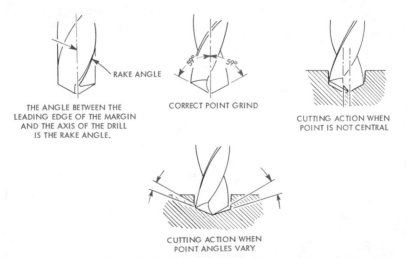

Fig. 15. The effects of improperly ground drill points are shown.

isfactory results. However, grinding drills by hand requires a great deal of practice.

Whether a drill is ground by hand or machine, three things must be considered: (1) lip clearance, (2) length and angle of the lips, (3) location of the dead center on the central axis of the drill.

Lip *clearance* is the relief given the cutting edges of the drill so that metal behind them will not rub against the bottom of hole.

The *angle* and *length* of the lips or cutting edges must be ground equal. Then the dead center will automatically be where it belongs on the axis. The drill point gage, Fig. 16, is used to check the angle and length of the lips.

Analyzing Drilling Difficulties

Improper Speeds and Feeds

The drilling of holes involves so many variables that no set rules can be established to cover all conditions.

One of the most common errors made in drilling practice is the use of too-high speeds and feeds. The speed of a drill is governed by two factors, heat generation and chip disposal. The drill life may be lowered considerably by using an excessive speed which tends to burn the drill either at the chisel edge or the lip edge. If the metal is removed too fast, the chips may clog in the flutes and cause drill breakage or abrasion on the drill margin. If too heavy a feed is used, the drill will twist under torsional stress and either break or lead off center. Too heavy a feed may also result in a chip too large to be disposed of through the drill flute clearance.

Too high a speed will rapidly wear away the outer corners of the drill. Too much lip clearance or too heavy a feed will chip or break the cutting edges. Too much feed or not enough lip clearance may cause the web to split.

Workpiece Support Considerations

It is *important* that the drill be started in the correct location if a good hole is to be produced. At times, a rigid support for starting the drill is also needed. To meet the above requirements, drilling guides (called *drill jigs*) can be used. They position and support the workpiece and guide the drill to the precise position for the hole. The drill jig *guide bushing*—through which the drill passes to contact the workpiece—should be located as close to the workpiece as possible.

Any hole over three times deeper than its diameter must be considered a deep hole and be given special consideration as such.

An irregular feed rate, eccentric or loose spindle, eccentric guide bushing or badly ground drill will cause trouble in obtaining a smooth and straight hole. Fig. 15 illustrates the results of drilling with an improperly ground drill.

In drilling through-holes, breakthrough lunge often causes considerable drill breakage. The lunge is caused by lack of rigidity. The feed mechanism may, due to torsional windup or, if hydraulic, due to compression of the fluid medium, slow down under load and then spring back to normal when the load is suddenly reduced. If not rigid enough, the holding fixture will also spring under load and release when the load is suddenly reduced. The workpiece may be inherently too weak to stand the thrust load and cause a spring action. The material may button out under thrust and cause a like spring action. Therefore, it is important to maintain the rigidity and that it be carefully controlled to obtain optimum results in drilling. Utilization of back-up or supporting materials behind workpiece in through-drilling operations is often necessary.

Coolant Considerations

Lubrication and coolant conditions should be carefully considered. Chips must be disposed of and heat that is generated dissipated.

The periphery of the drill in contact with the hole wall and the contact of the cutting edge as it bites into the work causes frictional heat, which can be reduced by a cooling medium that has lubricating value as well as heat exchanging characteristics.

The finish of a hole can be materially improved by the use of a proper lubricant. It is important that the coolant be supplied in sufficient quantity and properly directed to best serve the purpose of dissipating the heat and lubricating the frictional surfaces of the drill. A large quantity flooding the work may dissipate the heat but fail to lubricate the drill, while a small quantity in the drill hole may lubricate the drill but fail to dissipate the heat. Cutting fluid problems can best be solved by contacting a manufacturer of such fluids.

A good lubricant will also lubricate the rake angle cutting edge and lessen the chip friction both in cutting and sliding through the flutes for disposal.

As a heat exchanging medium, water is excellent. Water has no lubrication value, however, and is only used when mixed with soluble oil. This mixture suffices for many materials but is inadequate for materials requiring much lubrication in which cases a mineral oil, preferably sulphurized, is found to be much more satisfactory.

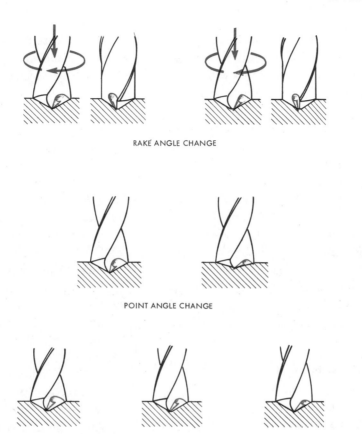

RAKE ANGLE CHANGE

POINT ANGLE CHANGE

HELIX ANGLE CHANGE

Fig. 17. This illustration shows the change in chip formation due to changes in rake, point, and helix angles.

Chip Considerations

Chip problems must be solved before a satisfactory hole is produced. The chips should be broken into small particles, the smaller the better. Small chips are easily removed either by the conveyor action of the helical flutes, gravity, coolant pressure, or withdrawal of the drill. Smooth polished flutes are very desirable especially for materials of high ductility (like lead).

Contrary to common belief, a long stringy chip is not a good condition as it is hard to dispose of. Stringy chips tend to remain wrapped around the drill, filling the flutes and rubbing on the flutes and hole wall.

Many factors affect the formation of chips. Chip formation is affected primarily by the ductility of the material being drilled, and the thickness of the chip made by the feed rate of the drill. When drilling materials with a low ductility (that is, hard materials) the chips will break up when standard rake angles are used. The higher the ductility, the more tendency there is for the chip to curl and remain in a continuous stringy ribbon.

Many methods are used to break up the chips. Changes in rake angle, point angle, and helix angle all affect chip formation. The most commonly used and probably the most satisfactory is the method of decreasing the cutting edge rake angle. Fig. 17 illustrates the effect of rake angle change on chip formation.

Check List for Drilling Difficulties

Drill Breakage—Causes

   (a) Feed too heavy.

   (b) Break-through lunge due to spring in work or machine.

   (c) Point not ground correctly.

   (d) Drill bends in hole due to margin wear.

   (e) Drill splits up the center due to insufficient lip clearance.

   (f) Cutting edges chip due to excessive feed or too much lip clearance.

   (g) Drill dull.

   (h) Chips packing in flutes.

   (i) No back-up or supporting materials behind workpiece.

Drilling Problems—Causes

   (a) Corners of lips wear or burn due to excessive speed.

   (b) Drill will not penetrate work due to dull drill or insufficient lip clearance.

   (c) Large chips in one flute, small or none in other flute. One lip doing all the cutting. Bad grind.

   (d) Walls of hole rough. Drill dull or improperly ground or chips wedging in flutes.

   (e) An unguided drill when entering the work, weaves around and does not produce a true circle or diameter until the full diameter is en-

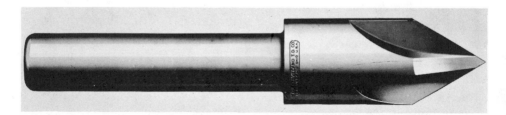

Fig. 18. A countersink is used to make a cone-shaped enlargement at the end of a drilled hole. (Cleveland Twist Drill Co.)

gaged. For this reason the guide bushing in a jig should be as close to the work as chip conditions will permit.

(f) Spring in the work causes an uneven cutting action, therefore, it is necessary that the work be so supported that the end pressure does not de-

flect the work.

Drill Sharpening Requirements

(a) Chisel point central.

(b) Cutting edges of equal length.

(c) Cutting edges equally spaced.

(d) Cutting edges of equal angle.

(e) Lip angles equal.

(f) Back taper minimum for material being drilled.

Other Types of Cutters Used on Drill Press

Countersinks

A countersink, Fig. 18, is a cutter that makes a cone-shaped enlargement at the end of a hole. A common example of work done with this tool is countersinking for a flathead screw to bring the head flush with the surface, Fig. 19. Countersinks

Fig. 19. Countersinking is most commonly performed so that a flathead screw will be flush with the surface after insertion. (South Bend Lathe, Inc.)

for machine screws and wood screws have an included angle of 82 degrees.

Counterbores

A counterbore, Fig. 20, is used to enlarge a portion of a cylindrical bore or hole. For example, a drilled hole may be enlarged for a little way to receive the head of a fillister-head screw. The pilot of the counterbore guides the tool so that the counterbore will be concentric with the hole in which it is used.

Spot-Facers

A spot-facing tool, Fig. 20, usually takes a light cut. The combination counterbore and spot-facing tool is useful either to enlarge a hole or to square a surface to furnish a bearing for the head of a bolt or a nut. This tool also has interchangeable pilots.

Core Drills

Most forgings or castings are, where possible, made with rough holes or cavities to conserve material. A cored hole saves metal which would otherwise be wasted when the hole is machined.

For enlarging cored, drilled, or punched holes, three- or four-flute core drills are employed. A core drill, Fig. 21, is similar to a counterbore except that it does most of its cutting by the chamfered edge terminating each flute. It thus machines the sides of the cored hole. A core drill is not designed to cut a hole in solid metal, as the cutting edges do not extend to the center of the drill.

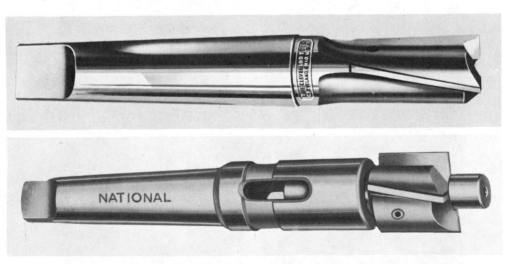

Fig. 20. A counterbore is used to make a cylindrical enlargement in a drilled hole. *Top:* A solid, combination counterbore-spot facer. (Cleveland Twist Drill Co.) *Bottom:* Replaceable counterbore is shown with pilot. (National Twist Drill and Tool Co.)

Fig. 21. A four-flute taper-shank core drill is used in enlarging pre-cast holes. (Cleveland Twist Drill Co.)

Reamers—Type and Selection

The purpose of reaming is to finish a previously drilled hole to exact size. Since drilled holes are not finished to close tolerance or fine surface finishes, it is essential that secondary operations follow drilling to meet these specifications.

A reamer may be defined as a rotary cutting tool with one or more cutting elements used for enlarging to size and contour a previously formed hole. See Fig. 22. Its principal support during the cutting action is obtained from the previously drilled hole. In using reamers there

are a number of factors which must be considered carefully in order to produce a properly finished hole.

Types of Reamers.   Reamers are classified in two ways: (1) on the basis of construction or design, and (2) by the method of holding or driving when in operation. Under these two systems of classification, the following types of reamers are commercially available.

REAMERS CLASSIFIED BY DESIGN.

Solid reamers are those made from a single piece of tool material. See Fig. 23.

Carbide-tipped solid reamers are

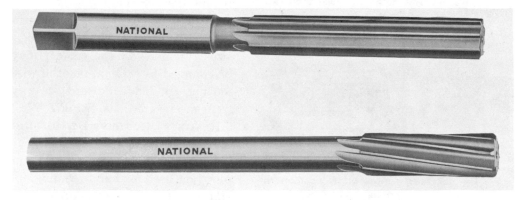

Fig. 22. Reamers come in many sizes and designs. They are used to accurately size and finish a drilled or bored hole, usually removing up to .015 inches in the process. (National Twist Drill and Tool Co.)

Fig. 23. Solid reamers are one piece reamers. They may have straight or taper shanks and straight or helical flutes. (National Twist Drill and Tool Co.)

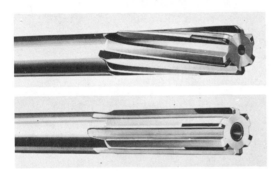

Fig. 24. Carbide is brazed to the cutting edges of a solid reamer to provide added hardness and durability. (National Twist Drill and Tool Co.)

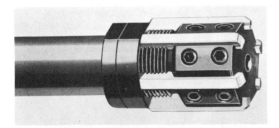

Fig. 25. Allen screws firmly position the inserted teeth on this inserted-blade, adjustable reamer. The diameter of the cutting head can also be changed. (Cleveland Twist Drill Co.)

solid reamers which have a body of one material with cutting edges of carbide brazed in place. See Fig 24.

Inserted blade reamers are usually solid. The cutting blades, however, are mechanically fastened to the tool, and may be adjusted or replaced. See Fig. 25.

Expansion-type reamers are those whose size may be increased by deflecting or bending segments of the reamer body. See Fig. 26.

Adjustable reamers are similar in purpose to expansion reamers. Their sizes (diameter of the cutter) may be changed by sliding, or otherwise moving, the blades toward or away from the reamer axis. See Fig. 25.

REAMERS CLASSIFIED BY HOLDING METHOD.

Hand reamers, as the name implies, are used for hand reaming operations. This type of reamer has a driving square at the end of the shank which is turned with a wrench. The cutting end is provided with a starting taper for easy entry into the drilled hole. See Fig. 27. Hand reamers are often used to finish a hole which has been drilled and machine-reamed to within a few thousandths of an inch of the desired size. Hand reamers are used for finishing operations where a high degree of accuracy and finish is required.

Machine reamers are used for production work on drill presses and

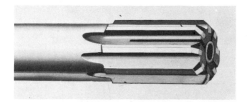

Fig. 26. The expansion-type reamer has slots which permit the head to vary in diameter as an adjusting screw is turned in or out at the end of the cutter head. (Cleveland Twist Drill Co.)

lathes for both roughing and finishing operations. There are two types of machine reamers: the rose reamer, used primarily for roughing operations, and the fluted reamer, used for finishing operations where extreme accuracy is not an important consideration.

Rose reamers, Fig. 28, have a 45° chamfer, permitting the reamer to take a heavier cut than a finishing reamer. The cutting action takes place at the front of the teeth. Rose reamers are used, for example, to take the heavier cut required to clean a cored hole in a casting. Rose reamers are usually made .003 to .005 inches under nominal reamer sizes.

Fluted machine reamers have more cutting teeth than the rose reamers. The teeth are relieved to allow efficient cutting along the side, whereas the rose reamer cuts primarily with the end. See Fig. 28.

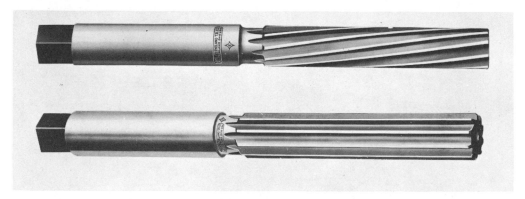

Fig. 27. The hand reamer is identified by the driving square at the shank end of the tool. The cutting end is tapered to allow easy entry into the hole. (Cleveland Twist Drill Co.)

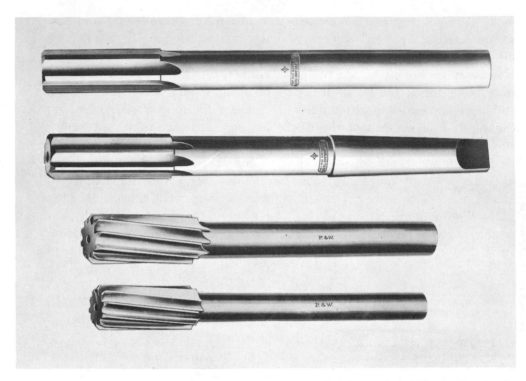

Fig. 28. *Top:* The rose machine reamer has a 45° chamfer at the cutting end and relatively fewer teeth. All cutting occurs at the leading edge of the rose reamer. (Cleveland Twist Drill Co.) *Bottom:* The fluted machine reamer has more teeth. The cutting action is on the side as well as the leading edge, and the side cutting teeth are accordingly backed off. (Pratt & Whitney Div., Colt Industries, Inc.)

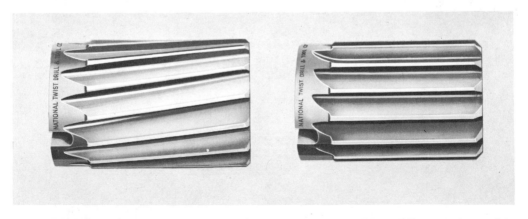

Fig. 29. Shell reamers are readily interchangeable. They offer economy and complete flexibility in reaming operations. (National Twist Drill and Tool Co.)

Machine reamers can have straight or tapered shanks, and straight or helical flutes. See Fig. 28. Tapered shank reamers are usually mounted directly in the drill press spindle. Straight shank reamers, called *chucking reamers*, must be inserted in a chuck, which mounts in spindle.

Shell reamers are reamer heads which must be mounted on arbors to be used, Fig. 29. Shell reamers offer the advantage of simply having to replace the reamer head. When

the shell reamer is worn beyond re-use, it is simply discarded. A new shell reamer is then mounted on the arbor, which accommodates many different sizes of reamers.

Shell reamers are used mostly for roughing operations, although shell reamers are available for the range of operations performed by solid reamers. See Fig. 30. Arbors for shell reamers may be straight or tapered, Fig. 31, depending on the type of shell reamer used.

Fig. 30. Shell reamers are also available which are adjustable and which have replaceable teeth. (Cleveland Twist Drill Co.)

Fig. 31. Arbors, to which the shell reamers are attached for use, can be straight or taper-shank types. The shell reamers have slots which are engaged by the two fins seen on the arbors. (Cleveland Twist Drill Co.)

135

Reamer Selection. A number of factors influence the selection of a particular reamer for a particular job:

1. The material to be removed.
2. The amount of stock to be removed.
3. The diameter and depth of the hole.
4. The accuracy and finish required.

The material to be reamed has much to do with the type of reamer one selects. If the material is easily cut, reamers of fairly light construction can be used. If the material is hard, or difficult to machine, some provision must be made to meet this condition. This is normally done by selecting a reamer with a cutting edge material that will withstand the hardness and abrasion of the workpiece material. Reamers with carbide tips are usually chosen for this purpose.

The amount of stock to be removed has a direct influence on the necessary driving force and, in turn, on the strength and rigidity the reamers must possess. A rose reamer would be used to remove material where the maximum amount of stock is to be removed. Fluted machine reamers would be used where not much stock is to be removed and only reasonable accuracy is required. Hand reamers are primarily designed to produce a fine finish and accuracy of size. They should not be used where a heavy cut is required.

The size of the hole to be reamed is an important factor when selecting the proper design of reamer for the job. Correct selection will help to avoid chatter when reaming various materials. Chatter is vibration of tool or workpiece caused by the motion of the cutting tool. Chatter can be caused by improper fastening of the workpiece or improper mounting of the tool as well as improper selection of the type of reamer for the job. Accuracy and finish can be adversely affected when chatter occurs.

If a hole is fairly large in diameter and not very deep, the tendency to chatter can be overcome by using a short reamer. If the hole is small in diameter and fairly deep, it is more difficult to prevent chatter. In order to overcome this problem, reamers are designed with unevenly spaced flutes, spiral flutes, or with pilots to support one or both ends of the flutes. The purpose of uneven spacing is to prevent any tendency to synchronize slippage (set up tendency to chatter). Reamers with spiral flutes will accomplish this objective even more effectively than uneven spacing, and are widely used where chatter is likely to occur, as in deep, small diameter holes. The key point to remember is that reamers should have enough spiral so that two or more flutes overlap in the length of the hole being reamed.

Reamer Guiding and Driving Tools. Since the function of a reamer is to finish a hole to a given size, the reamer must produce holes that are parallel in alignment and free from bell-mouthing at either end. In order to accomplish this, it is sometimes necessary to provide a suitable guide and drive device for production operations. For example, if two holes are located at an exact distance from some point or from another hole, then the reamer must be guided through a jig. This is done by the bushings that are normally found in drill jigs. In production work, the reaming operation would immediately follow the drilling operation. On the other hand, if the reamer is to guide itself into a previously drilled hole such rigidity is not satisfactory, because the reamer must be free to float and be guided by the pre-drilled hole. To accomplish this, floating reamer drivers are available.

Speeds and Feeds for Reamers. Reamers should be run at slower speeds and higher feeds than those for the corresponding diameters of drills. Speed should be two-thirds to three-fourths the speed of drills. If the speed is too slow, reamer tool life is shortened. If the speed is too high, the material being reamed will tend to cling to the edges and lands of the reamer thereby causing dulling and roughing of the walls of the hole. Extreme care should be taken when reaming hard materials.

Proper selection of feed is very important when machining difficult-to-cut materials. For general work a feed of two to three times that for corresponding diameters of drills will be most satisfactory. Both feed and speed are affected by the amount of material to be removed. In other words, the more stock to be reamed, the lower the speeds and feeds.

Feeds are also governed by the finish desired. A coarse feed tends to produce revolution marks and rough walls. If the feed is too fine, it keeps the reamer idle in the cut, thereby causing it to wear.

Taps

Taps are frequently used in the drill press to cut threads. Taps have already been discussed in considerable detail in a previous chapter.

Holding Devices

Holding devices attached to or used in connection with drill presses are of two kinds: (1) those that se-cure the tool or cutter in a properly mounted position, and (2) those that secure the workpiece in place for drilling, boring, etc.

Cutter-Holding Devices

Various methods are used for holding drills and other cutting tools in the drill press spindle while different operations are being performed. The method used or hold-

TAPER
SHANK

TANG

FOR
USE OF
KEY

JAWS FOR HOLDING DRILLS

Fig. 32. A key-type drill chuck is used to hold small drills, usually under ½ inch in diameter. (Black & Decker Mfg. Co.)

ing device employed depends upon whether the cutting tool has a straight shank or a taper shank.

Straight-Shank Drill Holders. The straight-shank twist drills are the small-size drills under ½″ diameter used in drill presses and portable drills. A chuck, Fig. 32, is used to hold this type of drill. It has three small jaws which tighten simultaneously against the straight shank of the drill.

The chuck for the drill press is removable from the spindle of the machine. The chuck has a taper shank and tang that fit into the taper in the drill press spindle. The taper on the chuck shank is the common Morse taper.

Taper-Shank Drill Holders. Taper-shank twist drills are inserted in various holding devices having a taper hole, or are inserted directly into drill press spindles. As the taper shank wedges itself into the drill spindle or socket, it naturally lines up true with the center line of the spindle. Increasing pressure on the drill point tends to wedge the taper tighter into the spindle. This friction drive, however, is not enough to prevent the drill from slipping under a heavy cutting load. The flat tang on the end of the shank of the drill permits a positive drive.

The tang fits into a small slotted recess on the inside of the spindle, above the tapered hole, and keeps the drill from slipping. Taper-shank

Fig. 33. A taper-shank drill is removed from a drill press using a drift pin.

drills are removed from the taper hole by means of a drift, Fig. 33.

Drill Sleeve and Socket. Frequently the taper hole in the spindle is too large for the taper shank of the drill or chuck. In such case, the taper-shank tool is held in a *tapered sleeve*, Fig. 34. The sleeve has a

Morse taper on the outside so as to fit into the tapered hole in the spindle. The tapered hole on the inside of the sleeve fits the taper on the drill shank.

The drill socket shown in Fig. 35 differs from the sleeve in that a larger taper-shank drill may be inserted in the socket than would fit the spindle or sleeve.

Tapping Attachment. A tapping attachment must be used for tapping operations on the sensitive drill press. Without this attachment, it would be impossible to tap on a sensitive drill press, as the machine does not have a reversing mechanism. When the spindle is brought downward into the work, the tap rotates clockwise. To reverse or back the tap out of the hole, the machine spindle is raised. A tap used in a tapping attachment stops when it strikes an obstruction in the hole or hits the bottom of a hole. This is accomplished by an adjustable friction nut, which can be adjusted to suit the size of tap used. A device

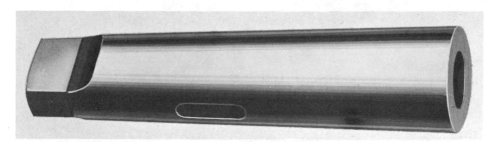

Fig. 34. When the spindle is too large for the taper shank of the drill a tapered sleeve is used. (National Twist Drill and Tool Co.)

139

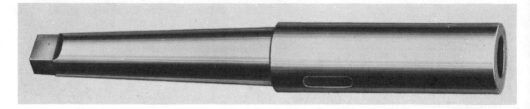

Fig. 35. When the spindle is too small for the taper shank of the drill, a drill socket is used. The drill is placed in the socket; the socket is then mounted in the spindle. (National Twist Drill and Tool Co.)

of this kind makes it possible to tap a hole at speeds considerably faster than those of hand tapping. The big advantage in tapping with this attachment is that it reduces tap breakage to a minimum.

Work-Holding Devices

Mounting and supporting the workpiece in the drill press are important. A workpiece improperly mounted or insecurely fastened will cause accidents and give poor results.

Vise. A drill press vise, like the one shown in Fig. 36, is used on a drill press table to hold and support the work to be machined.

Parallels. Parallels, Fig. 37, are accurately machined bars made in

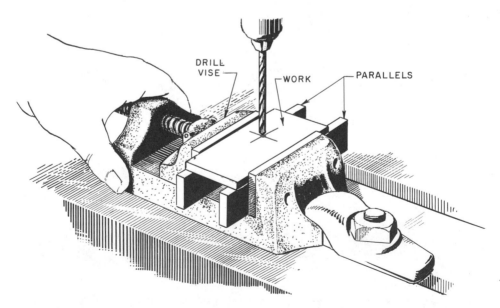

Fig. 36. A drill press vise is used to hold and support the work to be machined.

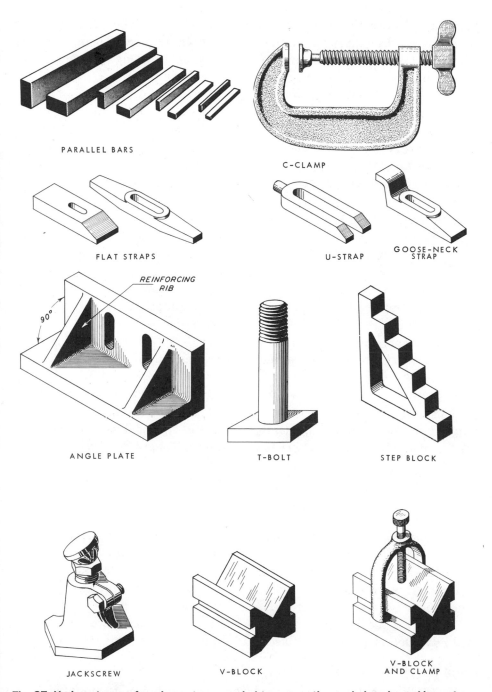

PARALLEL BARS

C-CLAMP

FLAT STRAPS

U-STRAP

GOOSE-NECK
STRAP

REINFORCING
RIB

90°

ANGLE PLATE

T-BOLT

STEP BLOCK

JACKSCREW

V-BLOCK

V-BLOCK
AND CLAMP

Fig. 37. Various types of equipment are needed to secure the workpiece in making setups.

pairs. They raise the work so the operator can drill through the part without damaging the table or vise. The parallels should be placed so that the drill will not damage them in passing through the work.

T-Bolts. Drill press tables are usually provided with T-slots in which T-bolts, Fig. 37, are inserted to fasten the work or a workholding device, such as a vise, to the table. T-bolts should have strong heads which fit the T-slots properly.

Clamp or Straps. Mounting work on the drill press table requires an ample assortment of clamps or straps, Fig. 37.

The C-clamp, Fig. 37, is convenient on small machines where a T-bolt cannot be used. The parallel clamp is also widely used.

Step Blocks. To properly support certain shapes of work, or to support the end of the strap clamp, one or more step blocks are useful.

Jackscrew. The jackscrew, Fig. 37, is an adjustable supporting device used in many different setups on the drill press.

Angle Plates. Another supporting device used in drilling and other operations is the angle plate. Holes and slots are provided to clamp or bolt the plate to the table and to secure the workpiece firmly to the angle iron.

V-Blocks. To hold round work, V-blocks are used, Fig. 37. The diameter of the workpiece to be drilled determines the size of V-block employed. Some V-blocks are provided with U-shaped clamps to hold the work securely.

Drill Jig. When several pieces of work have to be drilled alike, a drill jig is used, Fig. 38. With the aid of a drill jig, the hole or holes will be drilled at the same location on each piece of work. Thus, the pieces will be interchangeable with one another.

Setups

The correct procedure for making setups and for operating the drill press must be carefully followed in order to get good results. The layout work must be done properly first, of course, to insure that the hole will be drilled in the proper place.

Setup Involving Use of V-Block

Fig. 39 shows a V-block being used

to hold round stock. Care must be taken in this setup to see that the work is securely clamped in the V-blocks. Otherwise the work will swing around when the drill takes hold in the workpiece and possibly injure the operator.

Use of Drill Press Vise

The workpiece is mounted in the

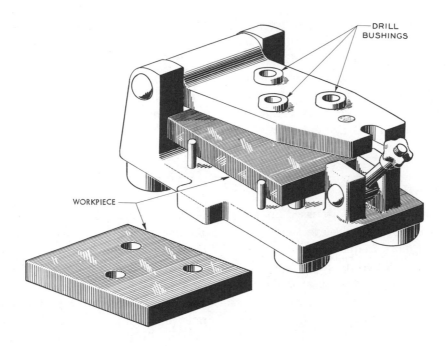

DRILL
BUSHINGS

WORKPIECE

Fig. 38. A drill jig is often used in production work where many identical pieces of work are to be drilled alike.

vise as shown in Fig. 36. Then the vise is aligned under the drill and securely fastened to the table by means of T-bolts.

Safety in Mounting Work

The importance of securely fastening the work to the table cannot be too strongly emphasized. Many injuries have been caused by holding the work in the hand instead of fastening it in a vise or to the table. Both the table and the table bracket must be tightly clamped to the column before the drilling is started.

Movement of the table after drilling has started will result in a broken drill and perhaps injury to the operator.

The bolts holding the work to the table should be placed as near the work as possible. Remember, the piece to be drilled must be held rigidly. Parallels should be well placed under the work. If they are too far apart, the work tends to spring and bind the drill.

A well-mounted job reflects the operator's mechanical skill and judgment. Drilling will go smoothly if he will (1) use T-bolts that are just long enough, (2) choose clamps of the right type and size, and (3) use V-blocks of a size appropriate to the job. If the operator will make sure that the work is securely clamped, he can avoid the drilling inaccuracies due to workpiece movement.

Adjustments Required in Setups

The arm supporting the table should be brought to the right height. The table should be clamped in position. The spindle travel is adjusted with stops to give the required range of movement.

Operations

The operations discussed in the following paragraphs are those performed most frequently on the drill press. Careful study should be given to the setup required and also to the procedure for each operation. The setup should be made and checked, and the operation should be performed under supervision until it is learned correctly.

Drilling

Drilling is the operation of pro-ducing a circular hole by removing solid metal. It is one of the major operations on the drill press. Drilling is often done on work held in a vise, Fig. 36. Work is sometimes clamped to V-blocks, Fig. 39, or to angle plates, Fig. 40.

Before drilling is done, the center of the hole is located by drawing two lines at right angles. A center punch is used to make an indentation at the center to help start the drill, Fig. 41A. Proof marks are made at four places on the circle

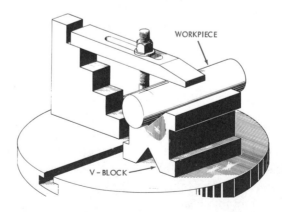

Fig. 39. V-Block, Step Block, and T-Slot Bolt and Strap used to hold round stock in preparation for drilling.

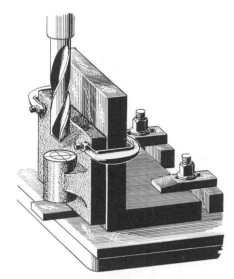

Fig. 40. An angle plate and C-clamps are used to secure a workpiece for drilling.

representing the hole to be drilled, Fig. 41*B*.

Precision Drilling. Precision drilling means drilling two or more holes which have been laid out in relation to each other. The holes must be drilled accurately to layout marks. Precision drilling also is performed by use of carefully prepared templates. Drilling with jigs and fixtures is classed as precision drilling, and so is drilling to a specified depth.

Suggestions on How To Drill in the Drill Press. Mount the workpiece in some suitable holding device. Adjust the worktable to a convenient height for drilling, taking into account the length of both the drill and the drill-holding device. Next, select the drill and drill-

holding device and mount them in the spindle. Adjust the height of the spindle to provide the shortest travel feed of drill to workpiece.

Select the proper speed. Start the machine and lower the drill to the work. Then feed the drill slowly into the work until the point of the drill has penetrated the metal. Check to see that the spot is centrally located.

Use a cutting lubricant where necessary. If the hole is to be drilled entirely through the piece, ease up on the drill as it breaks through the bottom.

Drawing the Drill. Very frequently the drill point fails to seat itself in the prick punch mark, and the desired start shown in Fig. 42*A* is not achieved. Instead, a small hole is made off center and close to the proof marks on one side, as shown in Fig. 42*B*. To determine whether the drill is going properly when a drill jig is not used, the operator should use layout lines as shown in Figs. 41 and 42. With layout circles or prick punch marks, he can readily see whether the circle made by the drill is concentric with the proof marks. If it is not, drilling should be stopped until the hole is prepared for drawing the drill.

The drill can be drawn back to the center of the circle by means of a groove or several grooves cut down the side of the shallow hole made by the drill point. For this, a round-nose chisel or a cape chisel is used, Fig. 42C. The cut is made in the drilled hole on the side farthest from the proof circle, that is, the side toward which the drill must be moved. The drill is then started again—this time in the groove—so as to work the drill back to the center of the hole or the center of the proof circle.

It is important that the operator check the drilling for proper location before the drill has cut too deeply. Once the drill begins cutting

its full diameter, it is too late to shift the tool without marring the surface of the workpiece.

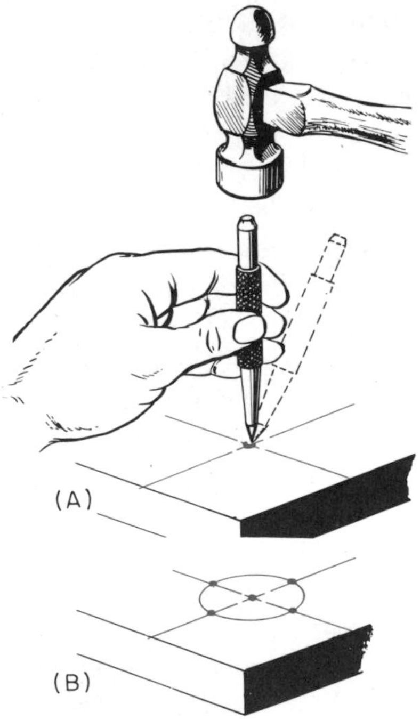

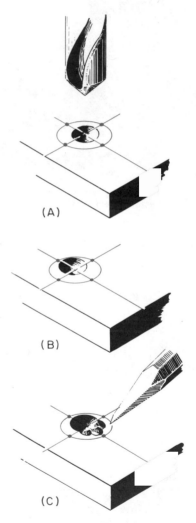

(A)

(B)

(C)

Fig. 41. A center punch is used to make an indentation to help start the drill at A. At B, four proof marks are made on the circle which represents the hole to be drilled. The proof marks and layout lines aid the operator in seeing that the drill is properly seated.

Fig. 42. At A, the drill point is properly seated. At B, the drill point enters the workpiece off-center. At C, an improperly seated drill can be drawn back to center by a series of grooves made with a round nose or cape chisel.

Countersinking

Countersinking means tapering or beveling the end of a hole with a conical cutting or reaming tool. Stated in another way, it is the operation of forming a cone-shaped enlargement of the end of a hole, as in making a recess for a flathead screw (see Fig. 18).

Suggestions on How To Countersink. Countersinks for machine screws and wood screws have an included angle of 82 degrees.

Rough countersinking can be caused by excessive speed of the spindle, a dull tool, or failure to clamp the work securely. The countersinking tool should be run slowly. For steel, use a cutting oil.

Good procedure includes fastening the workpiece in a workholding device, mounting the countersink in a drill chuck, selecting proper speeds, aligning the work with the countersink properly, and using care in feeding the tool to the work.

Counterboring

Counterboring means enlarging the end of a hole cylindrically, as in producing a recess for a fillister-head screw. Holes must sometimes be enlarged part way along their length to accommodate a bolt, stud, or pin of two or more diameters. This operation is performed with a counterboring tool or a conventional spot-facer.

Suggestions on How To Counterbore. The counterbore is guided in the original hole by a pilot on the end. The cutting edges on the counterbore produce the size of the enlarged hole. The "neck" of the counterbore is sometimes made long to enlarge a hole to a considerable depth. The shank is held in a tapered or straight shank-holding device.

Counterboring follows the drilling operation of the original hole. It is absolutely essential that the workpiece be lined up so the pilot will fit into the original hole.

Spot-Facing

Spot-facing is somewhat like counterboring. In this operation, just enough metal is removed at the

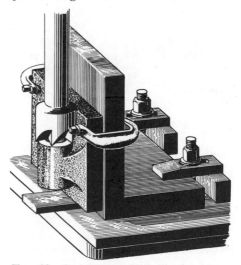

Fig. 43. Spot-facing is done to remove enough metal at the top of a hole to produce a bearing surface for a washer or cap-screw head.

top of the hole to produce a bearing surface against which a washer or nut or the head of a cap screw may be brought flat. See Fig. 43.

Suggestions on How to Do Spot-Facing. First, drill the hole as explained in the drilling operation. Then mount the spot-facing tool in the drill press spindle and align the pilot in the hole. The spot-facing is done on the surfaces specified on the blueprint.

Reaming

Holes do not always have to be made precisely accurate and smooth. A simple drilling operation, using ordinary care, does the trick in most cases. There are times, however, when the hole *must* be made to exacting requirements. To obtain a round and smooth hole of standard

Fig. 44. A machine reamer is shown mounted in the spindle of a drill press, ready to finish to size the hole drilled in the workpiece.

size, the hole should be drilled $\frac{1}{64}$ of an inch undersize. It should then be machine reamed. Sometimes the hole is finished with a hand reamer to the exact size. Reaming is defined, then, as the operation of sizing and finishing the inside of a hole, using a cutting tool (reamer) which has several cutting edges. See Fig. 44.

Suggestions on How to Ream in the Drill Press. The first operation is drilling the hole as near the reamer size as possible. Remember, removing metal with the reamer in excess of the necessary amount will shorten the life of the reamer. Therefore, holes should be drilled to a diameter $\frac{1}{64}$ of an inch under the reamer size. Usually the size of the hole is given on the blueprint, the dimension being followed by the word "ream." Thus, specification of a $\frac{3}{4}$-inch ream would mean that a drill $\frac{1}{64}$ of an inch under this size should be used.

Substitute the proper reamer for the drill in the press spindle without removing the work or changing its position.

Adjust the machine for the proper spindle speed. The reamer should revolve at a speed approximately half that of the drill.

Now proceed with the reaming. Use a cutting oil on steel and wrought iron. The automatic feed can be used to feed the reamer into the hole. Feeding may be done by hand, but the reamer should not be crowded. Be sure that the tool runs true, as otherwise a good finish cannot be expected.

Reamers have slight tapers at the end to facilitate their entry into the hole. Therefore, they should be run through the hole to a depth of at least an inch and a half to produce the true size.

After removing the reamer from the drill spindle, lay it upon a piece of cloth or waste to protect the delicate edges. A reamer should never be run backward, as this tends to break or dull the edges.

Tapping

After being drilled, holes may require tapping in the drill press. Tapping a drilled hole means cutting a thread with a tool called a tap.

Suggestions on How to Tap in the Sensitive Drill Press. The drill press, when used as a tapping machine, employs one of two methods. By the first method, the drill press is used merely as a guide for precision hand tapping. The workpiece is mounted on the drill press table or in a vise on the table. A lathe center is mounted in the hollow spindle of the drill press. The tap is prepared with a tap wrench mounted on the square shank. The tap is then placed in the workpiece hole, and the center point of the lathe center is placed in the center hole in the shank end of the tap. Enough pressure is maintained on the tap with the hand-feed lever of the drill press to hold it steady without forcing it. The tap is turned into the work with the tap wrench.

The second method employs a special tapping attachment which holds the tap and is rotated by the spindle of the drill press. The tap-holding device has a friction clutch built into it. If the tap sticks, jams on the bottom of a blind hole, or works under excessive pressure, the clutch will slip before the tap breaks. Tapping attachments have a reversing mechanism by which the tap may be backed out, simply by raising the spindle of the machine.

Safety Precautions—Drill Press

The principal hazard to the drill press operator is that of bodily injury. The injury may be caused by contact with moving machine parts or tools, flying chips entering the eyes, or material falling on fingers or toes. The drill press operator can avoid these injuries by observing the following safety rules.

1. The machine table should be equipped with a drill press vise, clamps, or other means for properly

holding the work in place while it is being machined.

2. Drill press operators should be provided with and required to wear goggles which will prevent chips and other flying particles from injuring their eyes.

3. Floors about drill presses should be maintained in good repair and, if necessary, should be covered with an anti-slip material to safeguard the operator from slipping and falling.

4. The operator should not attempt to oil the machine or make any adjustments to the work while the drill press is in motion.

5. Run the drill only at the proper speed; forcing or feeding too fast may result in broken or splintered drills and possible injury.

6. Only properly sharpened drills, sockets, and chucks in good condition should be used. Drills with battered shanks will not run true.

7. Drill press operators should not wear gloves, finger rings, aprons, long flowing ties, or loose or torn clothing. Female operators should wear hair nets.

8. Chips should be removed from the table or work by using a brush. A rag or the hands should not be used.

Courtesy of the National Safety Council

NOTE: Please see review questions at end of book.

ENGINE LATHE

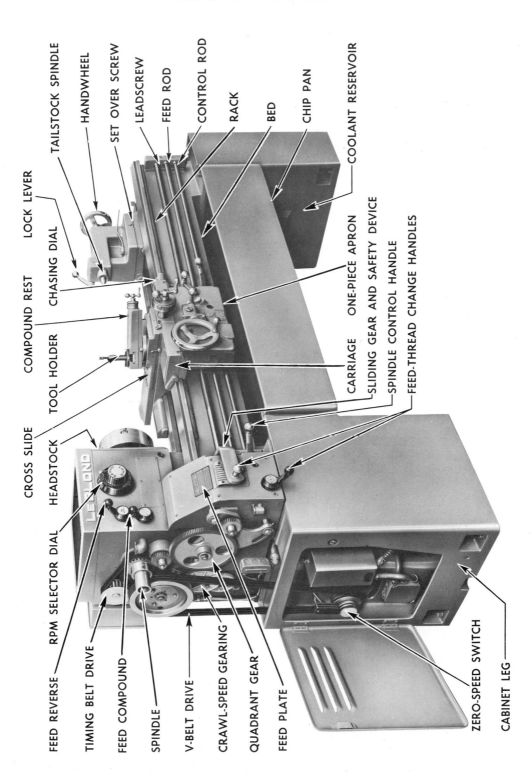

TAILSTOCK SPINDLE
HANDWHEEL
SET OVER SCREW
LEADSCREW
FEED ROD
CONTROL ROD
RACK
BED
CHIP PAN
COOLANT RESERVOIR

LOCK LEVER
CHASING DIAL
COMPOUND REST
TOOL HOLDER
CROSS SLIDE
HEADSTOCK

CARRIAGE
ONE-PIECE APRON
SLIDING GEAR AND SAFETY DEVICE
SPINDLE CONTROL HANDLE
FEED-THREAD CHANGE HANDLES

FEED REVERSE
RPM SELECTOR DIAL
TIMING BELT DRIVE
FEED COMPOUND
SPINDLE
V-BELT DRIVE
CRAWL-SPEED GEARING
QUADRANT GEAR
FEED PLATE
ZERO-SPEED SWITCH
CABINET LEG

PROBLEM: TURN A TWO DIAMETER SHAFT B FROM BAR STOCK A

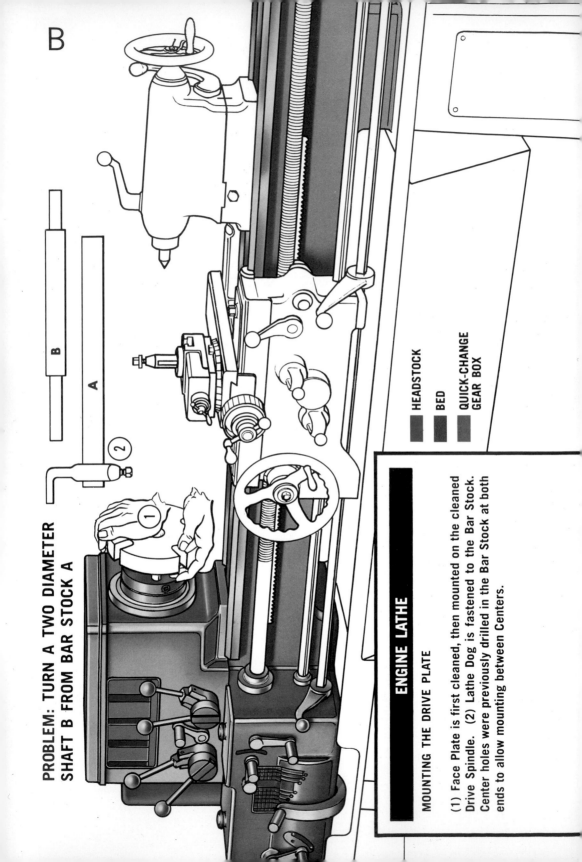

ENGINE LATHE

MOUNTING THE DRIVE PLATE

(1) Face Plate is first cleaned, then mounted on the cleaned Drive Spindle. (2) Lathe Dog is fastened to the Bar Stock. Center holes were previously drilled in the Bar Stock at both ends to allow mounting between Centers.

- HEADSTOCK
- BED
- QUICK-CHANGE GEAR BOX

IMPORTANT CONTROLS AND SETTINGS FOR THE LATHE

THE CARRIAGE HANDWHEEL

The carriage handwheel is a manual control located on the lower half of the carriage apron. This handwheel is used to move the carriage longitudinally between the headstock and the tailstock.

The primary function of the handwheel is to align the cutting tool with any desired part of the workpiece before using the power feed controls.

THE TAILSTOCK HANDWHEEL

The tailstock is also manually operated. The tailstock handwheel is turned to secure the tailstock dead center in the center-drilled hole in the right end of the workpiece. Turning the handwheel clockwise advances the spindle (which holds the dead center) toward the workpiece.

THE FEED CONTROLS

The power feed lever is located on the carriage apron. The power feed lever activates either the longitudinal power feed of the carriage or the cross feed of the cross slide.

In addition to the power feed lever, the cross feed can be manually set with the cross-feed screw handle. The micrometer dial on the cross-feed screw handle permits precise settings. The power feed must first be disengaged and then the micrometer dial is set to remove the remaining stock to obtain the finished workpiece.

CROSS-FEED SCREW CRANK

POWER FEED

THE COMPOUND REST SCREW

The compound rest screw is a manual control which positions the compound rest on the cross-slide. It is located just above the cross-feed screw, and also has a micrometer dial for precise settings.

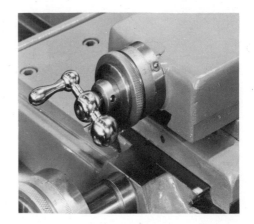

The compound rest screw is used to obtain the fine settings required in finishing and thread-cutting operations. For operations such as roughing and heavy turning, the compound rest screw is seldom employed.

THE RPM SELECTOR DIAL

The speed selector dial is usually found on the headstock. The speed settings are in RPM. Lathe speeds are determined in accordance with the size and type of workpiece, and the type of cutting tool to be used. The lathe speed setting is usually made after the workpiece and tool have been mounted.

THE FEED SELECTOR
(QUICK-CHANGE GEAR BOX)

The quick-change gear box is located just below the headstock. A variety of gear ratios, are possible which permit a feed range of approximately .002″ to .130″. A chart shows the operator how to select the feed or gear ratio he desires.

In the type of gear box shown, the feed is made by setting the lower lever to the letter position indicated by the chart. The upper tumbler lever is then moved to the proper numbered slot, also indicated by the chart. Each feed setting requires the proper positioning of both levers.

Engine Lathes:

Types, Accessories, and Attachments

Chapter

6

The variety of operations the engine lathe can perform make it one of the most useful and necessary machines in the shop. The major function of the engine lathe is to change the size, shape, or finish of a revolving workpiece by one cut or a series of cuts into the workpiece with an adjustable cutting tool. The type of engine lathe and its auxiliary equipment will determine the extent or variety of operations for which the machine is adapted. With proper attachments and adjustments, a lathe can also drill, ream, tap and thread.

Lathe Fundamentals

The metal-cutting action of the lathe is the basic principle in any machining operation. This action involves the formation of a chip from the metal workpiece by the tool and the movement of the chip across the face of the tool. This chip-producing occurs in all metal-cutting operations, whether it is turning, drilling, milling, or sawing. When the tool cuts into the metal, it executes a pressure of about twenty tons to the square inch. This heavy force stretches and deforms the material and in turn creates heat. The movement of the chip across the face of the tool creates friction. Friction, along with the heat created when the metal is deformed, are the common factors in all chip-removal processes.

Basic Forces. The three basic forces which act upon the cutting tool are identified as: *longitudinal feed force, radial force,* and *tangential force.* These forces are all influenced by the cutting velocity. See the insert illustrating these three forces on Trans-Vision, Page E.

The *longitudinal feed force,* F_{long} on Page E, is parallel to the axis of turning and in the direction of the feed movement. The direction is from tailstock to headstock. The feed force and feed velocity determine the power required for the feed drive. This force affects the chip formation. Feed is expressed in inches per revolution (*ipr*).

The *radial force*, F_{rad}, is most important to tool life. This force is radial from the turned surface in the direction of the depth move-ment (Toward the center of the stock).

The *tangential force*, F_{tan}, is in the direction of the turned surface and at a right angle to the axis of turning in the direction of the cutting speed. F_{tan}, is most important where power is concerned.

The cutting velocity determines the workpiece rotation and is expressed in feet per minute (*fpm*).

Lathe Types and Construction

To meet the demand of the metal-working industries, different types of engine lathes are available to fit various operations. Engine lathes may vary in size, appearance, and the relative position of their operating elements, but all engine lathes are similar in operation.

A lathe must have the mechanism for applying power or force in the form of motion. It must have mechanisms for controlling the speed of that motion and for holding the material or workpiece. A lathe must also be equipped to hold the cutting tool or tools and to control the tool movements so that the material can be machined to close tolerances.

Types of Engine Lathes

Engine lathes may be grouped into four classifications: light-power machine lathes, tool room lathes, standard lathes, and large swing or long bed lathes.

Light-Power Machine Lathes. This type is designed for training purposes and general machining of light work. These machines are available in two styles: *bench models* and *floor models*. They have all the parts and features of the larger and heavier lathes, but are usually not as expensive. Fig. 1 illustrates both bench and floor models.

Tool Room Group. The difference between this group and the others is that the tool room lathes are more accurately constructed and may be equipped with special attachments and accessories, thereby providing for a *greater variety* of precision work. Their primary function is to provide tooling and gages for production machine tools. These machine tools are available in bench and floor models. The bench-type lathe is primarily used for turning short, small-diameter work.

The floor-type machine is a preci-

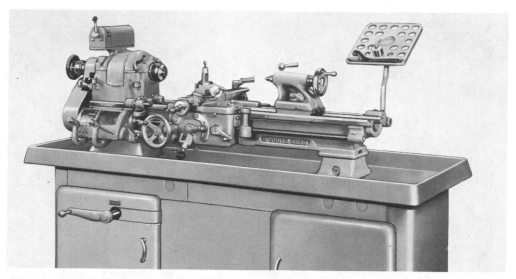

Fig. 1. Light-power machine lathes have all the parts and features of larger lathes. (*Top:* South Bend Lathe, Inc. *Bottom:* Clausing Div., Atlas Press Co.)

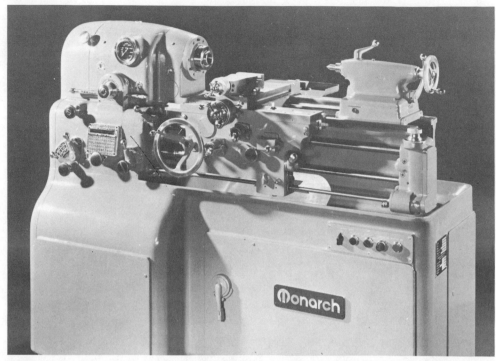

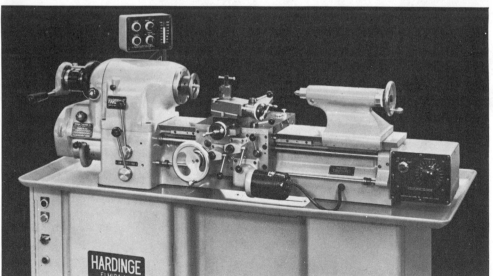

Fig. 2. The tool room group of lathes varies in design to meet almost any type of tool and die work. (*Top:* Monarch Machine Tool Co. *Bottom:* Hardinge Bros., Inc.)

sion machine equipped with variable speeds, micrometer stops, and various attachments required for tool room work. See Fig. 2.

Standard Group. These lathes are frequently referred to as standard manufacturing lathes, or all-purpose lathes. The machines are built a little heavier, and have more horsepower, wider ranges of speed and feed, and a more adequate capacity for handling larger work than the other groups. The lathe selected for the Trans-Vision is one of the standard types. Rigidity of design adapts this group of machines to all types of general purpose machining and to situations involving small lot production. See Fig. 3.

Large Swing and Long Bed Group. Engine lathes in this group are designed for machining long shafts of large diameter for almost any special job. See Fig. 4. Usually this group of machines is not equipped for high speeds. Rigidity and horsepower are more essential for this group in order to remove a considerable amount of metal per cubic inch.

Major Lathe Elements and Their Functions

Certain principal parts are common to all lathes. An understanding of the parts, their location, and their use in machining enables one to do more accurate work with less spoilage. The lathe is composed of six

principal parts: the bed, headstock, tailstock, carriage, quick change gear box, and the main motor drive.

Bed. The lathe bed, identified by a *russet* color in the lathe Trans-Vision, is the foundation on which the lathe is built. It must, therefore, be accurately designed and solidly constructed. The V-ways machined in the bed's surface are precision finished and hardened to insure proper alignment of all working parts mounted on the bed. Another view of the bed is seen in Fig. 5; other principal lathe parts are also identified.

The *ways* of the lathe, Fig. 6, serve as a guide for the saddle of the carriage as it travels along the bed and guides the cutting tool in a straight line. Usually, the two outer V-ways guide the lathe carriage. The inner V-ways and flat way together provide a permanent seat for the headstock and a perfectly aligned seat for the tailstock no matter where it is positioned. The ways should be oiled so the saddle can move easily along the bed.

The lathe must be carefully leveled both lengthwise and crosswise, Fig. 6. A slight twist in the bed would cause the machine to produce imperfect work. It is occasionally necessary to relevel the lathe after long periods of use.

Headstock. The headstock of the lathe, identified by the color *red* in the Trans-Vision, is mounted

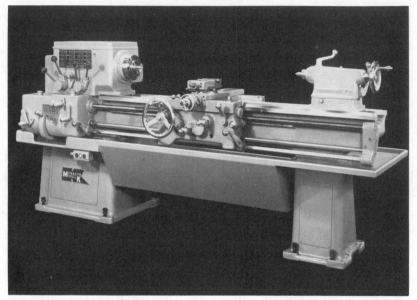

(Monarch Machine Tool Co.)

(American Tool Works Co.)

Fig. 3. Standard manufacturing lathes, like those shown on these pages, are adapted to all types of general purpose machining.

(R. K. LeBlond Machine Tool Co.)

(South Bend Lathe, Inc.)

Fig. 3 (cont.)

Fig. 4. Large swing and long bed lathes are used for work of unusually large size. (American Tool Works Co.)

Fig. 5. The principal parts of the standard lathe are identified. (R. K. LeBlond Machine Tool Co.)

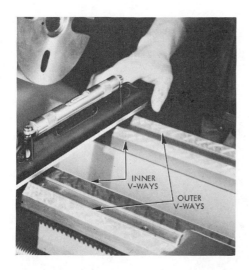

INNER
V-WAYS

OUTER
V-WAYS

Fig. 6. The lathe must be level lengthwise and crosswise. A slight twist of the bed will cause imperfect work to be produced. (South Bend Lathe, Inc.)

at the left-hand end of the machine. It is the unit which turns the workpiece. It contains the spindle and a series of gears or a combination of gears and different-sized pulleys (*steps 1, 2, 3,* and *4* of *cone pulley,* Fig. 7) by which the spindle is rotated at various speeds. Where gears are employed, the headstock unit includes control levers for selection of proper speed, Fig. 8. Power is supplied to the headstock by an electric motor that usually is built into the machine. Fig. 5 shows another view of the headstock.

Spindles and Spindle Noses. The hollow spindle is built into the headstock. The nose of the spindle

CONE PULLEY

BACK GEAR
LEVER

BULL GEAR
LOCK

FEED REVERSE
LEVER

1 2 3 4

Fig. 7. A cone pulley headstock. (South Bend Lathe, Inc.)

Fig. 8. A geared headstock. (American Tool Works Co.)

projects out from the headstock housing. The common screw nose is shown in Fig. 9, *lower left*. A faceplate, drive plate, or chuck can be screwed on these threads to support and rotate the workpiece.

The new lathes, especially the larger models, have different spindle noses. Fig. 9, *top*, shows a standard key-drive taper nose. This is primarily a locating and seating taper for guiding and seating chucks, faceplates, and fixtures. See Fig. 10. It has a substantial driving key and a large locking collar which screws onto the faceplate or chuck as they are slid onto the taper.

A spindle nose of the cam-lock type may be seen in Fig. 9, *lower right*. The chuck or faceplate has projections which fit into holes in the nose plate. A clamp holds them securely.

The end of the hollow spindle has an internal tapered hole into which tapered centers can be placed.

Tailstock. The tailstock, identified by the color *orange* in the Trans-Vision, is movable on the bed ways. It contains a taper-bored spindle in which the dead centers for turning work are inserted. See Fig. 11. Taper-shank drills and reamers can also be mounted in the tailstock spindle when required. Centers or tools mounted in the tailstock arc

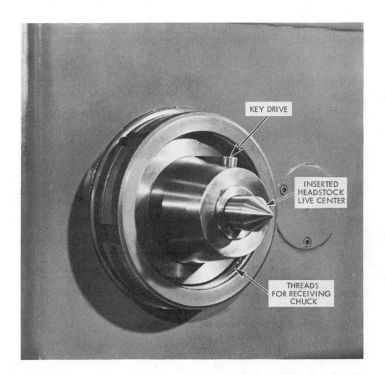

KEY DRIVE

INSERTED
HEADSTOCK
LIVE CENTER

THREADS
FOR RECEIVING
CHUCK

Fig. 9. *Top:* A key-drive tapered spindle nose. (The Lodge & Shipley Co.) *Lower Left:* A threaded spindle nose. *Lower Right:* A cam-lock spindle nose. (R. K. LeBlond Machine Tool Co.)

Fig. 10. A three-jaw chuck is being mounted on a tapered spindle nose. (R. K. LeBlond Machine Tool Co.)

Fig. 11. The tailstock can be moved along the ways to hold shorter or longer work pieces. (R. K. LeBlond Machine Tool Co.)

removed by turning the hand wheel counterclockwise.

Carriage. The carriage, identified by the color *yellow* in the Trans-Vision, supports the *cross slide, tool post*, and the *cutting tool*. It runs lengthwise along the bed between the headstock and the tailstock. The carriage, Fig. 12, is made up of two major parts: the *saddle*, which straddles the lathe bed and rests on carefully machined ways; and the *apron*, attached to the front of the saddle.

The carriage is moved by gears and controls housed in the apron. Carriage movement can be accomplished by hand through use of the handwheel seen on the apron in Fig. 12. The carriage can also be moved along the ways by engaging the power feed lever on the apron.

Cross Slide. The cross slide is mounted on the saddle and supports the compound rest, Fig. 12. The cross slide can be moved toward or away from the operator (across the ways) by turning the cross-feed screw crank. The cross-feed crank has a micrometer dial, or *adjusting collar*, which permits metal removal in thousandths of an inch. *Power* crossfeed can be had by shifting the *feed-change lever* to "cross-feed" and depressing the *power-feed lever*.

Compound Rest. The compound rest, Figs. 12 and 13, is mounted on the cross slide and supports the tool post. The compound may be swiveled horizontally through 360 de-

Fig. 12. The carriage of the lathe moves along the bed between the headstock and tailstock. Its principal function is to position the cutting tool. (R.K. Le Blond Machine Tool Co.)

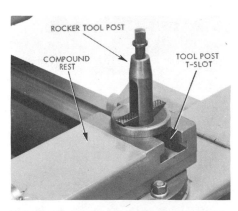

Fig. 13. The tool post assembly is used to hold cutting tools. The rocker post shown mounted on the compound rest can be used for almost all cutting tools except carbide-tipped tools.

grees and clamped at whatever angle desired. It can be moved forward and backward in this position (at the angle set) on the cross slide by turning the compound rest feed screw. The compound rest feed screw, like the cross slide feed screw, has a micrometer collar graduated in thousandths of an inch for precision settings.

Tool Post and Tool Supports. The tool post assembly is mounted on the compound rest, Fig. 13, and is used to hold the cutting tool. The rocker tool post shown is used to support a wide range of ordinary cutting tools, but is not suitable for

163

Fig. 14. The quick-change gear box changes the rate of movement of the carriage, or feed of the tool. (R. K. LeBlond Machine Tool Co.)

carbide-tipped tools. Types of tool posts are discussed in detail in the section which follows on Lathe Accessories and Attachments.

Quick-Change Gear Box. The quick-change gear box, identified by the color *gray* in the Trans-Vision, is located directly below the headstock on the operator's side of the lathe. It is the unit through which the lead screw and feed shaft are rotated to move the carriage along the bed ways at a desired rate of travel. Fig. 14 is a close-up view of a quick-change gear box. The table above the slot settings indicates proper settings for slot pin and letter lever to obtain desired feed.

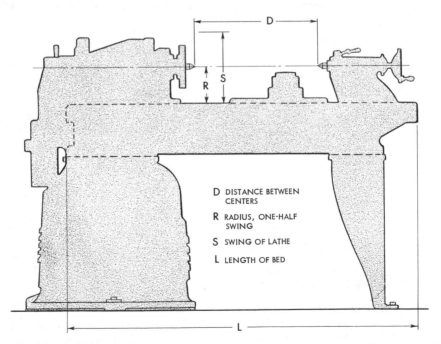

D DISTANCE BETWEEN CENTERS

R RADIUS, ONE-HALF SWING

S SWING OF LATHE

L LENGTH OF BED

Fig. 15. The maximum swing and the maximum distance between centers are two criteria used to designate the size and capacity of a lathe. (South Bend Lathe, Inc.)

The feed and thread-cutting mechanism includes gears and other parts which are necessary to transmit motion from the main spindle of the headstock through the gears and lead screw. This motion causes the carriage and thread-cutting tool to move. The same gears may be used in the feed mechanism.

Some lathes are built with a lead screw and a feed rod. On these machines the lead screw is used exclusively for thread cutting, while the feed rod, used for drive power, feeds for turning operations. Some-

times the lead screw is used for both power feed and thread cutting.

Size and Capacity of Lathes

Engine lathes vary in size from the small bench lathe to the large type many feet in length. The size of the engine lathe is based upon two measurements: (1) the approximate diameter (*maximum swing, S*) in inches of the largest piece of work that can be revolved over the ways of the lathe, and (2) the *maximum distance between centers (D)* in inches, study Fig. 15.

Lathe Accessories

There are numerous accessories used on the lathe. Included among these accessories are work-holding and work-supporting devices, tool-holding devices, and attachments for special operations.

Lathe work-holding devices are used to hold the workpiece securely in position on the lathe while one or several successive operations are performed. Chucks, faceplates, and lathe centers are examples of such devices.

Chucks

Some workpieces, because of their size or shape or the character of the operation to be performed, must be chucked in the lathe. This means that they are gripped and rotated

by one of several types of devices called chucks, which are attached to the headstock spindle of the lathe.

Three-Jaw Universal Chuck. A three-jaw, or universal chuck, Fig. 16, has three jaws that move simultaneously an equal distance toward or away from center by turning a single chuck key. Since the jaws all move toward the center at the same time, the workpiece is automatically brought to an on-center position.

Four-Jaw Independent Chuck. A four-jaw, or independent chuck, Fig. 17, has four jaws each of which can be adjusted independently. Thus, by adjustment of the jaws a workpiece of cylindrical, square, or irregular contour can be held on-

Fig. 16. A three-jaw or universal chuck automatically brings the workpiece on-center. (R. K. LeBlond Machine Tool Co.)

Fig. 18. A cam-lock chuck is being mounted on the spindle. The cradle is used to support the weight of the chuck and acts as a safety device if the chuck should slip while being installed.

Fig. 17. A four-jaw independent chuck allows a workpiece to be positioned on- or off-center. (R. K. LeBlond Machine Tool Co.)

center, or at a predetermined distance off-center.

Three- and four-jaw chucks are made to fit the screw-on, key-drive, and cam-lock types of spindles. Fig. 18 shows the cam-lock chuck ready to be mounted onto a cam-lock spindle. Note the cradle of wood holding the chuck at the correct height so that it can be slid into the holder. A wooden cradle is a handy safety device to use with any type of large chuck. Should the chuck slip while being mounted, the cradle will catch the chuck, preventing possible damage to the ways and possible hand injuries to the operator.

Fig. 19. By lengthwise movement of the collet on the draw-in collet chuck, the diameter of the gripping surfaces of the chuck can be expanded or contracted. (R. K. LeBlond Machine Tool Co.)

Collet Chucks. A *draw-in collet chuck*, Fig. 19, has a hollow, split, and tapered head which can be adjusted to grip small-diameter workpieces and bar stock. The collet fits into a sleeve of corresponding taper. A lengthwise movement of the collet causes contraction or expansion of the gripping surfaces of the chuck. The lengthwise movement is obtained by a *drawbar*, Fig. 19, which is inserted through the hollow spindle of the lathe headstock. When the collet head is drawn into its closing sleeve, the workpiece is automatically gripped on all sides and held in an on-center position.

Draw-in collet chucks are made in sets, since different collet chucks are needed for different diameters of work. The range of work which a single collet can be expected to hold and grip securely is small.

Newer designs in collet chucks have eliminated the draw-bar and have made it much easier for the operator to mount and clamp work. The modern *Jacob collet chuck*, Fig. 20, does not have the draw-bar. Instead of the draw-bar, it features a handwheel on the spindle of the lathe which is an impact-tightening wheel with a set of eleven rubber flex collets. Each collet is able to accommodate a workpiece size up to $\frac{1}{8}$ of an inch larger than could be held by the next smaller collet size in the set. These collets firmly grip the workpiece, which may vary in diameter from $\frac{1}{16}$ to $1\frac{3}{8}$ inches— the range of the collet set. See Fig. 20.

167

Fig. 20. The Jacobs collet chuck grips the workpiece in a manner similar to the draw-in collet chuck. The Jacobs chuck, however, is turned at the *front* of the headstock to open or close the collets. (The Jacobs Mfg. Co.)

Another collet chuck is the *Sjogren collet chuck*. To secure the workpiece in this chuck, Fig. 21, a clockwise motion of the handwheel will tighten the collets. Spring steel collets are used with a wide range of fractional and decimal sizes. Collets are available for holding round, square, and hexagonal shapes.

Drill Chucks. The drill chuck is one of the most useful tool-holding devices. It is used for clamping straight-shank twist drills and is usually inserted in the tailstock spindle of the lathe. See Fig. 22. Drill chucks are available with a capacity of up to ½ of an inch. Twist drills ranging from decimal sizes up to ½ inch diameters are clamped in this type of chuck.

Faceplate

If a workpiece cannot be ma-

Fig. 21. The Sjogren collet chuck grips the workpiece with steel spring collets. This type is also adjusted at the front; note the adjusting handwheel. (Hardinge Bros., Inc.)

Fig. 22. The drill chuck can be inserted in the tailstock spindle of an engine lathe. The drill remains fixed while the work is turned. (The Jacobs Mfg. Co.)

chined in a lathe chuck of the three- or four-jaw type, it can be fastened to a faceplate with bolts, studs, or clamps. The faceplate is mounted directly to the spindle. Most face-

Fig. 23. A faceplate is used on the headstock spindle to hold a workpiece that cannot be machined in a lathe chuck. (Monarch Machine Tool Co.)

plates are made with a flange around the outer rim and with radial slots equally spaced around the plate. See Fig. 23.

Lathe Centers

Lathe centers support the workpiece between the headstock and the tailstock. The center used in the headstock spindle is called the *live center*. It rotates with the headstock spindle. See Fig. 9, *top*.

The *dead center* is located in the tailstock spindle, Fig. 24. This center usually does not rotate. However, some manufacturers are making a roller-bearing or ball-bearing center in which the center point can revolve with the work. A stationary tailstock center must be hardened to stand the wear of the work revolving upon it.

Fig. 24. The stationary tailstock center is called a *dead center* because it does not revolve with the workpiece. (R. K. LeBlond Machine Tool Co.)

Fig. 25. Heavy-duty live centers can be installed in the tailstock spindle for high-speed work which would create friction problems if stationary centers were used. (The Ready Tool Co.)

Fig. 26. A revolving pipe center is used to support hollow cylindrical shapes such as tubing or pipe. (The Ready Tool Co.)

The hole in the spindle into which the center fits is usually of a Morse standard taper. It is important that the hole in the spindle be kept free of dirt and also that the taper of the center be clean and free of chips or burrs. *If the taper of the live center has particles of dirt or a burr on it, it will not run true.* The centers, remember, play a very important part in lathe operation. Since they give support to the workpiece, they must be properly ground and in perfect alignment with each other. The workpiece must have perfectly drilled and countersunk holes to receive the centers. The center must have a 60-degree point.

Engine lathes designed to operate at higher spindle speeds will have

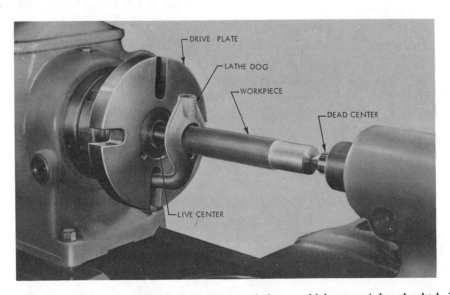

Fig. 27. A lathe dog is used to drive long workpieces which cannot be chucked. The lathe dog is fastened to the workpiece; its tail is then inserted in a slot in the drive plate. (J. H. Williams & Co.)

Fig. 28. A steady rest is used to support long cylindrical pieces. (R. K. LeBlond Machine Tool Co.)

Lathe Dog

A lathe dog, Fig. 27, is used to provide a simple and flexible means of driving a shaft between centers. The dogs are mounted near the end of cylindrical work, mandrels, and other pieces. The tail of the dog is inserted in the slot of the face or drive plate.

Steady Rest and Follow Rest

A steady rest, Fig. 28, is used to support long, slender, cylindrical

Fig. 29. A follow rest is used to support the workpiece from behind and above. (R. K. LeBlond Machine Tool Co.)

heavy-duty live centers mounted on the tailstock spindle. Note Fig. 25. Revolving tailstock centers are inserted in the tailstock spindle, thereby eliminating the dead center. These are very useful for revolving the workpiece at higher speeds, since lubrication of the dead center is eliminated.

In supporting cylindrical shapes such as tubing or pipe, it is necessary to use an inserted revolving pipe center to support the workpiece. This center will support work with large-diameter holes and offers all the advantages of a revolving center. See Fig. 26.

Fig. 30. The follow rest is attached to the carriage and moves with the tool along the workpiece. (South Bend Lathe, Inc.)

workpieces. It is attached directly to the ways of the lathe bed.

A follow rest, Fig. 29, supports the workpiece from behind and above. It is attached to the carriage when in use, Fig. 30, and is moved as the tool and carriage move.

These detachable devices prevent the cutting tool from bending slender workpieces. A steady rest is frequently used to support the "right" end of long pieces of material held in the chuck.

Mandrel

Mandrels are cylindrical workholding devices which have a slight taper over their entire length. The workpiece to be turned has a hole in its center. The workpiece is slipped onto the small end of the mandrel and moved along the length of the mandrel until the workpiece seats itself on the progressively increasing diameter of the mandrel.

Fig. 31 shows a pulley casting mounted on the mandrel. The mandrel is rotated by a bent-tail lathe dog. The casting is held firmly in place, since the cutting tool maintains pressure on the casting, holding it to the tapered mandrel.

Tool Posts

Cutting tool holders vary in design and shape. Because of this it is necessary to have a variety of tool posts which can be mounted on the

Fig. 31. A mandrel is mounted between centers. It is tapered along its length and is used with work which has a center hole. (R. K. LeBlond Machine Tool Co.)

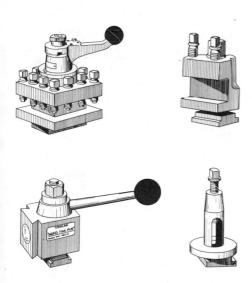

Fig. 33. The rocker tool post is the tool post generally used for most operations. (R. K. LeBlond Machine Tool Co.)

Fig. 32. Four common types of tool posts.

Fig. 34. The open-side tool post is used with carbide cutting tools. (R. K. LeBlond Machine Tool Co.)

Fig. 36. The Aloris quick-change tool post cuts tool-changing time to a minimum. (Aloris Tool Co., Inc.)

Fig. 35. The turret tool block saves time which might be wasted in changing tools. (R. K. LeBlond Machine Tool Co.)

compound rest. See Fig. 32. The four most common tool posts include:

Rocker Tool Post. The rocker tool post is standard equipment for all lathes. It clamps to the compound "T" slot and holds one tool at a time. An adjustable wedge or rocker permits positioning the top of tool on the work center line, Fig. 33. This type of tool post is best suited for tool holders that clamp high-speed cutting tool bits.

Open Side Tool Block. The open side tool block, Fig. 34, has more rigidity than the post type and should be used for exceptionally heavy cuts when using carbide cutting tools. Carbide cutting tools are always set on dead center and do not require adjustment for positioning the tool at the work center line.

Turret Tool Block. This tool block, Fig. 35, is designed to accommodate four tools at one setting and provides quick tool changes for work requiring multiple operations. Twelve station indexing is possible with this tool block.

Aloris Quick-Change Tool System. For multiple turning operations a number of tool blocks with preset tools can be quickly and accurately slid into place and locked with a clamping lever. This type of tool post, Fig. 36, provides for heavy cutting without chatter or vibration. Shims under tools are eliminated, since each tool holder has a separate height adjustment and a knurled knob.

Lathe Attachments

Taper Attachment

Fig. 37 shows a special device for taper turning, boring, and thread cutting. This taper attachment saves setting over the tailstock, and provides greater accuracy in taper cutting. The taper attachment is fastened to both the cross slide and the lathe bed. When not in use, it is detached from the cross-slide movement, but remains mounted on the saddle of the lathe.

Turret Attachment

The turret attachment, Fig. 38, is frequently added to the lathe to increase production. Various cutting tools are mounted in the revolving turret. Different machining operations such as drilling, reaming, tapping, threading, countersinking,

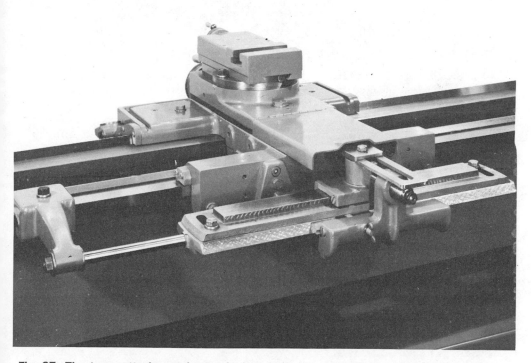

Fig. 37. The taper attachment is attached to both the cross slide and the lathe bed. (R. K. LeBlond Machine Tool Co.)

Fig. 38. The hexagon bed turret attachment can accommodate six cutting tools which can be used for sequential operations without time lost for tool-changing operations. (South Bend Lathe, Inc.)

CARRIAGE STOP

Fig. 39. The micrometer carriage stop is used to bring the carriage to exactly the same position for repetitive cuts. (R. K. LeBlond Machine Tool Co.)

counterboring, etc., may be performed without the usual delay caused by changing tools and toolholders.

Fig. 38 shows the lathe equipped with a hexagon bed turret operated by a hand lever. The lathe shown is also equipped with a double-tool cross slide.

Micrometer Carriage Stop

The micrometer carriage stop fastens to the front bedway and is equipped with a knurled micrometer barrel graduated in thousandths of an inch, Fig. 39. It permits the operator to bring the carriage to precisely the same position on repetitive cuts. Total adjustment of the stop is 1″.

Adjustable Thread Cutting Stop

The stop bracket clamps to the cross slide dovetail and the knurled stop screw extends through the bracket and fastens to the bottom slide. As successive thread cutting passes are made, the cross slide is

retracted, the carriage is repositioned, the depth of additional cut is fed in using the compound slide, and the bottom slide brings the stop screw against the stop bracket. This stop may also be used to gage the total depth of cut. See Fig. 40.

Fig. 40. The adjustable thread-cutting stop. (R. K. LeBlond Machine Tool Co.)

Maintenance of the Lathe

General

Care and maintenance are two very important items that must be attended to in the use of a lathe. Good care and maintenance make it possible for the operator to get the best results and lengthen the life of the machine. A neglected machine soon wears out if not properly cared for and maintained. To keep the lathe in the best operating condition, it is necessary to make frequent inspections and various adjustments. The periodic check should include such things as levelness of the machine, condition of the spindle bearings, clutches, gibs, cross feed and lead screws, gearing, and lubrication.

Gib Adjustments

Gibs may be either tapered or flat metal bars for taking up wear between bearing surfaces, such as the dovetailed surfaces of the cross slide, compound rest, or carriage. Gibs are provided with thrust screws by means of which the necessary adjustments are made. When making gib adjustments, (1) loosen the lock-screw; (2) tighten the gib screw until a smooth snug fit is obtained; and (3) lock the adjustment. If gibs are adjusted too tightly, binding will result. Gibs on the compound slide, Fig. 40, should be fairly tight when the compound is not being used for cutting angles.

Periodic Oil Changes

When a lathe is run daily, the oil should be changed in the headstock reservoir about every six months. A good grade of machine oil of SAE 20 or 30 should be used. The operators' instruction manual will state the grade of oil that should be used for the various machines. When changing oil in the reservoir, the plugs should be removed and the reservoir flushed with kerosene before refilling. The machine should be left running during the flushing process. All bearings fitted with oil cups should be oiled daily or as often as necessary. The performance of a lathe depends on the attention it receives. During the first three or four days, or the *breaking-in period*, all bearings should be carefully oiled and checked to see that none run hot.

Care and Storage of Accessories

The various attachments for the lathe should have a definite place for storage when not in use. When stored, they should be coated with a film of oil to prevent rusting. Care should also be taken in the use of accessories. Do not abuse them.

Safety Precautions—Lathe

The operator of the engine lathe will be exposed to certain hazards. Hazards can be minimized, however, if safe practice is followed in making the setups and performing the operations. The following rules of safety should be studied and followed:

1. Loose clothing, neckties, long sleeves, rings or wrist watches may

become entangled in revolving lathe parts. Remove these before starting to work.

2. All machine shops should require that goggles be worn at all times while working in the machine shop.

3. Chucks should always be started on the lathe spindle *by hand* instead of power. Remove the chuck wrench immediately after adjusting the chuck.

4. The operator should wait until the lathe stops of its own accord after the power has been shut off.

5. The lathe should be maintained in good condition at all times. Any defects should be reported to the foreman or instructor for correction as soon as they are discovered.

Courtesy of the National Safety Council

NOTE: Please see review questions at end of book.

Engine Lathes:

*Cutting Tools, Setups,
and Operations*

The engine lathe is an important machine shop tool because it is capable of performing many varied operations in job shops and tool rooms. The modern lathes are flexible and accurate. They are capable of working at high speeds using a variety of cutting tool materials. The cutting tool materials and the tool geometry selected for each cutting tool form the basis for correctly performing the many machining operations the engine lathe is capable of doing.

Our approach to the machining of metal on the engine lathe will be through a study of the operations involved. All work in industry, all parts made of metal, are manufactured in a given operational sequence. On the lathe, operations fall into two general classifications: *external operations* (facing, turning, knurling, external thread cutting, etc.) and *in-ternal operations* (drilling, boring, internal threading, etc.).

Importance of Proper Setups

Whatever the operation to be done, it cannot be executed until the proper machine setup has been made. The workpiece must be put into the machine properly. The machine itself must be prepared for the particular operation. To make the proper setup, the operator must rely on his knowledge of the lathe's major parts and their functions. In addition, he must be able to select work-holding devices of a suitable kind.

The word *setup*, sometimes referred to as *mounting the job*, means a lot more to the experienced operator than to the beginner, who is not fully aware of the elements involved. In some industries, the setup is made by one person and the operation carried out by another.

To mount the job properly, the operator must know the correct answers to these basic questions:

1. What kind of working-holding device should the workpiece be mounted on or in?
2. How should the workpiece be mounted in the work-holding device?
3. When is the workpiece accurately mounted?
4. What kind of cutting tool is needed, and how should it be mounted?

In the preceding chapter, the most common work-holding devices were illustrated. Their basic functions were described to give you the knowledge necessary to answer question number one correctly.

In order to answer question number two intelligently, you must consider the size, shape, and proportions of the workpiece, as well as the location of the surface upon which the cutting operations are to be performed. In this chapter, the ways in which the workpiece may be mounted in work-holding devices and afterward checked for accuracy will be discussed.

After mounting the workpiece properly in the machine, the next important step in making the setup is selecting and mounting the right cutting tool. In doing this, the kind of metal being machined and the grinding of the cutter to the correct shape and angles for the material and job operation are important. Next, the cutter must be mounted in the toolholder and the holder mounted in the machine at a suitable height and position.

Finally, you are urged to keep in mind that the various lathe operations are performed *only after the proper setup for the operation has been made.* Many times the same setup can be used in performing several operations merely by making a slight change—like changing the tool position or using a different tool. First, think through the operations needed to complete the job at hand; then arrange the operations in a workable sequence.

Lathe Cutting Tools

In order that metal may be machined effectively and accurately with an engine lathe, the correct type of lathe cutting tool, called a *tool bit*, must be employed. The cutting edge must be keen and well supported. The tool bit used should be suited to the kind of metal being machined and set at the proper position in relation to the work. Thus, the cutting angles will be correct.

The cutting tools widely used with

the engine lathe are further defined as single-point cutting tools. A single-point cutting tool is a relatively small piece of high speed steel ground to form the cutting edges essential in removing metal efficiently. The point of the tool may be integral with the shank, tipped with hard metal, or consist of interchangeable or replaceable bits of hardened cutting material attached to the end of the shank.

The cutting tool must be ground in precise form for two reasons. First, the cutting edge must be shaped and located properly in relation to the shank of the tool (which supports or supplies backing to the cutting edge). Second, the shape should allow the cutting edge to penetrate the work in the most efficient manner to provide maximum efficiency in removing metal.

Cutting Tool Surfaces

As the lathe tool bit is ground to form the cutting edges, various faces or surfaces are produced. You should become familiar with these surfaces in order that you may use the tool bit properly, and, if necessary, be able to grind a tool bit to the various cutting angles required for a particular job. The most important surfaces of a tool bit are the *top*, (or *face)*, the *end*, and the *side*. Also of importance is the *nose*, which is the cutting edge of the turning tools shown in Figs. 1 and

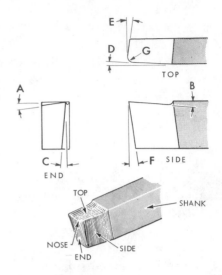

Fig. 1. The important surfaces and angles of a tool bit are shown. The tool has positive side and back rake.

2. The nose is formed by the meeting of the tool's principal surfaces (top, end, and side) along the ra-

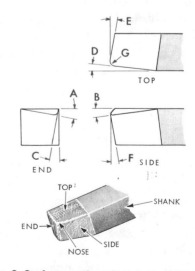

Fig. 2. Surfaces and angles of a tool bit with negative side and back rake are shown.

dius G. The unground part of the tool bit held by the tool holder is called the *shank*.

Function of the Cutting Tool Angles

The tool geometry, or the cutting angles ground on the tip of the cutting bit, play an important part in machining. There are six basic angles and a nose radius that must be clearly understood in order to see the function each angle plays in the chip cutting process. See Figs. 1 and 2.

Side Rake Angle. The term *side rake* means that the surface that forms the top (or *face*) of a tool has been ground back at an angle sloping from the side cutting edge. See Angle A, Figs. 1 and 2. The extent of side rake determines the angle at which the chip leaves the workpiece as it is directed away from the side cutting edge. Angle A, Fig. 1, shows a *positive* side rake angle. Angle A of Fig. 2 shows a *negative* side rake angle.

Back Rake Angle. The term *back rake* means that the surface that forms the top (or *face*) of the tool has been ground back at an angle from the nose. See Angle B, Figs. 1 and 2. However, when a tool is held by a tool-holder, the holder also determines the *overall* back rake angle. This angle normally is 16½ degrees. The amount of back rake affects the angle at which the chip leaves the workpiece as it is directed away from the nose of the tool. Angle B, Fig. 1, shows a *positive* back rake angle; Angle B of Fig. 2 shows a *negative* back rake angle.

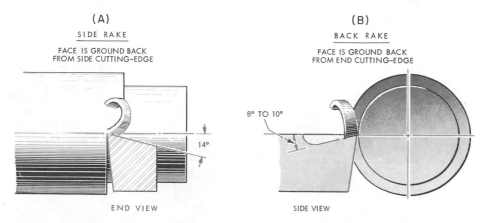

(A)
SIDE RAKE
FACE IS GROUND BACK
FROM SIDE CUTTING-EDGE

14°

END VIEW

(B)
BACK RAKE
FACE IS GROUND BACK
FROM END CUTTING-EDGE

8° TO 10°

SIDE VIEW

Fig. 3. Rake angles permit a smooth chip flow over the face of the tool and also provide a means of controlling the angle of shear.

183

The rake angles have a two-fold function: (1) to direct and facilitate the flow of the chip over the face of the tool, Fig. 3, and (2) to modify and control the cutting forces, which are also affected by the shear angle.

The greatest amount of force exerted on the top, or *face* of a turning tool, is concentrated in the area where the side and end cutting edges meet to form the tool nose. This force must be distributed along the entire lengths of the side and end cutting edges by one of two methods: (1) by grinding the face back at an angle sloping from the side cutting edge (*side rake*) as shown by the 14° angle in Fig. 3A or (2) by grinding the tool back at an angle sloping from the end cut-

ting edge (*back rake*) as illustrated by the 8° angle in Fig. 3B.

Side Clearance Angle. The term *side clearance* (or *side relief*) means that the surface forming the side, or flank, of a tool has been ground back (in) at an angle sloping from the side cutting edge. Side clearance concentrates the thrust exerted on the flank of a tool in a small area near the side cutting edge. This is shown by Angle C in Figs. 1 and 2.

Front Clearance Angle. The term *front clearance* (or *end relief*) signifies that the end surface of a tool has been ground back at an angle sloping down (in) from the *nose* of the tool. Front clearance concentrates the thrust exerted on the end surface of the tool in the area about the nose. This is repre-

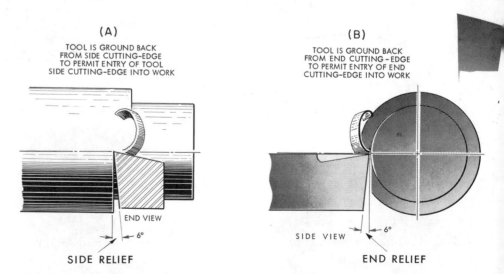

<table>
<tr><td>(A)</td><td>(B)</td></tr>
<tr><td>TOOL IS GROUND BACK FROM SIDE CUTTING-EDGE TO PERMIT ENTRY OF TOOL SIDE CUTTING-EDGE INTO WORK</td><td>TOOL IS GROUND BACK FROM END CUTTING-EDGE TO PERMIT ENTRY OF END CUTTING-EDGE INTO WORK</td></tr>
</table>

END VIEW — 6°   SIDE VIEW — 6°

SIDE RELIEF   END RELIEF

Fig. 4. Clearance angles permit the cutting edges to enter the workpiece material with a minimum of friction between cutter and workpiece.

sented by Angle F in Figs. 1 and 2.

The clearance (relief) angles, Fig. 4, permit entry of the tool into the workpiece so that the tool can cut, yet prevent excessive rubbing of the work on the tool. The size of the clearance angles depends to a great extent upon the amount of feed used and the diameter of the work for the particular cut. Clearance angles should be kept as small as possible since any metal ground away to form these angles *weakens support* to the cutting edges, and decreases the tool's ability on conduct heat away from the critical cutting area. If, however, clearance angles are too small, the tool will not cut freely; the side and end of the tool will bind on the workpiece. Fig. 4 illustrates side and front clearance angles of 6°.

Side Cutting-Edge Angle. The term *side cutting-edge angle* (Angle D of Figs. 1 and 2) signifies that the surface which forms the side of a tool has been ground back at an angle to the side of the shank. This angle establishes the tool's side cutting edge in relation to the shank.

End Cutting-Edge Angle. The term *end cutting-edge angle* (Angle E of Figs. 1 and 2) signifies that the surface that forms the end (or front) of a tool has been ground back at an angle sloping from the nose to the side of the shank. This establishes the angle of the end cutting edge of the tool in relation to

the workpiece.

Fig. 5 identifies the location of the side and end cutting-edge angles. The side cutting-edge angle establishes the angle of the side cutting-edge in relation to the shank, while the end cutting-edge angle establishes the angle of the end cutting edge of the tool in relation to the work surface. Fig. 5 illustrates a typical side cutting-edge angle of 15° and an end cutting-edge angle of 20°.

The side cutting-edge angle has an important but often overlooked effect on the metal cutting operation. This angle allows the tool to

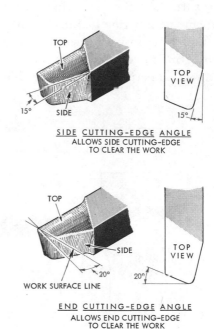

SIDE CUTTING-EDGE ANGLE
ALLOWS SIDE CUTTING-EDGE
TO CLEAR THE WORK

END CUTTING-EDGE ANGLE
ALLOWS END CUTTING-EDGE
TO CLEAR THE WORK

Fig. 5. The cutting-edge angles allow the cutting force to be distributed along the entire cutting edge.

enter the work so that the initial load is taken some distance *away* from the tool nose, Fig. 6, *right,* which is the weakest area of the tool. Likewise, this angle acts to gradually release the load on the tool when running out of the cut, Fig. 6, *left.* Both considerations are of special importance with carbide tools.

For efficient cutting action, the side cutting-edge angle should be large, Fig. 7. A study of this illustration, however, shows that there is a practical limit to the size of this angle. Here, in the left hand view, a thin, unsupported workpiece is being turned. A slender boring bar is supporting a boring tool in the view at the right. In both cases, the tool is exerting force against the work in a direction perpendicular to the side cutting edge. Of course, an equal and opposite force must be exerted by the work against the tool. This force tends to push the tool and work apart if the side cutting-edge angle is too great. The magnitude of this force can be measured in terms of the amount of deflection of the work with consequent chatter, as in view A of Fig. 7, or with resultant chatter or bell-mouthing by the boring bar in view B.

Radius. The term "radius" applies to the radius of a circle which the rounded part of a tool's cutting edge would form if the circle were completed. This arc of a circle is generally on the nose of a tool. It may be ground to any radius required to accomplish a desired cutting result. Letter *G* in Figs. 1 and 2 indicates the nose radius of tools

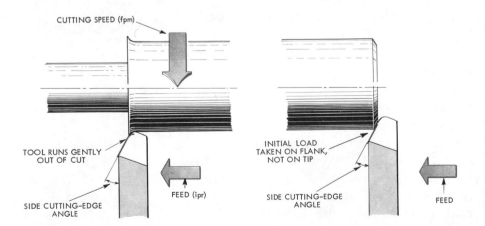

Fig. 6. *Right:* The side cutting-edge angle allows the initial load to be placed on the side rather than on the nose of the cutting tool. *Left:* This angle also permits the tool to run gradually out of the cut, avoiding sudden springing of the work when the load is reduced.

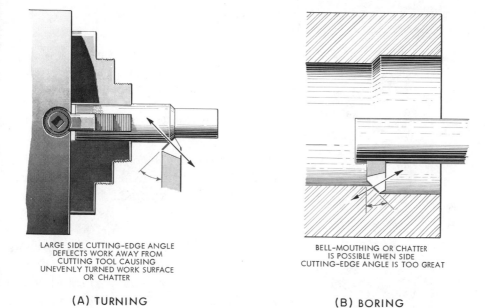

LARGE SIDE CUTTING-EDGE ANGLE
DEFLECTS WORK AWAY FROM
CUTTING TOOL CAUSING
UNEVENLY TURNED WORK SURFACE
OR CHATTER

(A) TURNING

BELL-MOUTHING OR CHATTER
IS POSSIBLE WHEN SIDE
CUTTING-EDGE ANGLE IS TOO GREAT

(B) BORING

Fig. 7. If the side cutting-edge angle is too large, the force normal to tool and workpiece surfaces will push the surfaces apart, deflecting either the workpiece or the cutter.

whose cutting edge is located at the corner tip where the top, end, and side surfaces meet.

Naturally, it is the proper grinding of the right amount of angle in each case that makes the tool a cutting tool. Some tool bits are ground to cut on the right side, some on the left side, and some are ground to cut on both sides.

Chip Types and Their Formation

By understanding the functions of the basic single-point tool angles, it is easier to visualize what happens when a lathe tool enters a workpiece and removes unwanted metal. This unwanted metal is removed in the form of chips. You will recall that chip formation and removal played an equally important role in drilling operations (Chap. 5). There are three basic types of chip formations.

Type 1 is the *Discontinuous or Segmental Chip*, produced in machining cast iron or other materials with deliberate inclusions intended to produce relatively brittle shear planes. Other conditions generally favoring formation of the Type 1 chip are: large chip thickness, small rake angle, and low cutting speeds. See Fig. 8.

The *Continuous Chip* is the Type 2 Chip. It does not have the built-up

187

Fig. 8. Section view of a segmental or discontinuous chip.

Fig. 9. Section view of a continuous chip.

Fig. 10. Section view of a continuous chip with a built-up edge.

edge and is commonly considered the ideal chip in turning. Conditions favoring its formation are: ductile workpiece material, small chip thickness, large rake angle, high cutting speed, and keen tool edge. See Fig. 9.

The Type 3 Chip is the *Continuous Chip with the built-up edge.*

This chip is commonly produced from very ductile materials (especially when having high work-hardening properties) under heavy cuts. See Fig. 10.

Right-Hand and Left-Hand Tools

Tools are often named from the manner in which they are used. This is done by prefixing the name of the tool with the qualifying term *right-hand* or *left-hand*, as the situation requires. A roughing tool, for example, is given the prefix right-hand (or left-hand) to identify it. This is the case with many kinds of tools. Figs. 11 and 12 will help you understand the prefix terms right-hand and left-hand. Tools are called right- or left-hand from the location of the tool's side cutting edges, but not in the manner you might expect.

The prefix *left-hand* is applied to tools which have their side-cutting edges on the right, and *move to the right*. The prefix *right-hand* is applied to tools which have their side cutting edges on the left, and *move to the left*. The reason for this is not clear. Perhaps the important point was the starting position of the cut. Thus, a tool that began cutting on the operator's left was called a left-hand tool even though it moved to the right and had its side cutting edge on the right.

Holding the Tool Bit

The small square tool bit is

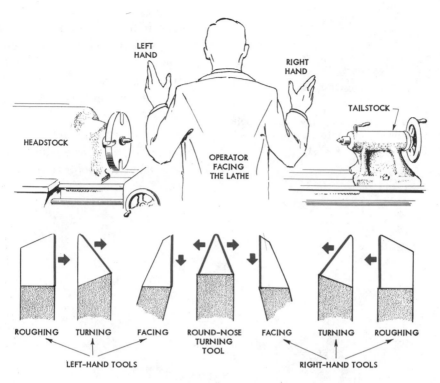

LEFT HAND

RIGHT HAND

TAILSTOCK

HEADSTOCK

OPERATOR FACING THE LATHE

ROUGHING   TURNING   FACING   ROUND–NOSE TURNING TOOL   FACING   TURNING   ROUGHING

LEFT–HAND TOOLS

RIGHT–HAND TOOLS

Fig. 11. Right-hand tools cut in the direction of the headstock. Left-hand tools cut in the direction of the tailstock.

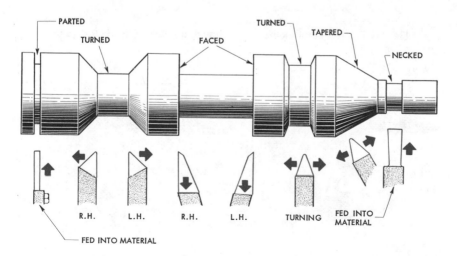

PARTED

TURNED

TURNED

FACED

TAPERED

NECKED

R.H.   L.H.   R.H.   L.H.   TURNING   FED INTO MATERIAL

FED INTO MATERIAL

Fig. 12. This illustration shows a variety of right- and left-hand tools and the cuts they produce.

Fig. 13. This left-hand forged steel tool-holder is used with interchangeable tool bits. (J. H. Williams & Co.)

clamped in a forged steel toolholder, Fig. 13, which makes a strong holding device. There are times when the toolholder must be bent to the side, as shown in Fig. 14, so the tool can be held properly without the holder getting in the way. This toolholder is called a left-hand toolholder. Other toolholders are shown

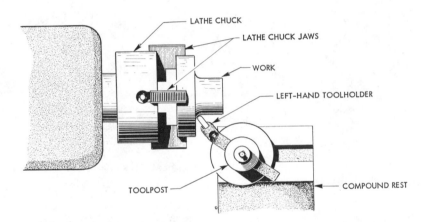

Fig. 14. The left-hand toolholder is shown in position in the tool post.

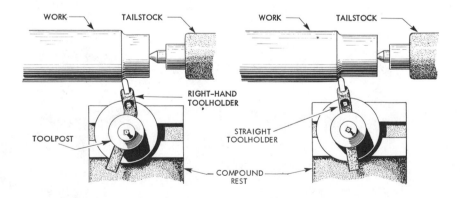

Fig. 15. *Left:* A right-hand toolholder can be used where the tool post might otherwise strike the tailstock. *Right:* A straight toolholder.

in Fig. 15. The holding device called a right-hand toolholder, shown in Fig. 15, *left*, permits work to be placed so that the toolholder support (tool post and compound rest) will not bump into the tailstock. It is commonly used in facing operations. Fig. 15, *right*, shows a straight toolholder being used.

Types of Cutting Tools

Facing Tool. When facing the end of a workpiece that is supported between centers of the lathe, the tool must be ground so it can cut close to the center of the workpiece without striking the dead center of the lathe, Fig. 16.

Some machinists use a special milled dead center, as shown in Fig. 17 *A*. Note that the dead center shows only about a fourth to a third of the point milled away. If less

than half the center remains to support the end of the piece, it will act as a countersink and cut into the end of the workpiece. A center of this kind should be mounted in the tailstock spindle so the milled portion is toward the cutting tool. The facing tool can be placed within the center hole and the facing operation performed. The workpiece will be smooth across the entire end, Fig. 17 *B*.

The end surface of a workpiece cannot be completely faced unless the tool is set at the same elevation as the workpiece center. If the tool is set below (or above) the work center line, the center portion of the workpiece will not be faced. Fig. 18 shows the height at which the tool should be set.

Round-Nose Tool. The round-nose tool may be employed in the

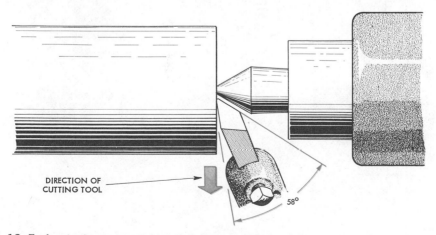

DIRECTION OF CUTTING TOOL

58°

Fig. 16. Facing tools are ground so that they can be brought to the center of the work-piece without striking the dead center.

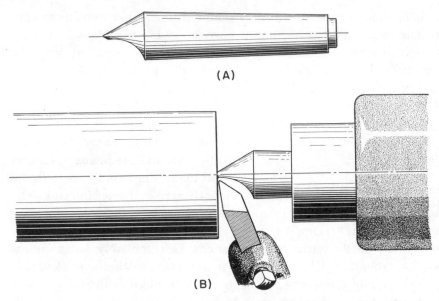

Fig. 17. When the dead center is milled as shown at A, the facing tool can be brought right to the center-drilled hole in the workpiece as shown at B.

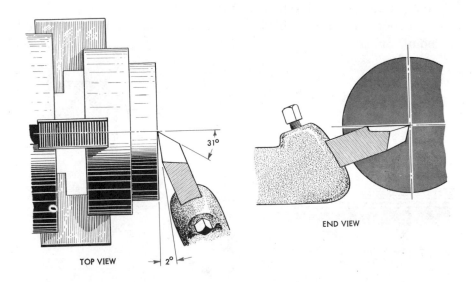

Fig. 18. The facing tool must be set to the correct height (on the axis of the workpiece) if the workpiece is to be completely faced.

greatest variety of lathe operations. This is true because (1) the tool may be ground so its side cutting edge is at any desired angle to the shank; (2) the nose may be ground to any desired radius; and (3) the tool may be set at different angles. When it is ground without side rake, as in Fig. 19, a cutting edge is established on each side and on the nose. The tool can, therefore, be fed to the right or left, or directly into the workpiece.

Heavy-Duty Roughing Tool. In taking "heavy" cuts, a heavy-duty roughing tool is used to remove a large amount of metal as rapidly as possible. Consequently, the tool is subjected to very high pressures and temperatures. Because of this, the tool is ground so that the cutting edge is brought parallel with the shank, or nearly so. The nose is rounded, but at a small radius. The front (end) and side clearances, as well as side rake, are reduced to a minimum. Even the front (end) cutting-edge angle is slight. All clearance angles on a heavy-duty roughing tool are reduced to a minimum to provide support for the cutting edge of the tool. If this were not done, the cutting edge would break down due to the high pressures developed when taking heavy cuts. Ordinarily, the tool is set slightly above center.

Finishing Tool. A finishing tool takes a light cut. It produces a smoother and more accurate finish than is ordinarily possible with a rough turning tool.

A finishing tool is usually ground with a larger radius on the nose

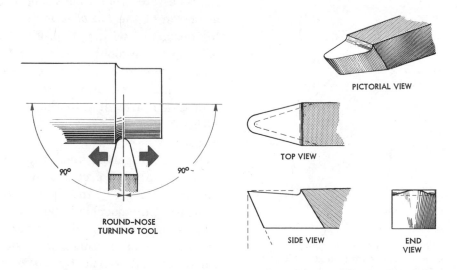

PICTORIAL VIEW

TOP VIEW

ROUND-NOSE
TURNING TOOL

SIDE VIEW

END
VIEW

Fig. 19. The round nose tool is an all-purpose tool which can cut right, left, or straight in.

of the tool than that of a roughing tool. As the tool cuts on the circumference of the rotating workpiece, it moves sidewise at a given rate per workpiece revolution—that is, at a given feed. Thus, the broader the tool's nose, the more it will overlap cuts previously taken by the roughing tool. A rough cut reduces the workpiece to a size approaching the shape and proportions desired in the shortest period of time, regardless of finish.

A relatively smooth finish is the product of many factors—all combined. Among them are the *kind of metal* in the workpiece, the *surface*

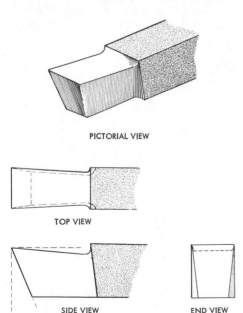

PICTORIAL VIEW

TOP VIEW

SIDE VIEW   END VIEW

Fig. 20. The necking tool cuts only with its nose, or end cutting-edge. This tool is used for cutting a groove having parallel sides.

speed of the piece, the *ratio of tool feed to rotating speed*, the *depth of cut* taken, and the *shape of the tool* used. As a general rule, the smoothest finish is produced by taking a light cut at a fine feed and high r.p.m. speed. The r.p.m. speed is increased not to produce a better finish but to save time.

Necking Tool, Cutoff Tool. You should also be familiar with the necking tool and the cutoff or parting tool. Both are square-nose tools; the width of the nose determines the width of the cut. The term *necking* refers to the operation of cutting a groove of predetermined depth in a workpiece. The term *parting* refers to the operation of cutting a workpiece in two.

To prevent binding as a square-nose tool penetrates the workpiece, the sides are ground back slightly, as shown at the *top view* in the illustration of a necking tool, Fig. 20. A square-nose tool has no side rake because the cutting edge is confined to the nose, and also because the tool is fed straight into the workpiece. It may, however, have a back rake.

A cutoff or parting tool is similar to a necking tool except that it is narrow—usually $\frac{3}{32}$ to $\frac{1}{4}$ of an inch in width. The tool should be as narrow as the work permits. This tool is usually manufactured with side clearance and is held by a special toolholder, Fig. 21. Back rake is confined to an area close to the nose since shallow cuts are taken.

Carbide and Oxide Cutting Tools

Besides the high speed steel cutting tools, tungsten-carbide and ceramic or oxide cutting tools are playing an important part in increasing productivity. Today's production machines have more horsepower and speed, and are capable of handling cutting tool materials that will withstand high speeds and temperatures for economical machining.

Carbide cutting tools are available in a wide variety of grades and styles in order to meet a variety of machining conditions. These cutting tools are made in blanks, or disposable inserts, in a variety of shapes, Fig. 22. These shapes include round, triangular and square designs, and are constructed to resist shock, wear, and abrasion.

Disposable inserts require special tool holders in order to mount these tips for use. Typical holder designs are illustrated in Fig. 23. Holders are available with either positive or negative rake angles, with a wide variety of adjustable chip breakers and clamping devices, as shown in Fig. 23. You will also note that the side cutting-edge angle is pre-ground on the tool holder.

Cemented carbide cutting tools are also available with the carbide tip brazed to the shank, which is usually plain steel. Brazed cutting

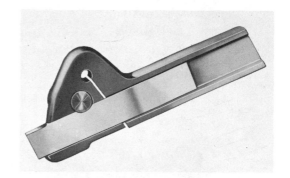

Fig. 21. The cutoff tool is a narrow tool used for "cutting off" a portion of the workpiece or for cutting narrow grooves. It must be supported with a special toolholder.

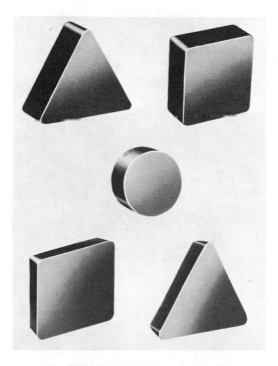

Fig. 22. Replaceable carbide cutting tool blanks come in a variety of shapes. (Metallurgical Products Dept., General Electric Co.)

195

tools are also available in a variety of sizes and shapes. See Fig. 24. These tools are more suitable for production turning operations, but are not as flexible as the disposable-tip tools described above.

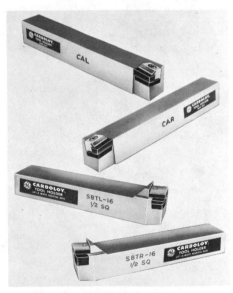

Fig. 23. Adjustable toolholders are available with a variety of pre-ground rake and cutting-edge angles. (Metallurgical Products Dept., General Electric Co.)

When selecting carbide cutting tool materials, careful consideration should be given to the manufacturer's grade classifications. The following information on the method for selecting carbide grade in cutting tools is reprinted from the *Carboloy Steel Tools and Blanks Catalog*, by permission of the Metallurgical Products Department, General Electric Company.

Selecting Carbide Grades. There are two main groups of carbides from which to select grades for most machining operations. The straight tungsten carbide grades commonly used for machining cast iron and non-metallics, and the crater-resistant grades normally used for machining steels.

The family of straight tungsten carbide grades, that is, those containing only tungsten carbide and cobalt are grade for grade, the strongest and the most wear resistant. The addition of anything to this basic composition serves to reduce the strength of the carbide and its abrasive wear resistance.

These grades, however, are generally not satisfactory for machining most steels because of their tendency to crater. The hot steel chip welds itself to the carbide and pulls out small particles of the cutting tool material forming a crater which causes rapid tool failure. To prevent this, titanium carbide and tantalum carbide are added to the basic composition of tungsten carbide and cobalt. These materials greatly reduce the cratering action but at the sacrifice of some abrasive wear resistance and strength.

In selecting a grade of carbide for any given job, proceed as follows:

Select the type of carbide most likely needed for the job. Straight tungsten carbide grades should be selected for cast iron, non-ferrous metals, non-metallics, series 300 stainless steels and many high temperature and high strength alloys.

Estimate the strength level needed in the carbide. A grade of carbide strong enough to handle the cut without chipping or breaking is required. The tougher the grade, the lower its wear resistance and vice-versa.

Take a trial cut. Run the tool for one or two minutes, or if machining time is short, machine one piece.

Examine the carbide. With a low-power magnifying glass, examine the cutting edge, top and flank of the carbide. Determine if the carbide is cratering, chipping or wearing.

Change grade of carbide as required. If a crater formed in the top of the insert where the chips flowed over it, change to a steel cutting grade, that is, a grade containing titanium and/or tantalum. If a steel cutting grade is being used and it still cratered, a change to a harder grade is indicated.

If the edge chipped or if the insert or tip broke, a change to a tougher grade of carbide may be required. Further checking should be done to make sure the chipping was not due to some other cause. If the flank of the cutting edge shows sign of excessive wear, a change to a harder, more wear resistant grade is indicated.

Recheck. Change only one condition at a time and recheck.

Is setup rigid? Is there chatter? That is if both cratering and wear occurred, change first from a straight tungsten carbide grade to a steel cutting grade of comparable hardness.

Was edge honed? If the cratering is eliminated but if rapid wear still occurs, change next to a harder steel cutting grade.

No matter what the particular problem, the basic approach to grade selection is the same. The machinist must decide two things: what strength level (or transverse rupture strength) is necessary, and what type of wear it is necessary to combat. Is it the abrasive-type wear of the non-metallics and non-ferrous materials? Or is it cratering, seizing, welding and galling that is encountered on ferrous metals? Or is it a combination of both?

There are four guideposts to carbide selection:

1. Always use a grade with the lowest cobalt content and the finest grain size whose strength is sufficient to prevent breakage.
2. Use straight tungsten carbide grades to combat straight abrasive (not cratering) wear.
3. Use grades containing titanium carbide to combat cratering, seizing, welding and galling.

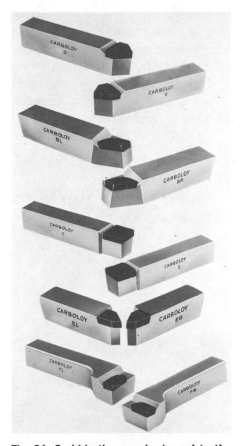

Fig. 24. Carbide tips can be brazed to the shanks of ordinary cutting tools. (Metallurgical Products Dept., General Electric Co.)

4. Use grades containing tantalum carbide for extremely heavy cuts in steel where the heat and pressure tend to deform the cutting edge.

Oxide or ceramic cutting tools are finding their way into industry on many turning jobs because they can be operated at higher speeds than cemented carbides. This cutting tool material is primarily composed of aluminum oxide or corundum. Corundum is the second-hardest natural mineral and is the material used in most grinding wheels.

Individual crystals of aluminum oxide are tightly bonded together to produce the cutting tool material. This material is usually stronger than the natural substance and just as hard. This cutting tool material

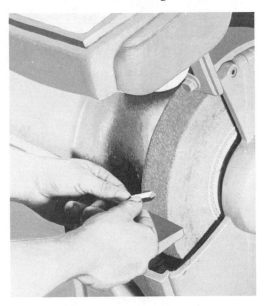

Fig. 25. The high-speed steel tool bit may be ground by hand on a conventional grinding wheel.

has a high hot-hardness and remains hard at high temperatures generated by high cutting speeds. Aluminum oxide tools will produce up to twenty times more pieces than cemented carbide tools before requiring replacement. Oxide tools are available in various sizes and shapes as disposable inserts, which must be mounted in tool holders similar to those for carbide blanks illustrated in Fig. 23.

Grinding the High Speed Steel Cutting Tool

While hand-grinding on the emery wheel is satisfactory, best results are obtained by the use of a regular tool-grinding machine.

The steel tool bit will need to be ground with regard to the duty expected of it. Its shape and angles must be right for the work to be done — facing, rough turning, or shouldering.

Steps in Grinding and Sharpening Tool Bits. Tool bits are generally "roughed out" while held by the toolholder. A generous portion of the tool bit should project out of the toolholder. Be careful not to grind into the holder itself.

After roughing out, the tool bit should be removed from the toolholder and finished to shape by holding it firmly in the fingers, Fig. 25.

Do not apply too much pressure on the grinding wheel. There is danger that the tool bit will be torn from your grasp, injuring you or damaging the tool. Also, too much pressure will burn the tool and fingers. A hot tool bit may lose its temper. In addition to grinding, a tool used for fine finished cuts should be whetted carefully on a fine oilstone.

Carbide cutting tools must be ground on a special type of grinding wheel. They are so hard that they cannot be ground satisfactorily on the ordinary corundum wheel. Retaining the proper cutting edge support requires accurate grinding of clearance angles; this should not be attempted on the corundum wheel. Generally, diamond wheels of fine grit are used to grind finishing tools. Vitrified silicon carbide wheels are used for roughing tools, or where the shape of the tool is to be substantially changed.

BASIC EXTERNAL MACHINING OPERATIONS

Facing

One of the first operations performed on a piece of work in the lathe may be that of facing and squaring the ends. Facing is done on a workpiece to produce a smooth and true face on the piece, and also to bring it to the desired length. Facing is very often done on the workpiece while it is held and revolved in the three- or four-jaw chuck. Another means of holding and revolving the workpiece for the facing operation is between centers. It is recommended that the student review the Lathe Trans-Vision at this point. The student will find that, while the operations performed on the lathe may vary, many of the setups are similar, if not identical, to the setup for the turning operation shown in the Trans-Vision.

Setup for Facing in the Chuck

Fig. 26 shows the operation of facing a piece of stock held in a three-jaw chuck mounted on the headstock spindle and rotated by the headstock power unit. The workpiece should not project from the chuck a distance greater than two and one-half times the diameter of the workpiece.

Mounting a Chuck on the Lathe Spindle. The first step in setting up the job is to mount the chuck on the spindle. Thoroughly clean and oil the threads of the spindle and the chuck before mounting the chuck on the lathe spindle. When removing the chuck from the spindle, a block of wood should be placed across the ways of the lathe. This prevents the chuck from damaging

Fig. 26. A facing operation is shown being performed on a workpiece mounted in a three-jaw chuck.

Fig. 27. Checking the workpiece setup for trueness with a piece of chalk.

the lathe ways when the chuck drops off the spindle nose.

Mounting Work in Four-Jaw Independent Chuck. With the chuck in position on the spindle of the lathe, the next step is to mount the work in the chuck. Notice the concentric grooves in the face of the chuck. These grooves will aid you in adjusting the movable, independent jaws to an approximate position for centering round stock. The grooves in the chucks serve as a guide. Before starting the machine, each of the jaws should be tightened against the work with an even pressure. Then remove the key.

As the chuck turns slowly, you can see whether the work is mounted in the center. A good way to check trueness is with a piece of chalk, Fig. 27. The chalk, held close to the work, will mark a "high" spot. Loosen the jaw on the side opposite the chalk mark, then tighten the jaw opposite the one loosened. This will center the work more accurately. Do not loosen more than one jaw at a time.

For very accurate centering, use a dial indicator mounted in the tool post, Fig. 28. If the workpiece is truly centered, the dial needle will almost stand still when you turn the chuck by hand.

Suggested Method for Facing in Chuck. Put a properly ground *facing* tool in the toolholder. Place the holder in the tool post on the rocker, with the tool post in the left-hand side of the slot on the compound. Run the cross-slide in until the point of the tool is even with the center of the workpiece. Be sure to check the height of the tip of the tool bit. It must be just even with the center, Fig. 29.

The tool is positioned so it looks like the setup in Fig. 30. Now lock the carriage to the lathe ways by tightening the carriage lock screw. This should be done to prevent the carriage from moving away from the work as the facing cut is taken.

If the facing is done well, there will not be a burr or fin remaining at the center of the workpiece.

Removing the Chuck from the Lathe Spindle. The first step in removing the chuck from the spindle is to stop the machine. Have the power switch button in the "off" position. Next, place the wooden lathe-chuck cradle or board on the bed of the lathe under the chuck. If necessary, lock the back gear and release the chuck by breaking it loose on the spindle. Now reverse the spindle by hand while holding the chuck steady until it is freed and resting on the wooden support. Special spanner wrenches are sometimes used in removing the chuck. The lathe spindle is locked and the wrench is placed in the T-slot provided. As pressure is applied to the handle of the wrench, the chuck is loosened.

Storing the Chuck. When the chuck is not in use, it should be placed where it will not be in the way of the operator. Place it where chips and dirt will not collect in the threads and where it will not be

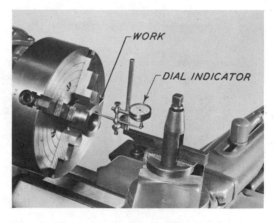

Fig. 28. A dial indicator can be used to check the workpiece setup for trueness. The workpiece is held in a four-jaw independent chuck.

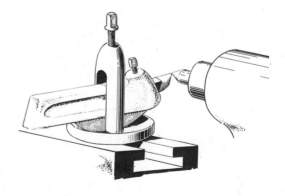

Fig. 29. This is one method of setting the cutting tool bit for the correct height.

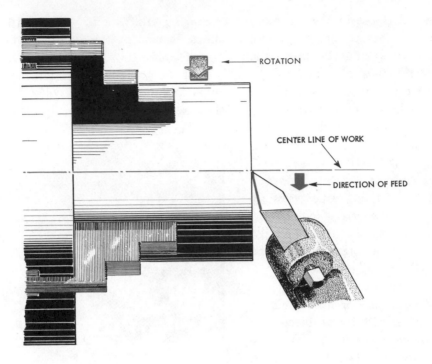

Fig. 30. The facing tool must be placed on the centerline of the workpiece.

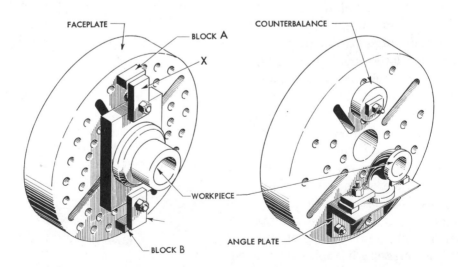

Fig. 31. Workpieces of irregular size or shape which are to be faced can be mounted on a faceplate in a variety of ways.

damaged by tools and parts that may be knocked against it.

Setup for Facing Work on the Faceplate

The faceplate is a very good work-holding device for certain shapes and sizes of workpieces to be faced. Once the workpiece is properly mounted on the faceplate, the setting of the tool and the facing operation itself are much the same as for facing in the chuck. See Fig. 31.

Mounting and Removing the Faceplate. Before mounting or removing the faceplate, place a board or cradle across the ways of the lathe to protect these parts. Dirt and chips should be removed from the threads of both spindle and faceplate. If the faceplate screws on with difficulty, look for dirt and burrs. When these are removed, try again. The hub of the plate should screw tight against the shoulder of the spindle, but the plate should not be spun up with a fast turn. Do not run the spindle by power while attaching the faceplate. It will be too hard to remove the plate. When the job is finished, the faceplate should be carefully removed and properly stored.

Mounting the Work on the Faceplate. The setups shown in Fig. 31 show workpieces clamped to the faceplate. For ordinary turning, boring, and facing operations, the work is bolted or clamped as shown at

the left. Notice the position of the workpiece and the clamping straps *X* and *Y*. The blocks *A* and *B* in this illustration are used for proper positioning of the clamping straps. A counterbalance is sometimes needed to balance the workpiece. Odd-shaped work is mounted on an angle plate. See Fig. 31, *right*.

Setup for Facing between Centers

The facing operation is occasionally performed on the right-hand end of a workpiece mounted between centers. However, before the workpiece can be mounted between centers, it is necessary to prepare the ends of the workpiece by center-drilling holes into which the points of the lathe centers can be inserted to support the work. See Fig. 32.

Setup for Drilling Center Holes. The workpiece is accurately mounted in the three- or four-jaw chuck. A combination drill and countersink is mounted in the Ja-

Fig. 32. Preparing the workpiece for turning between centers by center-drilling holes.

cobs chuck held in the spindle of the tailstock. The center drill is then fed into the end of the workpiece as the workpiece is revolved by the chuck.

Mounting Centers in Spindles. Before mounting the lathe centers in the headstock and tailstock spindles, thoroughly clean the centers and the tapered holes. Use a cloth and a stick to clean the holes. (CAUTION: Do not attempt to clean the spindle of the headstock while the spindle is revolving.)

The live center, Fig. 33, is held in the main hollow spindle of the headstock. It must run perfectly true. If the center does not run true, the work will run out of true.

The dead center in the tailstock must be in perfect line with the live center, Fig. 33. If the tailstock center is found to be out of alignment, loosen the tailstock clamp bolt. Then set over the tailstock top in the proper direction by adjusting the setover screws. Retighten the clamp bolts.

Lathe centers soon become worn or damaged, and it becomes necessary to true the point to restore it to proper condition. The point is either ground or turned in the lathe to the 60-degree included angle.

The dead center is usually hardened, and sometimes both live and dead centers are hardened. Hardened centers are usually marked by grooves near the conical end. The best way to restore hardened centers to good condition is to grind them. The points of soft centers are trued by a turning tool.

Mounting Drive Plate. To drive the lathe dog, mount the face or drive plate on the spindle of the headstock, Fig. 34. The drive plate is screwed onto the nose of the headstock after the screw threads in the plate have been cleaned. The nose on some spindles requires a special locking drive plate.

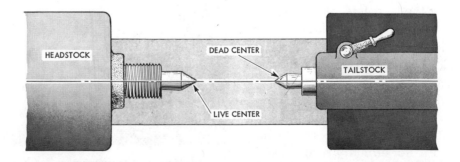

Fig. 33. The live center revolves with the workpiece. The dead center usually does not revolve with the workpiece, which explains its name.

Mounting Workpiece between Centers. Place the bent-tail lathe dog on one end of the workpiece. The tail should project enough over the end of the work to engage the slot in the drive or face plate. See Page B of the Lathe Trans-Vision.

Fill the dead center hole in the workpiece opposite the lathe dog with white lead or special lube.

The distance between the live and dead center points should be slightly greater than the length of the workpiece. The tailstock spindle should be drawn into the tailstock before attempting to mount the workpiece.

Place the work on the live center, with the tail of the dog inserted into one of the slots in the faceplate. Run the dead center into the right-hand end of the work, Fig. 34. The tail of the lathe dog should move freely in the slot in the plate. Adjust the tailstock spindle and clamp it as explained on Page D of the Lathe Trans-Vision.

Suggested Method of Facing Work Mounted between Centers. Mount the facing tool in the proper position at a height even with the center points or center line (axis) of the workpiece, Fig. 35.

Use the manual cross-feed and carriage feed, Page E of the Lathe Trans-Vision, to move the facing tool in as near the dead center as possible without bumping the point of the facing tool, Fig. 17.

Lock the carriage to the lathe bed

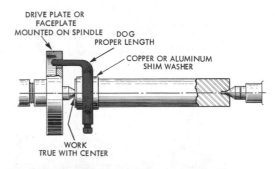

DRIVE PLATE OR FACEPLATE MOUNTED ON SPINDLE

DOG PROPER LENGTH

COPPER OR ALUMINUM SHIM WASHER

WORK TRUE WITH CENTER

Fig. 34. A lathe dog is shown inserted in one of the slots in the drive or faceplate mounted on the headstock spindle.

as explained in the chuck facing operation. Then start the lathe and take a cut using either a hand feed or a power feed. Return the cutting tool to the center of workpiece using the hand cross-feed.

If a power cut is to be taken, set the feed selector levers on the gear box and apron to obtain a medium power feed of the cross slide. Engage the friction feed clutch so the cutting tool faces the end of the workpiece.

Repeat the procedure until the end is square, or until the facing operation has faced the surface to spec-

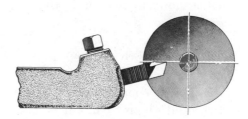

Fig. 35. Moving the facing tool into the proper position.

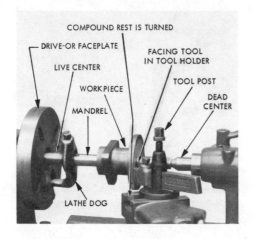

Fig. 36. A facing operation is shown being performed on a workpiece mounted on a mandrel. The mandrel is tapered and is driven between centers. The workpiece is fitted over the small (tailstock) end of the mandrel and forced on as far as it will go. (R. K. LeBlond Machine Tool Co.)

ifications, or the work is the proper length.

Setup for Facing a Workpiece on the Mandrel

Another setup for facing is shown in Fig. 36. Here the workpiece is mounted on a *mandrel* placed between centers. Cylindrical work that has been bored and reamed in a chuck is usually further machined on the outer surfaces by holding it on a mandrel. The surface to be faced in Fig. 36 is the flat surface on the sides of the workpiece. The facing tool is shown in position in the tool post.

How to Use the Mandrel. The mandrel is slightly tapered. It must be pressed into the hole in the workpiece tightly enough to prevent the work from slipping while it is being machined. Standard hardened and ground mandrels may be purchased or they may be made in various sizes.

They are usually ground to about .005 taper per inch of length.

The small end of the mandrel is placed into the hole of the workpiece after first coating the mandrel and hole with oil or a mixture of oil and white lead.

The larger end of the mandrel is mounted toward the headstock. When the pressure of the cutting tool is exerted against the workpiece, the piece will be forced more tightly onto the mandrel.

Suggested Method of Facing Work Mounted on the Mandrel. The procedure for facing work mounted on the mandrel is similar to that described in the preceding discussion on facing work held in the chuck, faceplate, or between centers. CAUTION: Be careful not to bring the facing tool in contact with the mandrel at any time, or the tool will be dulled and the mandrel scored or nicked.

Straight Turning Between Centers

Straight cylindrical turning, Fig. 37, removes metal from a piece of stock to form a cylinder. The cutting is done along the entire length of the shaft. Thus, the piece has the same diameter at all points along its length. This operation involves reversing the piece end-for-end on the centers in order to cut down to size the portion that previously lay under the lathe dog.

In machine shop work, two kinds of cuts are taken: a *roughing* cut and a *finishing* cut. When a piece of work is roughed out, it is nearly the finished size. However, enough material is left on the surface—usually about ¹⁄₃₂ of an inch (.0312)—to finish the work smoothly and to exact size.

Steel bars, forgings, and castings are made in shapes and sizes that allow easy machining. The pieces are given one rough cut and one finish cut. At times, however, more than one rough cut is necessary.

Setup for Straight Turning between Centers

Because straight turning is required in so many jobs, the setup is a common one. Every lathe operator must know how to make a setup for straight turning. While the operation to be performed in the Lathe Trans-Vision is a *shoulder* turning operation, the setup is basically the same as that for a straight turning operation. The operation of turning between centers was chosen for the Trans-Vision because setups for turning between centers are frequently made, and are basic setup steps for many related operations on the engine lathe .

Preparing Centers of Workpiece and Mounting the Work. The workpiece is faced and prepared

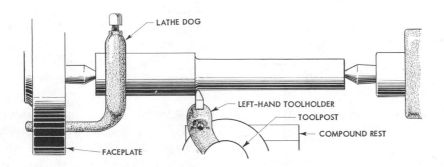

LATHE DOG

LEFT-HAND TOOLHOLDER

TOOLPOST

COMPOUND REST

FACEPLATE

Fig. 37. To turn the entire workpiece, the workpiece must be reversed on the centers to finish the part originally held and shielded by the lathe dog.

with accurately located and properly drilled and countersunk holes. Then the piece can be mounted between centers. (See the sections on Facing, Drilling, and Countersinking.) Check headstock and tailstock centers for alignment as shown in Fig. 33. It is recommended that the

Fig. 38. The lathe dog must be firmly secured on the workpiece with a wrench. (J. H. Williams & Co.)

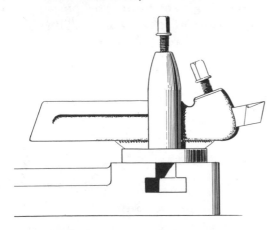

Fig. 39. The toolholder is mounted with just enough overhang to allow clearance to use a wrench for removing the cutting tool.

student again review the Lathe Trans-Vision before reading on.

The next step is to select and mount a lathe dog on one end of the workpiece. A driving plate is placed on the spindle and the work is mounted between centers, Fig. 38.

Suggested Method of Rough Turning. Grind a roughing tool and mount it in the toolholder. The toolholder should project no more than necessary. In Fig. 39, the toolholder is shown properly mounted. Tighten the tool post with the wrench. (Correct position of the cutting tool when viewed from the top is shown in Fig. 40.) The tool bit should be set slightly *above center*, rather than on-center as in facing operations.

Set the lathe for the suggested cutting speed and feed for the metal being turned. Start the lathe. Using the carriage feed, take a cut from the right-hand end of the work. Be careful to get the point of the tool under the scale on the work when turning cast iron.

The feed selector lever on the carriage can be set for a longitudinal feed. When the friction feed clutch is engaged, the carriage will move under power.

CAUTION: The proper lubricant and coolant should be applied at the cutting point. If the machine labors and starts to stall, stop the lathe and reduce the amount of feed.

To cut away most of the excess

material, the roughing cut should be as heavy as the machine and tool will stand. The aim is to remove the excess stock as rapidly as possible without leaving a badly torn and rough surface. CAUTION: Too heavy a cut may damage the centers or bend the workpiece.

Measuring Rough-Turned Diameter Using Calipers. To find the size of the turned piece after the cut is taken, it must be measured. The outside caliper is set to the measurement specified on the blueprint. The caliper is held over the work; see Fig. 41. When the workpiece is the proper size, the caliper will fit over the workpiece. The caliper is not a precision measuring tool; therefore, the measurement obtained will be reliable only within .005 of an inch. Because the workpiece has been only rough turned and is not machined to finish size, the measurement taken with the caliper is close enough.

Reversing Ends. After the right end of a workpiece has been rough turned to the proper diameter, allowing $\frac{1}{32}''$ or $\frac{1}{64}''$ for the finish cut, it becomes necessary to reverse the piece between centers in order to rough turn the part of the work held by the lathe dog.

The lathe dog is removed from the workpiece and is mounted on the turned end. If the material is soft, it may be necessary to protect the work from being dented by the screw of the dog as it clamps onto the work-

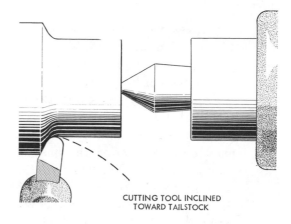

CUTTING TOOL INCLINED TOWARD TAILSTOCK

Fig. 40. Rough turning tool in correct position on the workpiece.

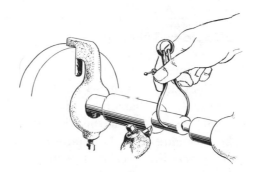

Fig. 41. Checking the diameter of the already-turned part of the workpiece with a caliper.

piece. A strip of metal can be placed under the end of the screw. White lead is applied in the center hole to be mounted on the dead center. Then the work is mounted between centers as before. In the final operation, the end, which originally was held by the lathe dog, is rough turned to size.

209

Suggestions on How to Do Finish Turning. Mount the finishing tool in the toolholder. Position the tool to take the finishing cut.

From the measurement taken with calipers, you can determine the depth of the finishing cut. Using the hand cross-feed, bring the nose of the tool up until it just touches the work lightly. Take the reading on the micrometer collar of the cross-feed ball crank. Then turn the hand carriage wheel to the right until the tool is clear of the work. Now feed the tool in with the hand cross-feed a distance equal to half the differ-

ence in diameter between the rough-turned piece and the desired finish size. The micrometer collar will help in determining this new setting.

Select the proper speed and take a cut. Stop the machine and measure the diameter with a micrometer caliper. If the work is oversize, adjust for a slightly greater cut. Now proceed to let the carriage feed as close up to the dog as possible. Then stop the machine and remove the dog from the workpiece and place it on the opposite end of the work. Now finish the end of the workpiece which originally was held by the dog.

Shoulder Turning

Work that is turned to more than one outside diameter, with steps from one size to another, is said to be "shouldered." As mentioned in the preceeding section, the shoulder turning operation is featured in the Lathe Trans-Vision. Discussion of the setup for a straight turning operation applies equally to the shoulder turning operation. The additional setup requirements for shoulder turning are discussed below.

Setup for Turning Shoulders

The setup for turning shoulders is similar to that for straight turning, Fig. 38. The type of tool used, Fig. 42, will depend on the type of shoul-

der desired—for example, a sharp corner or one with a radius. The filleted corner, it should be noted, gives greater strength at the shoulder.

Suggestions on How to Turn Shoulders. Before starting to turn a shoulder, lay off the required distance as specified on the blueprint or drawing. This is usually done by measuring from the faced end at the tailstock center.

Laying Off Lengths with Hermaphrodite Caliper or with Scale. To lay off lengths, the hermaphrodite caliper is helpful. This instrument is most useful in making marks fairly close to the ends of the work. After the ends have been faced, you

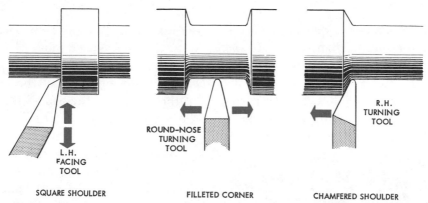

SQUARE SHOULDER   FILLETED CORNER   CHAMFERED SHOULDER

Fig. 42. Three common types of shoulders are shown, with the turning tools used for each.

can lay off lengths on the work while it is mounted in the lathe.

Hold a piece of chalk against the work as it revolves. Chalk the area to be marked. With the rule, set the hermaphrodite calipers to the length to be laid off. Hold the bent leg of the caliper against the end of the work and revolve the workpiece slowly while the scriber makes a mark on the chalked part of the workpiece.

Lengths on the workpiece can also be measured off simply by using the machinist's rule and scribe. After the desired length is measured, turn the workpiece by hand to scribe the line around the piece.

Knurling

Knurling is the process of checking the surface of a piece of work by rolling depressions into it, Fig. 43. This operation of embossing a pattern on the surface of cylindrical work is done with a knurling tool as the work revolves in the lathe. Knurling on metal shapes provides a better grip for the hand and improves the appearance of the surface.

Setup for Knurling

Fig. 43 shows the knurling operation being performed on a cylindrical workpiece mounted between centers. Here the lathe dog is used to rotate the workpiece. Therefore, the setup is similar in many ways to that for straight turning. Occasionally the piece to be knurled is held in the chuck.

Knurling Tool. The tool-post type of knurler with a single pair of hardened rolls is used more commonly than any other knurling tool, Fig. 43. This tool is known as a *knuckle-joint* knurler.

The knurls or rolls are made in

211

two patterns—the *diamond* and the *straight line*. Each of these patterns can be made in three pitches: 14 (coarse), 21 (medium), and 33 (fine); see Fig. 44. The pitch is the number of teeth contained per linear inch. The size and nature of the work

Fig. 43. Diamond-point knurling of a cylindrical workpiece. (South Bend Lathe, Inc.)

determine the pitch and the pattern to be used for a given job.

Suggestions on How to Knurl. First, locate the limits of the knurl on the surface of the work—that is, the beginning and end of the knurled portion.

Set the knurling tool so that the top roller is the same distance above the center of the work as the bottom roller is below the center of the work. The working faces of the rolls should be set parallel to the surface of the work.

Set the lathe to run at the slowest speed of the back gear. Then set the feed selector levers to move the carriage a distance approximately half the width of the knurling rolls per revolution of workpiece.

Move the cross-slide and carriage to position the knurling tool at the

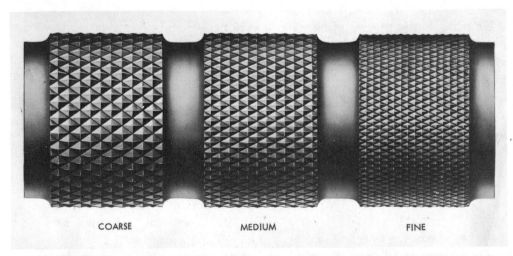

COARSE   MEDIUM   FINE

Fig. 44. Coarse, medium, and fine diamond-point knurling rolls. (J. H. Williams & Co.)

right edge of the section to be knurled. Start the lathe and force the knurls into the work with the hand cross-feed control to a depth of about $\frac{1}{64}$ of an inch.

At this point it is well to take a test cut to make sure the tool is mounted square with the work. A perfect diamond pattern should be produced. The diamond marking must in no case be split by one of the wheels. The finished job should show a clean, sharp diamond. If the wheels do not track properly, the knurling tool may have to be raised or lowered to produce a good clean knurl. Then, when the tool makes the proper knurl, go over the whole surface.

Oil is used to lubricate when knurling material of any kind.

Cutting Off

Many parts made on the lathe are machined out of stock that originally was cut a little longer than the finished dimensions. This allows for the center holes drilled in the ends to be cut off, leaving a finished end. The cutoff tool, Fig. 20, serves this purpose. Frequently, several parts are made out of one piece and later cut apart.

Setup for Use of Cutoff Tool

In cutting off, the setup of the tool should be made rigid. The tool should be fed into the work at right angles to the center line of the work-piece, Fig. 45. The carriage is clamped to the bed to keep the tool from moving to the right or left.

When cutting off steel, keep the work flooded with oil. No oil is needed, however, in cutting off cast iron. Be sure to set the cutoff tool so that the point of the tool is on the center line of the workpiece. This operation should be performed at slow speed, using the back gears.

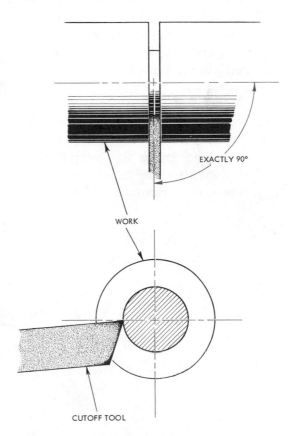

Fig. 45. The cutoff tool is always set at a right angle to the workpiece.

213

Filing Work in the Lathe

Workpieces are filed in the lathe for a number of reasons. Burrs and sharp corners on the piece can easily be removed with a file. Tool marks can also be removed by filing. By doing careful work, one piece can be made to fit another more accurately. Filing in the lathe, however, is generally not considered good practice. Much of what is done by the file might be done better with an accurately ground tool bit and careful workmanship. Excessive or careless filing will destroy the accuracy of the turned piece.

Setup for Filing

Many of the setups previously described can be used in filing. The reason for this is that filing usually is done after completing turning operations. Performance of the filing operation while the workpiece remains in the chuck, on the faceplate, or between centers merely requires that the proper speed be selected and the correct file for the job be chosen.

Fig. 46 shows a typical setup for filing. In this illustration the workpiece is mounted between centers.

The recommended procedure for this operation is to grip the file handle in the left hand and use the right hand to hold the tip of the file. This procedure *(suggested by the National Safety Council)* keeps the operator's arm out of range of the driver plate and lathe dog or chuck. When filing in the lathe, the important thing to keep in mind is that every precaution should be taken to avoid accident. In the shop, the safe way is the best way.

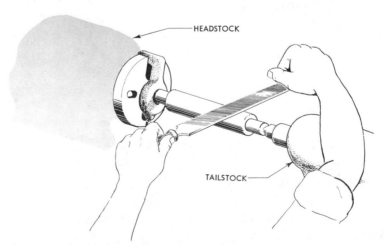

HEADSTOCK

TAILSTOCK

Fig. 46. Filing a workpiece mounted on the lathe.

Suggested Method of Filing in the Lathe. An 8-inch or 10-inch mill file is the best tool to use for filing in the lathe. Filing is done dry —without oil or other lubricants.

To avoid a hot dead center when filing, loosen and oil the dead center before, and occasionally after, increasing the speed. The lathe is run at a speed that will turn the work two or three revolutions for each stroke of the file. If the work is re-volved too fast, the file will not "bite" but only graze the work.

Long, slow strokes should be used in filing. The strokes are crossed while the operation is in progress. This crossing of the strokes is called cross filing. It is done by moving the file from left to right and right to left. Crossing the strokes eliminates the tendency to file more material from one part of the workpiece than from another.

Polishing Work in the Lathe

Work is frequently given a polishing while it remains in the lathe, the purpose being to improve the appearance of the piece. Whenever possible, a speed lathe rather than an engine lathe should be used in performing this operation, Fig. 47. The gritty spatterings produced in polishing will injure working parts of an engine lathe. An exception may be made when the workpiece is machined in a chuck in an engine lathe; in that case, the polishing is done before the workpiece is removed from the chuck.

Setup for Polishing

Fig. 47 shows the polishing operation. In this kind of polishing, a piece of emery cloth is held in the hand. Do not bring the hands too close to the lathe dog. The emery strip should be of adequate length.

Suggested Method of Polishing. Speeds for polishing operations in the lathe should be about 5,000 to 6,000 f.p.m. surface speed.

Various kinds and grades of emery cloth are used. Grade 60 to 90 will give a satisfactory polish. For a bril-

Fig. 47. Using a strip of aluminum oxide (emery) cloth to polish the turned surface of a workpiece. (The Carborundum Co.)

215

liant polish, used grade 120 emery cloth and "flour." First, apply grade 60 emery with hard pressure until tool marks and scratches in the metal are eradicated and the pores in the metal have nearly disappeared. Now apply grade 90 emery with a lighter pressure until all scratches from the coarser emery have been removed.

Use lard oil on the emery cloth or on the workpiece, distributing it with the fingers. Use the oil sparingly to avoid spattering. If sharp corners are desired, do not let the emery cloth slip off the edges of the work.

Grinding

Grinding can be done in the lathe if the machine is equipped with an electric grinding attachment similar to that shown in Fig. 48. This permits the grinding of lathe centers and the sharpening of reamers and some milling cutters. The grinder is mounted directly on the compound rest of the lathe. The unit consists of a grinding wheel powered by its own electric motor.

Fig. 48. Grinding can be done on the lathe using an electric grinding attachment mounted on the compound rest. (R. K. LeBlond Machine Tool Co.)

Setup for Grinding

The V-ways of the lathe bed should be covered with a heavy cloth to protect them from the dust and grit that collects in grinding.

When grinding the hardened centers of the lathe, Fig. 48, the center is mounted in the headstock spindle. The grinder is mounted on the compound rest and set at an angle of 30 degrees with the axis of the headstock spindle. By moving the slide of the compound rest, the grinding wheel is traversed across the center.

Honing

On certain precision tools and on gages and metal parts required for accurate machine construction, abrasive wheels do not give a fine enough surface. A higher refinement in surface finish is accomplished with metal disks, rings, or sleeves whose surfaces have been charged with a fine "flour" abrasive. A tool used for this purpose is called a *hone*, and its use is referred to as *honing*.

Common Uses of Honing

Small tool parts which are commonly honed include micrometer spindles and anvils, plug, ring, and thread gages, holes in jig bushings, and die and punch work of fine character.

Fig. 49 shows the operation of honing the inside of a hardened steel bushing with emery dust and oil. The hone consists of a strip of emery cloth attached to a shaft that is held in the chuck.

Hones are generally made of a material soft enough to permit the abrasive substance to be pressed into the surface. So treated, the tool is said to be *charged*. Soft, close-grained cast iron, copper, brass, or lead may be used for the hone. Cast iron is the material generally preferred.

The ideal abrasive for honing is one which will break down or become finer as the honing proceeds. Two general types of abrasives are available: *natural* abrasives and *manufactured* abrasives. In the first group

Fig. 49. A honing sleeve is shown in position in front of a hole in a hardened steel bushing. The operator is applying a fluid which will help to obtain a smooth honed surface. (South Bend Lathe, Inc.)

are emery, corundum, rouge, oxide of chromium, oxide of tin, quartz, and diamond dust. Compounded abrasives are available in powder and paste forms; very satisfactory results can be obtained by suitable selection of these compounds.

Thread Cutting

The screw thread is an important mechanical unit. There are few assemblies that do not require threads for fastening, adjusting, or transmitting motion. *Thread cutting in the lathe, Fig. 50, is one of the most exacting lathe operations. It requires a thorough knowledge of thread-cutting principles and procedures. It is the operation most frequently per-*

Fig. 50. Thread cutting in the lathe. (R. K. LeBlond Machine Tool Co.)

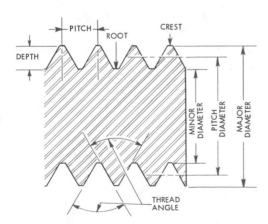

Fig. 51. The major dimensions and terms used to describe screw threads are illustrated.

formed and, without doubt, is one of the most difficult.

The operation of threading means cutting a helical groove of a definite shape or angle, with a uniform advance for each revolution, on the surface of a round piece of material or in a cylindrical hole.

Thread-Cutting Terms and Parts

Certain thread-cutting terms and the names of parts of an internal and external thread will be used in the discussion of threads and thread cutting. If these are learned at the outset, the setups and operations will be more readily understood.

Major Diameter. The major or outside diameter *(O.D.)* of an external thread is the diameter of the piece on which the thread is cut. It is the largest diameter of the thread, Fig. 51.

Depth of Thread. The depth of a thread is the distance from the top or crest of the thread to the root measured vertically.

Minor Diameter. The minor diameter is the smallest diameter of the thread of the screw. It is sometimes called the root diameter *(R.D.)* and can be found by subtracting twice the depth of the thread from the major diameter.

Crest. The crest is the top surface joining the two sides of a thread.

Number of Threads per Inch. The number of threads per inch *(N)* can be counted by placing a rule against the threaded part, as shown in Fig. 52, and counting the threads in one inch. The first thread is not counted since, in reality, not the crests but the spaces between the crests are what is being counted.

A second method is to use a screw-pitch gage, Fig. 53. This method is more suitable for checking the finer pitches of screw threads.

Pitch. The pitch *(P)* of a thread is the distance from a point on a screw thread to a corresponding point on the next thread, measured parallel to the work axis. The pitch

of a thread in inches can be found by dividing the whole number 1 by the number of threads per inch *(N)*.

$$P \text{ (in inches)} = \frac{1}{N \text{ (per inch)}}$$

To determine the number of threads per inch *mathematically* is simple. By dividing 1 by the pitch, we get the number of threads per inch. For example, if the pitch is .0625, the number of threads per inch is:

$$1 \div .0625 = 16 \text{ threads per inch}$$

Pitch Diameter. On a screw thread, the pitch diameter is that of an imaginary cylinder. The outer surface of this cylinder would pass through the threads at such points

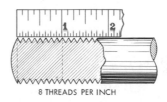

8 THREADS PER INCH

Fig. 52. The simplest method of finding the number of threads per inch is to use a rule as shown.

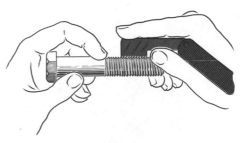

Fig. 53. Using the screw-pitch gage to find the number of threads per inch.

as to make equal the width of the threads and the width of the spaces, Fig. 51. On a 60-degree V-type thread and on National form threads, the pitch diameter can be found by subtracting the single depth of the thread from the major diameter of the thread.

Lead. The lead of a thread is the distance a screw will advance into a nut in one complete revolution. The lead is the same as the pitch on a single-thread screw, Fig. 54, *top.*

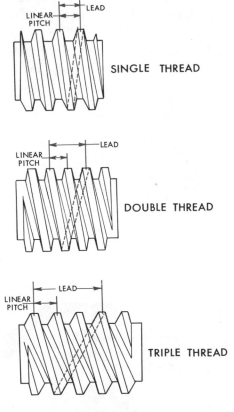

SINGLE THREAD

DOUBLE THREAD

TRIPLE THREAD

Fig. 54. Relation of linear pitch to lead in single- and multiple-thread screws.

The formula for the lead of a single-thread screw, then, is:

$$\text{Lead} = \frac{1}{\text{number of threads per inch}}$$

The lead for a screw with 9 threads per inch is .1111, which is the same as the pitch in inches (.1111).

On a double-thread screw, Fig. 54, *center,* the lead is twice the pitch. On a screw with 9 threads per inch with a double-thread screw the lead would be doubled or .2222. On a triple-thread screw, Fig. 54, *bottom,* the lead is three times the pitch, and so on.

Single Screw Threads. Most screw threads are single. The single screw thread has a single ridge and groove, Fig. 54, *top.*

Multiple Screw Threads. Multiple screw threads, Fig. 54, *center* and *bottom,* are used to obtain an increase in lead without weakening the screw itself. Multiple screw threads are referred to as double, triple, and so on. A double thread, Fig. 54, *center,* has two threads and two grooves starting on opposite sides of the screw. The lead of a double thread is twice that of a single thread having the same pitch.

A triple thread, Fig. 54, *bottom,* has three threads and three grooves. The lead of a triple thread is three times that of a single thread of the same pitch. There also are screws made with four or more threads starting in equal spaces around the end of the screw.

Angle of Thread. The thread angle is the angle included between the sides of the thread, Fig. 51. The thread angle of the Unified National Form is 60 degrees, for example.

Forms of Screw Threads

Screw threads are so widely used to connect parts that they are of prime importance when produced in the machine shop. Great amounts of study and research have been devoted to the standardization of screw thread forms. Tables I and II in the Appendix list the basic dimensions for Unified and National Coarse, Fine and Extra Fine threads.

The basic form of the Unified

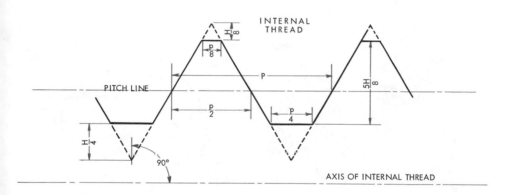

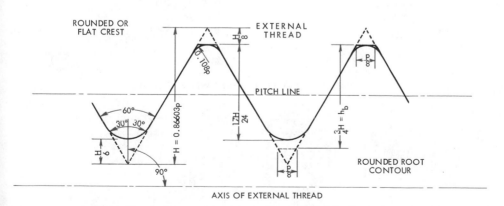

Fig. 55. The basic form of internal and external Unified screw threads.

screw and nut, with a 60-degree angle of thread, is shown in Fig. 55. The crest of the thread may be flat or rounded. The flat crest is preferred in American practice, and the rounded crest is given preference in British practice.

Over a period of many years, different screw thread forms and standards have been adopted in the United States. These different forms originated chiefly because of special requirements or because they were considered superior to other forms. Fig. 56 shows several screw thread forms which have been widely used. In addition to the American Standard Unified thread, they are the Acme, Square, and the 29-Degree Worm threads. These four threads are to be discussed individually in the paragraphs which follow.

American Standard Unified Thread Form

The American Standard Unified

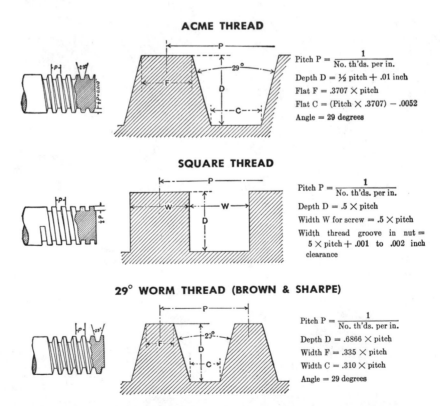

ACME THREAD

Pitch $P = \dfrac{1}{\text{No. th'ds. per in.}}$

Depth $D = \frac{1}{2}$ pitch $+ .01$ inch

Flat $F = .3707 \times$ pitch

Flat $C = (\text{Pitch} \times .3707) - .0052$

Angle $= 29$ degrees

SQUARE THREAD

Pitch $P = \dfrac{1}{\text{No. th'ds. per in.}}$

Depth $D = .5 \times$ pitch

Width W for screw $= .5 \times$ pitch

Width thread groove in nut $= 5 \times$ pitch $+ .001$ to $.002$ inch clearance

29° WORM THREAD (BROWN & SHARPE)

Pitch $P = \dfrac{1}{\text{No. th'ds. per in.}}$

Depth $D = .6866 \times$ pitch

Width $F = .335 \times$ pitch

Width $C = .310 \times$ pitch

Angle $= 29$ degrees

Fig. 56. Standard screw-thread forms other than the basic V-thread are shown, with general dimensions.

thread form is used as the standard locking thread form in the United States. This thread form is used on practically all mating parts in modern machine construction. The Unified thread form is essentially identical to the former standard, the American National thread form. The two forms are interchangeable for most diameter-pitch combinations. For example, a 1-64 National Coarse thread is interchangeable with a 1-64 Unified National Coarse thread.

Fig. 55 illustrates the American Standard Unified thread form and dimension formulas.

The Unified thread form is most widely used in six series of pitches as follows:

Unified National Coarse (UNC)
Unified National Fine (UNF)
Unified National Extra-Fine
(UNEF)
Unified National 8-Pitch (8 UN)
Unified National 12-Pitch
(12 UN)
Unified National 16-Pitch
(16 UN)

In the coarse, fine, and extra-fine series, the number of threads per inch increases as the diameters decrease. The coarse and fine threads are widely employed for general use.

Eight-pitch is used for bolts, for high-pressure pipe flanges, cylinder-head studs, and so on. Twelve-pitch is used in modern machine and boiler construction for thin nuts or for shafts and sleeves. Sixteen-pitch

is intended for adjusting collars, bearing retaining nuts, or any part requiring a fine thread.

In order to produce an American Standard Unified thread on the engine lathe, the ratio of tool feed to workpiece speed must be adjusted so the width of the crest will correspond with the width of the tool's nose when the thread is cut to its full depth. Thus, the root and crest are the same width.

Thread-Cutting Tools and Thread Forms

For each of the types of thread shown in Figs. 55 and 56, a tool bit to form that particular thread must be ground. Since the form of thread is the end product, the grinding of the tool bit must take into consideration the clearance angle, side rake, form, and other factors to achieve the result desired.

Clearance. Because of the rapid advance of the cutting tool across the workpiece, the clearance angle is an important factor. Clean, accurate threads are impossible unless both sides and front of the tool are given enough clearance to permit the tool to move freely in the groove being formed. A cutting tool ground with proper clearance is shown in Fig. 57.

When the tool is fed into the work at an angle, as for Unified form threads, the tool should have 3 to 5 degrees of side clearance, as shown in Fig. 57.

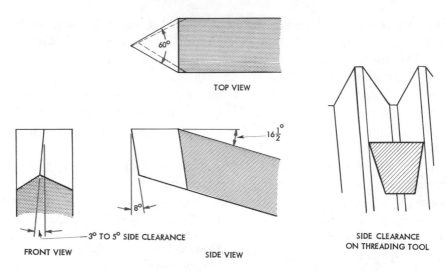

Fig. 57. The side clearance on a V-thread cutting tool is shown.

V-thread tools are ground flat across the top, Fig. 57, with about a 5-degree side-clearance angle. If the top of the tool is at an angle, the thread will be incorrect.

Center Gage for Checking Cutting Tools. The center gage is a tool used for checking the points of cutter bits for the American Standard screw threads which must be ground to an angle of 60 degrees. The 60-degree included angle on one end of the gage is used when grinding the tool to the exact angle. This same tool is useful in checking the angle of lathe centers. The V on the edge of the gage is used for setting the tool in the tool post of the lathe so that the center line of the V-point is perpendicular to the axis of rotation of the cylinder to be threaded,

Fig. 58, *left*. This setting will give the correct thread form.

The two opposite sides of the gage are parallel and have graduations on the edges for checking the number of threads per inch.

One side of the gage at the point is engraved to show the sizes of tap drills for 60-degree V-threads (internal threads), and to show, in thousandths of an inch, the double depth of commonly used threads. See Fig. 58, *right*.

The 60-degree angle at the point is useful to gage the thread after it is cut.

Size of Workpiece before Threading. Before the threading operation is begun, the workpiece needs to be turned to the maximum or major diameter of the thread spec-

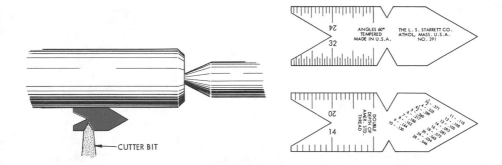

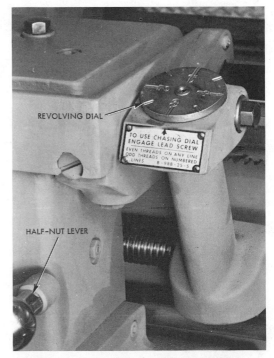

Fig. 58. *Left:* Center gages are used to check the angle ground on the thread-cutting tool, and to align the cutting tool with the workpiece. *Right:* They are also used to find the number of threads per inch. Thread information may be found on one or both faces of the gages. (The L. S. Starrett Co.)

ified on the blueprint or other speci-fications. This size is usually given as *O.D.* (outside diameter) ex-pressed in inches and decimal parts of an inch. When cutting V-type threads, it is also good practice to chamfer or bevel the ends of the shaft to be threaded. The chamfer can be at any angle from 30 to 45 degrees. For instructions on straight turning, see the section on that operation.

Setup for Cutting an External Unified (V) Thread on the Lathe

The setup for external threading will depend largely upon the shape and size of the workpiece. Threads are often turned on shafts mounted between centers and revolved by a lathe dog. After this setup is made and the work is turned to the proper outside diameter, the workpiece is left right in the lathe and the thread-

Fig. 59. The thread-chasing dial is used in conjunction with the half-nut lever to align the cutting tool with the workpiece for increasingly deeper cuts in a thread-cutting operation.

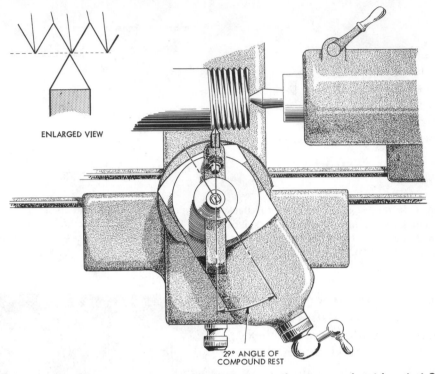

ENLARGED VIEW

29° ANGLE OF
COMPOUND REST

Fig. 60. To cut a right-hand external Unified (V) thread, the compound rest is set at 29° as shown. For left-hand threads, set the compound rest 29° to the left of zero.

ing operation is performed. Other jobs are threaded while held in a three- or four-jaw chuck in which they were mounted to be turned to size.

Because of the pressure exerted by any threading tool, it is necessary, when threading long, narrow shafts, to use a steady rest.

Thread-Chasing Dial. The thread-chasing dial is an indicator with a revolving dial. It is either fastened to the carriage of the lathe or built into it. The dial of the indicator, Fig. 59, serves as a guide to the operator. It tells him when to close the half-nut in the apron onto the lead screw so he may take successive cuts in the same groove, or to space grooves properly for multiple threads. When this is done, the lead screw and carriage bear the same relative positions as before. In other words, when the operator engages the half-nut as the proper mark on the dial comes into position, the threading tool moves into the same groove made on previous cuts.

The dial is engaged to the lead screw, as the operator desires, by a feed lever cn the apron. This lever is called a half-nut lever, Fig. 59. When the lathe is running but the half-nut lever is not engaged, the dial revolves. When the half-nut lever is engaged, the carriage moves but the dial remains motionless.

Before the operator can start the thread-cutting operation, he must decide, from the number of threads required per inch, at what point on the dial to engage the half-nut. For *chasing* all even numbers of threads per inch, such as 4, 6, 8, 10, etc., the half-nut is engaged for the first, and for all successive cuts, at any of the eight graduation marks on the face of the dial. For an odd number of threads per inch, engage the half-nut at any quarter-turn or numbered line on the dial (the main graduation marks).

For half-threads, such as 3½ threads per inch, the half-nut must be engaged at any odd-numbered line on the dial.

For quarter-threads, such as 2¼ or 3¼ threads per inch, the half-nut must be engaged at the same point on the dial each time a cut is started.

Setting the Compound for Right- and Left-Hand Threads. To cut right-hand external threads on the lathe the compound rest is turned at a 29-degree angle, Fig. 60. This prevents tearing of the thread and makes it easier to rechase the thread if the tool must be reset. The carriage is made to travel from right to left or, in other words, toward the headstock. Before each successive cut, the tool is fed in with the compound rest. To cut a left-hand thread, the compound is turned at a 29-degree angle toward the headstock, and the carriage is made to travel from left to right.

Setting Threading Tool. The threading tool is mounted in the tool post as shown in Fig. 60. Adjust the cutter point vertically to the exact center of the work. Then place a center gage with its back edge in contact with, or parallel to, the work or the tailstock spindle. Now adjust the tool horizontally by fitting the cutter point exactly into the 60-degree angle notch in the front edge of the center gage. Tighten the tool post screw. Be sure not to change the position of the holder. Recheck the tool setting after tightening the tool post screw.

Suggestions on How to Cut an External V-thread. After setting the compound rest and positioning the threading tool properly in relation to the work, it is necessary to select the proper speed and feed. Good thread-cutting practice requires that the back gears be engaged for this operation. This reduces the r.p.m. or speed to a minimum, and is necessary if best results are to be obtained. The correct selection of feed is determined by a gear box on all modern lathes. Directions are shown on the gear box regarding the setting of levers to obtain the correct feed, depending upon the threads per inch to be cut. See Trans-Vision Page G.

Now the compound feed-screw graduated collar is set to zero and the tool point is brought into contact with the work by turning the cross-feed screw. The tool point should contact the work lightly. Then run the carriage to the right, using the carriage hand wheel, until the tool clears the end of the work. Notice what the setting is on the cross-feed collar. This adjustment must be remembered so that, at the end of each cut, the cross-feed is always brought back to the same number or setting. Feed in on the compound .002″ and then start the machine and take the first trial cut. If using the thread dial, be sure to engage the half-nut lever at the correct line on the dial, depending upon the threads per inch you are cutting. This causes the carriage to start in motion.

A check should be made after this first trial cut to see that the correct pitch of thread is being machined. This is done by using a thread-pitch gage or a rule, Fig. 53.

To determine the total number of thousandths of an inch that the threading tool must be fed in by the compound feed screw, and in order to cut the thread to the desired depth, a simple formula is used. This formula is: divide the constant .750 by the number of threads per inch. For Example: If it is required that 8 threads per inch are to be cut, then divide .750 by 8, which gives a result of .0937″ (.094″ can be used).

This is the total number of thousandths of an inch the tool is to be fed into the work to cut 8 threads per inch. However, the compound feed is used until the tool has been fed in .090″. Then the cross-feed is used to remove the final four thousandths of an inch from the thread, making four cuts and feeding the tool in .001″ on each cut. This will help to polish the right side of the thread.

A good grade of lubricant should be used on the tool when threads are being cut, *except* when cutting cast iron, brass, or babbitt metal.

Mineral lard oil is a very good lubricant for threading. It is made of white lead, graphite, and fatty oil.

Resetting the Tool. If it is necessary for any reason to remove the tool before the thread is finished, reset the tool to gage regardless of the part of the thread already cut. Having the compound rest at an angle of 29 degrees makes it easy to reset the tool if it needs regrinding. The tool is clamped in the tool post after it is reground. Then it can be set with the center gage as before.

In resetting the tool, proceed as follows: first, reset the tool to the gage; then back the tool away from the workpiece. Start the machine and engage the thread-chasing lever as before. Let the workpiece make two or three turns and shut off the power with the threading lever still engaged. Adjust the tool into the thread previously cut by moving the compound and cross-feed until the tool is lined up properly in the thread groove. Back the tool out from the workpiece slightly and turn on the power. When the tool is aligned properly, proceed as before.

Thread Fits and Allowances

Before starting to cut threads on parts which are to be assembled, the operator must know the relationship between the two mating parts when they are assembled. The term for this relationship of mating parts is called *fit*. The fit refers to the allowance (looseness or tightness) between the mating parts when they are fastened together. The *class of fit* depends upon both the relative size and the quality of finish of the mating parts, since the classes of thread fit are distinguished from each other by the amount of tolerance or the amount of tolerance and allowance as applied to pitch diameter.

There are three classes of fit designated by American Standards now being used. Classes 1A, 2A, and 3A apply to external threads only, and class 1B, 2B, and 3B apply to internal threads only. The three classes of fit are:

Class 1. Loose Fit
Class 2. Medium Fit
Class 3. Close Fit

Loose Fit. This class possesses the largest allowance and is used where rapid assembly of parts is required

and looseness is not objectionable.

Medium Fit. This is used on the bulk of standard screws, bolts and nuts. A very small amount of looseness or shake may be present or if the parts are carefully made no movement can be noted, yet the nut can be screwed on by hand.

Close Fit. This is used on fasteners where accuracy of fit is highly important, and where no looseness is permitted. A wrench or some other tool must be used to force the nut onto the bolt.

Tighter fits than those mentioned require special fabrication, and are carefully specified for the job being assembled.

The note designating thread size and fit on the blueprint will specify the proper combination of thread classes for the components. For example, a Class 2A thread (an external thread) may be used with a Class 3B thread (an internal thread), thus ensuring that the mating parts will fulfill the exact requirements for the end use of the assembly.

Measuring External Screw Threads

The fit of the thread is determined by its pitch diameter. The pitch diameter is the diameter of the thread at an imaginary point on the thread where the width of the space and the width of the thread are equal. The fact that the mating parts bear on this point or angle of the thread, and not on the top of it, makes the pitch diameter an important dimension to use in measuring screw threads.

Screw Thread Micrometer Caliper. The thread micrometer, Fig. 61, is an instrument used to gage the thread on the pitch diameter. The anvil is V-shaped to fit over the V-thread. The spindle, or movable point, is cone-shaped (pointed to a V) so as to fit between the threads. Since the anvil and spindle both contact the sides of the threads, the pitch diameter is gaged and the reading is given on the sleeve and spindle, where it can be read by the operator.

Thread micrometers are marked on the frame to specify the pitch diameters which the micrometer is used to measure. One will be marked, for instance, to measure from 8 to 13 threads per inch, while others are marked 14 to 20, 22 to 30, 32 to 40, and so on.

This method of thread inspection is used by the inspection department and the lathe operator. The procedure in checking the thread is first to select the proper micrometer. Then the operator calculates or selects from a table of threads the correct pitch diameter of the screw Lastly, he fits the thread into the micrometer and takes the reading.

Ring (or Female) Gage. The ring gage, Fig. 62, is screwed onto the thread after it is cut. It checks the completed thread against a standard mating part. By means of a ring gage, the oversize and undersize sections of a thread can be found.

Ring gages are made in a variety of sizes and fits. The gage must be carefully chosen to control the desired limits for each thread, as determined by the design engineer. Overly close limits mean high manu-

Fig. 61. Checking thread size with a screw-thread micrometer. (Brown & Sharpe Mfg. Co.)

Fig. 62. Ring gage used for checking external thread sizes. (Taft-Peirce Mfg. Co.)

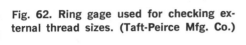

facturing costs, while overly wide limits cause assembly failures. Gages, then, should be selected in accord with the class of fit for which they are used. To give an example: if the blueprint calls for $\frac{9}{16}$—12UNC—2 thread, it means that the ring gage needed will have a major diameter of $\frac{9}{16}$ inch, with 12 threads per inch of the Unified National Coarse type, and it should give a class 2 fit. (See classes of thread fit under heading "Thread Fit" given in this chapter.)

Three-Wire Method for Measuring V-Threads. The three-wire method of measuring threads is another method of measuring the pitch diameter of a screw thread. It is considered the best method for extremely accurate measurement.

Fig. 63 shows three wires of correct diameter placed in the threads with the micrometer measuring over

them. The pitch diameter can be found by subtracting the wire constant from the measured distance over the wires. It can be readily seen that this method is dependent on the use of the "best" wire for the pitch of the thread.

The "best" wire is the size of wire which touches the thread at the middle of the sloping sides — in other words, at the pitch diameter. A formula by which the proper size may be found is as follows: Divide the constant .57735 by the number of threads per inch to be cut. If, for example, 8 threads per inch have been cut, we would calculate: .57735 ÷ 8 = .072. The diameter of wire to use for measuring an 8-pitch thread would be .072.

The wires used in the three-wire method should be hardened and lapped steel wires. They should be three times as accurate as the accuracy desired in the measurement of the threads. The Bureau of Standards has specified an accuracy of .00002 inch.

Suggested Procedure in Measuring Threads with Wires. After the three wires of equal diameter have been selected by using the above formula, they are positioned in the thread grooves, as shown in Fig. 63. The anvil and spindle of an ordinary micrometer are then placed against the three wires, and the reading is taken. To determine what the reading of the micrometer should be if a thread is the correct finished size, use the following formula: (for measuring Unified National Coarse threads) *add three times the diameter of the wire to the diameter of the screw; from the sum, subtract the quotient obtained by dividing the constant 1.5155 by the number of threads per inch.* Written concisely, the formula is:

$$M = D + 3W - \frac{1.5155}{N}$$

where

M = micrometer measurement over wires

D = diameter of the thread

N = number of threads per inch

W = diameter of wire used

EXAMPLE: Determine M (measurement over wires) for ½", 12 pitch, UNC thread. We would proceed to solve as follows:

$$M = D + 3W - \frac{1.5155}{N}$$

where

$W = .04811''$

$D = .500''$

$N = 12$

then

$$M = .500 + .14433 - \frac{1.5155}{12}$$
$$M = .500 + .14433 - .1263$$
$$M = .500 + .01803$$
$$M = .51803'' \text{ (micrometer measurement)}$$

When measuring a Unified National Fine thread, the same method and formula are used, except that the constant should be 1.732 instead of 1.5155. Too much pressure should not be applied when measuring over wires.

Checking Threads by Use of Optical Comparator. The optical comparator, an instrument based on the magnification principle, is used for obtaining a very accurate comparison between an enlarged template and the threaded part.

Acme Screw Thread, Twenty-Nine Degree Worm Thread

The Acme screw thread was designed to overcome the difficulty of cutting square threads with taps and dies. It is much easier to produce this thread accurately than the Square thread.

Acme Screw Thread. The Acme thread form is classified as a power-transmitting type of thread. The Acme thread form, when used on lead screws or similar parts, has a distinct advantage. This is because the 29-degree included angle at which its sides are established (which is almost perpendicular) reduces the amount of friction when matching parts are under load. Further, because of the wide root and crest, this thread form is strong and capable of carrying a heavy load. The formulas used in determining the depth of Acme and worm threads are given in Fig. 56.

Twenty-Nine-Degree Worm Thread. The 29-degree worm and the Acme thread-cutting tools are both ground so that their side cutting edges are at a 29-degree in-

cluded angle. However, these two threads should not be confused. They are different in depth of thread, width of tooth at top, and width of tooth at bottom. The Acme tool has a wider nose, while the 29-degree worm will cut a deeper thread than the Acme.

Suggestions on How To Cut an External Acme Thread. The first step is to grind a threading tool to conform to the 29-degree included angle of the thread. The tool is first ground to a point, with the sides of the tool forming the 29-degree included angle. This angle can

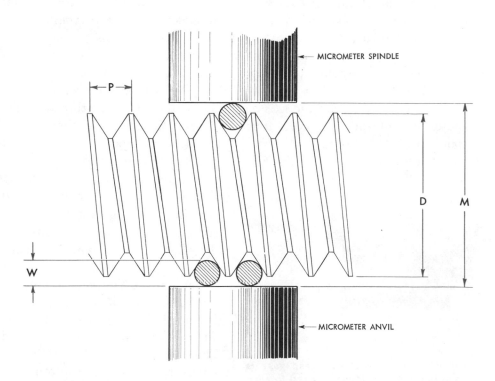

Fig. 63. The three-wire method for measuring the pitch-diameter of V-threads.

be checked by placing the tool in the slot at the right end of the Acme thread gage shown in Fig. 64. Be sure to grind this tool with sufficient side clearance so that it will cut. Now, depending upon the number of threads per inch to be cut, the *point* of the tool is ground flat so as to fit into that slot on the Acme thread gage that is marked with the number of threads per inch the tool is to cut, Fig. 64. You can·see that the size of this flat on the tool point will vary, depending upon the threads per inch to be machined.

After grinding the tool, set the compound one-half the included angle of the thread (14½ degrees) to the right of the vertical center line of the machine. Now mount the tool in the holder or tool post so that the top of the tool is on the axis or center line of the workpiece. The tool is set square to the work, using the thread gage as shown in Fig.

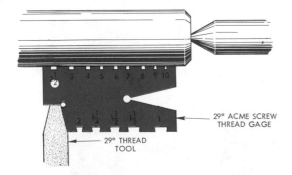

29° ACME SCREW THREAD GAGE

29° THREAD TOOL

Fig. 64. The Acme screw thread gage is used to check the cutting tool and to align it with the workpiece.

64. This thread is cut using the compound feed. The depth to which you feed the compound to obtain total thread depth is determined by the formula given and illustrated in Fig. 56. The remainder of the Acme thread cutting operation is the same as the V-threading operation previously described. The compound should be fed into the work only .002″ to .003″ per cut until the desired depth of thread is obtained.

Square Threads

Because of their design and strength, square threads are used for vise screws, jack screws, and other devices where *maximum transmission of power* is needed.

All surfaces of the square-thread form are square with each other, and the sides are perpendicular to the center axis of the threaded part. The depth as well as the width of the crest and root are of equal dimensions. Because the contact areas are relatively small and do not wedge together, friction between matching threads is reduced to a minimum. This fact explains why threads are used for power transmission.

Helix of the Thread. Before the square-thread cutting tool can be ground, it is necessary first to determine the helix angle of the thread. The sides of the tool for cutting the square thread should conform with the helix angle of the thread.

To determine the helix angle of a square thread, angle B of Fig. 65, draw the base line AC_2 equal in length to the circumference of the thread to be cut. Draw line C_2C perpendicular at C_2 and equal in length to the lead of the thread to be cut. Complete the triangle by drawing line AC. Angle B (angle CAC_2) in the triangle is the helix angle of the thread.

The tool bit should be ground at the same angle as angle B. See the end view of the cutting tool, Fig. 65. Note that the sides of the tool E and F have been ground to give clearance, while the helix angle has been maintained. The center line of the tool face is inclined as shown by center line KL, with clearance at each side as shown at I and J.

Width of Cutting Edge. For cutting the thread, the cutting edge of the tool should be ground to a width exactly one-half that of the pitch. For cutting the nut, it should be from one thousandth to three thousandths of an inch larger to permit a free fit of the nut on the screw.

Suggestions on How To Cut a Square Thread. The cutting of the square-thread form presents some difficulty. Although it is square, this thread, like any other, progresses in the form of a helix, and thus assumes a slight twist. Some operators prefer to produce this thread in two cuts—the first with a narrow

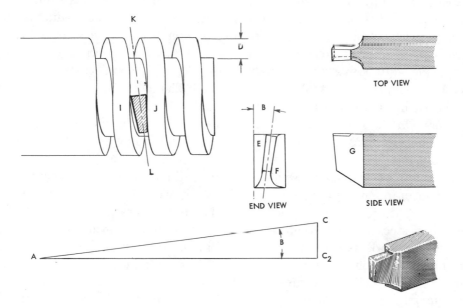

Fig. 65. Determining the helix angle for a square thread. The angle of the cutting tool will conform to this angle.

235

tool to the full depth, and the second with a tool ground to size. This procedure relieves cutting pressure on the tool nose and may prevent springing of the work.

The operation of cutting the square thread differs from cutting threads previously explained in that the compound is set parallel to the axis of the workpiece, and the feeding done only with the cross feed. The cross-feed is fed in only .002″ or .003″ per cut. The finish depth of the thread is determined by the formula:

$$\text{Depth} = \frac{1}{2} P.$$

The width of the tool point is determined by this formula also, and will depend upon the number of threads per inch to be machined. It is measured with a micrometer, as square-thread gages are not available.

Tapers and Taper Turning

One of the most important principles in machine shop practice is the principle of the taper, particularly the round taper shank and the round taper hole into which the taper shank fits. Practically all revolving spindles in various machines are provided with taper holes. The taper hole will receive and securely hold the taper shanks of various tools, such as drills, reamers, and centers. On the lathe there is a tapered hole in both the headstock and tailstock spindles. The tapered centers are inserted into these holes.

Taper in round work is defined as the difference in diameters for any length of stock, measured along the axis of the work. It is usually given in tables and on drawings, blueprints, and specification sheets as the amount of taper in inches per foot or in degrees.

Four elements are taken into account: (1) the amount of taper per foot; (2) the length of the taper; (3) the large diameter; and (4) the small diameter.

Taper turning as a machining operation is the gradual reduction in diameter from one part of a cylindrical piece to another part, Fig. 66. A single-point cutting tool is used. The reduction is at a predetermined rate.

Taper turning can be done in any of the following ways: (1) by offsetting the tailstock, thereby setting the lathe centers out of alignment; (2) by using the taper attachment, thereby duplicating on the workpiece the taper previously set on the slide of the taper attachment; and (3) by setting the compound rest at an angle. This last procedure furnishes a means of turning included angles. This operation could not be performed by either the first or second method. A typical example is the machining of the faces of bevel gears and lathe centers.

Kinds of Tapers

Morse Standard tapers are used for lathe and drill press spindles by most of the manufacturers of these machines. The taper is approximately .625 inch per foot. Table XII of the Appendix shows the principal dimensions of Morse Standard tapers.

Several other systems of tapers are widely used. The Brown and Sharpe tapers are used for milling-machine spindles and milling-cutter shanks. The taper is .5 inch per foot. The Jarno tapers are used for some makes of lathe and drill equipment. This taper is .6 inch per foot. Speci-

fications for these tapers are found in handbooks. The Sellers taper of .75 inch per foot is used on equipment made by William Sellers and Company. Taper pins of .25 inch per foot are used in assembly work.

Calculating Taper in Inches per Foot. Since taper attachments on lathes and grinders are usually graduated in degrees and taper in inches per foot (t.i.p.f.), it is frequently necessary to calculate one of these factors.

The formula for finding the taper in inches per foot is:

$$t.i.p.f. = \frac{\text{large diameter} - \text{small diameter}}{\text{length of tapered portion (in.)}} \times 12$$

To find the amount of taper per

Fig. 66. Taper turning with an offset tailstock (South Bend Lathe, Inc.)

inch (t.p.i.), use the formula as given but do not multiply by 12.

Calculating the Angle. To find the angle at which to set the compound or taper attachment when the t.i.p.f. is known, divide the t.i.p.f. by 24 and find (in the trigonometry tables) the angle whose tangent produces this result.

Setup for Taper Turning with Setover Tailstock

Setting over the top of the tailstock is a common method for turning tapers. This method is used when the workpiece is sufficiently long. The objection to turning tapers by the setover method, however, is that the lathe centers do not have full bearings in the ends of the work. The center holes are likely to wear out of their true positions.

The headstock center and tailstock are always set in perfect alignment for cylindrical turning. When, however, the dead center is set *out of alignment* with the live center, as in the setover method, the centers are at an unequal distance from the cutting tool on the carriage. Since the cutting tool is mounted on the carriage and moves with the carriage in a direction parallel to the axis of the lathe, the tool will take a deeper cut at the section nearer the tool. The workpiece will become tapered, Fig. 67. When the dead center of the lathe is offset, the center line of the work and the line of travel of the turning tool are no longer parallel. As the work revolves and the tool moves longitudinally, a taper is turned.

The amount of tailstock setover from the center line of the lathe depends on the t.i.p.f. and the length of the work. With the same amount of setover, pieces of different lengths will be machined with different tapers.

Provision is generally made on the tailstock for moving the dead center laterally toward the front or rear of the bed of the lathe, Fig. 68. The amount of taper to be cut will govern the distance the top of

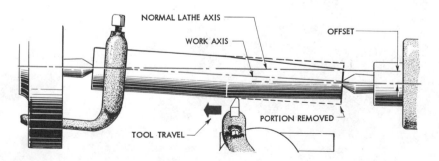

Fig. 67. Effect of offsetting the lathe dead center to obtain a tapered workpiece.

the tailstock is setover from the center line of the lathe. The tailstock top is adjusted by loosening the clamp nuts, shifting the upper part of the tailstock with the adjusting screws, and then tightening them in place.

Calculating the Amount of Setover. Divide the total length of

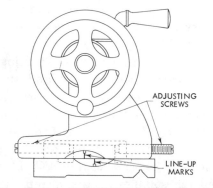

Fig. 68. How the tailstock is set off-center for taper turning.

the piece in inches by 12, and multiply this quotient by one-half the amount of taper per foot:

$$\text{Setover} = \frac{\text{total length in inches}}{12}$$
$$\times \quad \frac{\text{taper per foot}}{2}$$

Tapers are usually specified in inches per foot. For example, if a Morse taper of .625 inch (⅝″) per foot is to be machined on a shaft 1 foot long, the tailstock center should be setover .3125 inch (⁵⁄₁₆). If the shaft is only 9 inches long, the amount of setover would be ⁹⁄₁₂ of .3125 inch, or .2344 inch.

If the diameters at the ends of the taper are given, the rule is as follows: Divide the total length of the piece by the length of the portion to be tapered, and multiply this

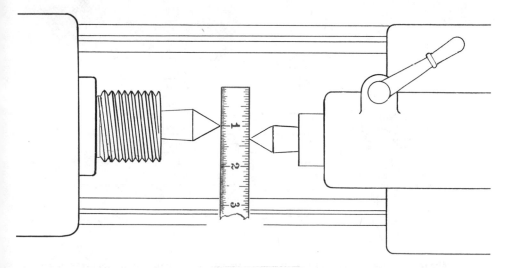

Fig. 69. A steel rule can be used to provide a rough measurement of tailstock setover.

quotient by one-half the difference in diameters.

Measuring the Tailstock Offset. A common but not very accurate method to measure tailstock offset is shown in Fig. 69. The steel rule is used to check the amount of offset when the tailstock is brought close to the headstock.

Where accuracy is required, the amount of offset may be measured by means of the graduated collar on the cross-feed screw. This is done as follows: First, compute the amount of setover; next, set the toolholder in the tool post so the butt end of the holder faces the tailstock spindle, Fig. 70. Using the cross-feed, run the toolholder in by hand until the butt end touches the tailstock spindle. The pressure should be just enough to hold a

slip of paper placed against the spindle.

The reading on the cross-feed micrometer collar may be recorded, or the graduated collar on the cross-feed screw may be set at zero. Next, move the cross-slide to bring the toolholder toward you to take up backlash. Now notice the reading on the cross-feed dial and bring the cross-slide toward you the necessary amount, which you have computed. The micrometer collar is used to set this distance. To bring the tailstock toward you and the toolholder, loosen the set screw on the back of the tailstock and tighten the set screw on the front of the tailstock until the piece of paper drags when drawn between the toolholder and the spindle. Lastly, clamp the tailstock to the lathe bed.

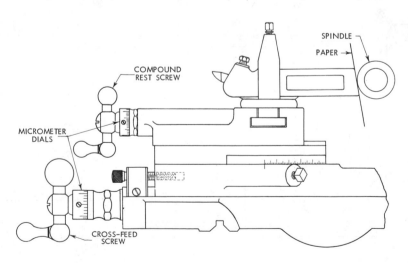

Fig. 70. The micrometer collar on the cross-feed screw can be used where accurate tailstock setover is required.

Setting the Cutting Tool for Taper Turning. The cutting edge of the lathe cutting tool should always be placed for height *exactly on center* when turning a taper. This rule applies to all taper turning and taper boring. If the rule is not followed, the work will not be truly conical, and the rate of taper will vary with each cut.

Checking Tapers. A taper may be checked with a micrometer by measuring the large and small diameters and the length of the taper. The taper is determined by using the formula discussed in the preceding material. Another way is to use standard taper gages. The taper can also be fitted to the spindle or sleeve and checked for fit. Still another method is to use a sine bar and Johansson gage blocks.

Fig. 71 shows Morse Standard taper plug gages and socket gages. They are often used to check tapers. For an outside taper, the socket gage is used; for an internal taper, the standard taper plug gage is used.

To Fit a Taper to a Gage. To test the taper, mark three equally spaced lines with chalk or Prussian blue for the full length of the taper. Then place the work straight in the taper hole it is to fit. Turn the work carefully one full revolution by hand. Remove the work, and the chalk-mark will show at what point the taper is bearing.

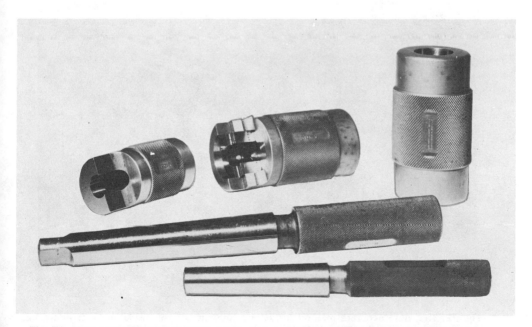

Fig. 71. Morse standard taper plug gages and ring gages (Brown & Sharpe Mfg. Co.)

If the taper is correct, chalk will be rubbed off the full length of the chalkmark. If the lines remain distinct at one end of the taper, the taper is not correct. Make the necessary adjustments on the tailstock. Take another light cut on the taper and test it again. Be sure that the taper is correct before turning it to the finished diameter.

Taper Turning with Taper Attachment

A taper attachment, Fig. 72, offers many advantages. For one thing, it is easily connected and set. The greatest accuracy is possible, since at one end the guide bar is gradu-

ated in taper in inches per foot and, at the other, in degrees of taper. Blueprints usually designate tapers in one of these ways. The graduations on the guide bar are given in both directions from zero. Matching tapers can be turned and bored readily by swiveling the guide bar to identical graduation points. A setting to one side of the zero mark is for the external taper, and to the other side for the internal taper. As a rule, the guide bar cannot be swiveled to produce a taper exceeding 3 inches per foot. The length of taper the guide bar will afford is usually not over 12 to 24 inches, depending on the size of the lathe. It is possible

Fig. 72. The most accurate method for setting taper is to use the lathe taper attachment. (R. K. LeBlond Machine Tool Co.)

to cut a taper longer than any the guide bar affords by moving the attachment after a portion of the desired taper length has been cut. Then the remainder of the taper can be cut. This operation, however, requires experience.

To cut matched tapers, the bored taper should be cut first. The taper should then be checked with a plug gage before the cuts are completed. In this way, inaccuracies can be corrected in the finishing cuts. The external taper should be checked with a ring gage. The two matching surfaces also should be checked when seated one in the other.

Suggestions on How To Turn a Taper, Using a Taper Attachment. The swivel bar of the taper attachment is bolted to the back of the lathe bed to form a guide. A slide moving with the guide is attached to the cross-feed slide of the carriage. This cross-feed slide is loosened; while the carriage is moved by the feeding mechanism, the tool is being moved in or out as the cross-feed slide of the carriage moves with the guide.

Set the guide bar so that the guide block does not project over the ends of the bar at the beginning or end of the taper.

Set the swivel bar for the correct taper in inches per foot. Set the compound rest at 90 degrees with the work. With the plain taper attachment, all tool adjustments will have to be made with the compound rest feed screw. Adjust the tool toward the work until contact is nearly established. Tighten the necessary lock bolts and binding screws.

Taper Turning with Compound Rest

The compound rest of the lathe, Fig. 73, may be used for turning and boring short, steep tapers and bevels, such as die and pattern work, bevel gear blanks, lathe cen-

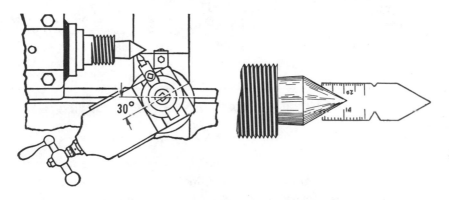

Fig. 73. Taper turning using the compound rest.

ter points, short taper gages, and so on. It will be recalled that the compound base is graduated in degrees of angle in order that it may be set to produce any angle of taper desired.

The angle at which the compound rest is set is computed in the following manner: (*a*) if the angle with the center line of the work is given, the compound rest is set to that angle; (*b*) if the included angle is given, the compound rest is set to one-half the given angle; (*c*) if the taper per foot, the diameters at the ends of the taper, and the length of the taper are all given, the angle for the compound rest setting is computed as in Fig. 74.

In Fig. 74, a taper of 2 inches per foot is to be turned. The procedure consists of subtracting the diameters and forming two identical right triangles on either side of the small diameter. Only one of these triangles is considered in computing the required angle of compound. The tangent of this angle is equal to one-half the difference of the diame-

ters divided by the length of the piece. The tangent of the angle equals $\dfrac{.125}{1.500}$ equals .08333. The angle whose tangent is .08333 is 4 degrees, 46 minutes (written 4° 46′).

Suggestions on How To Cut Taper, Using Compound Rest. The cutting tool should be set on center for height. Turn the compound rest feed screw by hand. The carriage is locked to the bed of the lathe when taking heavy cuts. After the first cut is made, run the tool back to the starting point by turning the compound rest handle. Then feed in to take the next cut by moving the cross-feed handle.

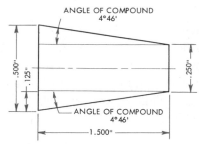

Fig. 74. The method for calculating the angle at which the compound rest is to be set is illustrated.

BASIC INTERNAL MACHINING OPERATIONS

Just as there are various lathe operations that can be performed on the outside surfaces of a workpiece, so likewise there are operations that

can be performed on the inside surfaces of the piece. To make a comparison, Table I is presented.

According to the data given in

TABLE I. EXTERNAL AND INTERNAL LATHE OPERATIONS

Operation	External	Internal
Center drilling..............	no	yes
Drilling....................	no	yes
Reaming	no	yes
Straight turning	yes	yes (boring)
Taper turning	yes	yes (boring)
Shoulder turning	yes	yes (boring)
Recessing.................	yes	yes (boring)
Threading	yes	yes
Tapping	no	yes
Knurling....................	yes	no
Cutting off.................	yes	yes

Table I, the operations classed as internal operations are: center drilling, drilling, reaming, boring, threading, tapping, and cutting off. These operations will be taken up individually in the pages which follow. As in the preceding section on external machining operations, the proper setups are first discussed before describing operations.

Center-Drilling in the Lathe

All lathe centers are made with a 60-degree angle on the pointed end. A workpiece supported between centers must have center holes drilled in the ends, Fig. 75. The conical points of the two lathe centers fit snugly in the center holes.

The center hole must have an accurate, smooth-finished conical-shaped surface to bear on the lathe center point. Do not feed the center drill into the work beyond the taper of the drill, as it will make a poor fit with the conical point of the lathe center. The bottom of the hole must be deep enough to keep the point of the lathe center from binding. It also serves as a reservoir for the lubricant used on the center. The center hole, then, consists of a plain drilled hole with a countersink following the drill. A combination drill and countersink is used, Fig. 75. It is a cutting tool having a short drill and countersink at each end. Center drills are made in different sizes. Care must be used in selecting the proper size for the job.

Setup for Drilling Center Holes

When the work permits, the most accurate method of drilling center

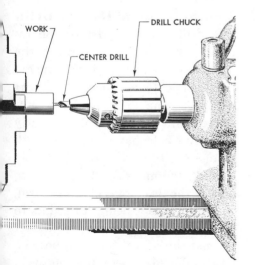

Fig. 75. Center-drilling in the lathe.

holes is with the workpiece held in the lathe chuck, Fig. 75.

Suggestions on How To Drill Center Holes. Before the center hole is drilled, the workpiece should first be faced smooth. Facing was ex-

plained in the discussion of external operations. A rough end may cause the drill to "walk" off a high spot and seat itself in a depression. With the combination drill mounted in a drill chuck, feed the drill from the tailstock.

Drilling in the Lathe

Drilling, Fig. 76, is one of the more common operations done on the lathe. The workpiece is held in the three- or four-jaw chuck for this operation.

Setup for Drilling

The twist drill can be held in the drill chuck mounted in the tapered

Fig. 76. Simple drilling operations can be performed in the lathe. The drill is held in the tailstock spindle. (R. K. LeBlond Machine Tool Co.)

hole of the tailstock spindle. Very frequently the drilling operation is done in this way, with the workpiece being rotated in the headstock. Only straight drills, however, can be mounted in the chuck. Other means must be used for holding taper-shank drills.

Taper-shank drills are held in a steel sleeve or socket. The sleeve, mounted in the tailstock spindle, has a taper shank that fits the taper hole in the tailstock.

Suggestions on How To Drill. The drill is advanced into the revolving workpiece by turning the handwheel of the tailstock, causing the spindle to advance out of the tailstock. The drill should be advanced into the workpiece slowly until the hole has reached its full diameter. Otherwise, the drill may "walk" to one side. One way to steady the drill point of a long drill, while the hole is being started, is to bring the butt end of the toolholder against the side of the drill, near its end, and hold it there. If a relatively large hole is to be drilled, a small lead hole is drilled first. This hole, which should be the size of the dead center of the drill, will reduce the feeding pressure required for the large-size drill.

Frequently, in drilling a hole in the end of a long piece of round stock, one end is chucked in the three- or four-jaw chuck. The other end is supported by the steady rest.

Reaming in the Lathe

Drilled holes are not always precisely round or straight. It takes a good lathe, a good setup, and a skillful operator to drill a hole to an exact standard diameter. When great accuracy is demanded, the hole is first drilled slightly undersize and then reamed, Fig. 77.

A reamer is a cutting tool used for perfecting holes. Many different types of reamers are in use, as you have already learned.

Setup for Reaming in the Lathe

Fig. 77 shows a typical setup. The reamer shown is held in a drill chuck placed in the tailstock. A machine reamer fitted with a taper shank can also be held in the tailstock spindle of the lathe. The work should rotate slowly and should be kept turning in the same direction until the reamer has been backed out. If necessary, the hole is finished to exact size with a hand reamer. With care in drilling and reaming, the hole should be perfectly round and accurate to within one thousandth of an inch. See also the material on reamers and reaming in Chapter Four.

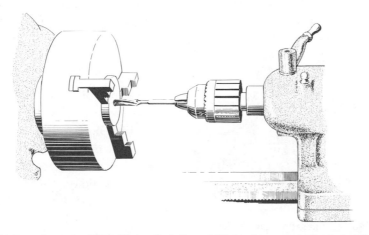

Fig. 77. Reaming operations normally follow drilling where accuracy is required.

Boring in the Lathe

Reaming was defined in an earlier chapter as finishing a previously drilled hole to exact size. However, the guide for the reamer is the drilled hole itself. Reaming, therefore, can only gloss over the irregularities of a hole which has been improperly drilled. To insure a properly located hole of finished size with perfectly straight sides, the drilled hole can be *bored* instead of reamed.

Boring is defined as the operation of enlarging and truing a hole with a single-point tool, Fig. 78. Unlike the reamer, the boring tool is independent of irregularities in the drilled hole. For example, the boring tool can take a heavier "bite" on one side to re-center a hole which has "wandered". Furthermore, the boring operation is not limited to drilled holes.

Boring is accomplished by either of two methods. By one method, the work revolves in a chuck or faceplate while the cutter moves along in a straight line. By the second method, the work is mounted in a fixed position on the carriage. The cutter bar is mounted between centers and revolves.

Drilled holes are seldom straight. Hard or soft spots and "blow holes" are faults in metal which cause drills to move out of alignment. A casting, for instance, may have a cored hole to help keep the drilling straight. Or if it has no cored hole, it would be drilled first-and then bored to get a straight hole of proper size.

Large holes cut with a welding torch or on a band saw may also be trued by boring, Fig. 79. Whenever a hole must be true, the best way to produce it is by boring.

Fig. 78. Boring operation and setup for the lathe. (R. K. LeBlond Machine Tool Co.)

Setup for Boring in the Lathe

The tool bit is usually held in a soft or semisoft bar called a *boring bar*. The boring bar is held in a toolholder mounted in the tool post, Fig. 78. Several types of boring bars are in use today. Fig. 80 shows the inserted-cutter type of bar which holds the tool bit in either a straight or an angular position.

The boring tool, like other cutting tools, must have side clearance, front clearance, and top rake. The size of the hole and its shape will determine the angle of the cutting edge and the position in which the tool is held, Fig. 81, *A*, *B*, and *C*.

The tool should be set for height on or a little above center, Fig. 82. The tool has a tendency to spring down if the cutting tool overhangs or extends out of the tool post too far.

A boring tool must be set properly before any cutting is attempted. The tool should be run in the hole to see that the end of the bar does not strike the side or end of the hole. This fault may be caused by too large a holder. An incorrect angle of the tool may also cause trouble.

Holding the Part To Be Bored. One of the principal work-holding devices for work to be bored is a four-jaw independent chuck, Fig. 79.

Fig. 79. In this illustration, the workpiece is held by a four-jaw independent chuck. (The DoALL Co.)

Fig. 80. A replaceable boring bit, bar, and toolholder. (J. H. Williams & Co.)

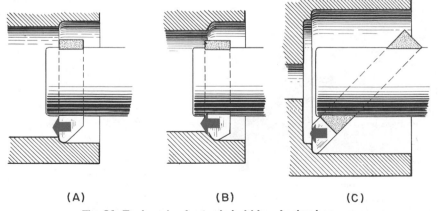

(A)   (B)   (C)

Fig. 81. Tool angles for tools held in a boring bar.

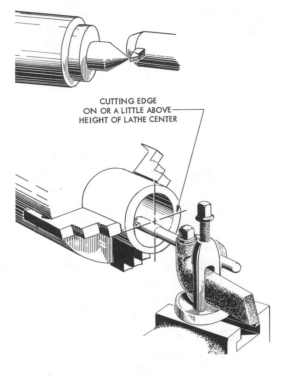

CUTTING EDGE
ON OR A LITTLE ABOVE
HEIGHT OF LATHE CENTER

Fig. 82. The tool bar and toolholder should both be set for minimum overhang. The height should be adjusted as shown.

Fig. 83. Work of irregular shape or size can be mounted on a faceplate, then bored on the lathe. (South Bend Lathe, Inc.)

The faceplate, Fig. 83, is used to hold various kinds of jobs to be bored. An assortment of bolts, angle plates, and straps is required because of the many different setups.

The angle plate may be clamped to the faceplate, Fig. 84, *left*. This presents a surface parallel to the lathe axis and permits the mounting of work for boring. The angle plate may be set at any angle to the faceplate, but 90 degrees is the one most commonly used. Each workpiece presents its own problem in mounting. Proper bolts and accessories must be available and used correctly.

Sometimes the weight of an angle plate or a workpiece of irregular shape may throw the job off balance. Then it is wise to use a counter-balance, Fig. 84, *right*.

Measuring the Size of Bored Holes

Several different instruments are used in measuring the size of the hole that is bored. They are (1) the telescopic gage, Fig. 85; (2) the small-hole gage, and (3) the inside calipers. Of these, the inside calipers are probably the least accurate. Their accuracy depends to some extent on the skill of the operator in using them and transferring the measurements to a precision measuring device.

Telescoping Gages. In using the telescoping gage, the operator places the gage in the part to be measured.

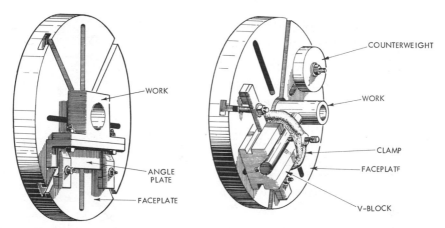

Fig. 84. *Left:* Method of holding workpiece with an angle plate. *Right:* Holding the workpiece with V-blocks.

He then releases the tightening device contained in the knurled handle to allow the plungers to extend against the sides of the hole. The plungers are locked in this position. After removing the gage, the operator determines the distance across the plungers by measuring it with a micrometer.

Taper Boring

Taper boring operations are usually performed with the boring bar and tool bit. Methods by which the tool is moved to cut the desired angle of taper are, however, re-

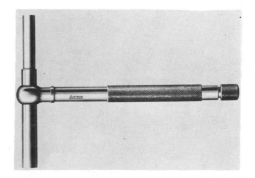

Fig. 85. Telescopic gage. (Lufkin Tool Co.)

stricted. The compound rest and taper attachment are the only means by which internal tapers may be cut.

Internal Thread Cutting

Internal threads are cut in nuts and castings. They are the mating parts into which bolts, threaded studs, threaded shafts, and other parts are screwed. Threaded holes up to about .5 inch are cut with the tap.

Size of Hole before Threading

Before cutting an internal thread,

251

the diameter of the hole to be drilled or bored is carefully worked out.

To determine the size of hole to drill for tapping or threading a Unified National thread, use the following full depth formula:

$$\text{Size of hole} = \text{major diameter} - \frac{1.299}{\text{number of threads per inch}}$$

For all practical purposes, a 75 per cent thread is as strong as a full thread. To cut a 75 per cent thread, the hole size is found by applying the formula given here:

$$\text{Size of hole} = \text{major diameter} - \frac{1.299 \times .75}{\text{number of threads per inch}}$$

EXAMPLE: Find the hole size to be drilled for a ¾-10 UNC 75 per cent thread.

$$.750 \text{ (major diameter)} - \frac{0.974}{10} = .750 - 0.097 = 0.653$$

Setup for Internal Threading

To hold a workpiece for internal threading, three devices are used: (1) draw-in collets; (2) three- or four-jaw chucks; and (3) faceplates.

Logically, threading follows the boring operation. To change from boring to threading, reduce lathe speed by engaging the back gears. Set the feed change levers on the gear box to correspond to the threads per inch to be cut. Of course, the boring tool is changed for a Unified (V) thread-cutting tool.

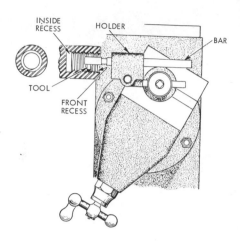

Fig. 86. To cut a right-hand internal 60° V thread, the compound rest is set 29° to the left of the cross-slide centerline.

In cutting a right-hand internal thread, the compound rest is swung around as shown in Fig. 86. By moving the compound rest toward the operator for each successive cut, the tool moves into the metal at the desired 29-degree angle.

A threading tool, Fig. 87, is used for cutting internal 60-degree form threads. It is mounted directly in a boring bar which in turn is mounted in the tool post of the lathe.

The point of the tool is set exactly on the center line of the work for height. The center gage is used this time, as shown in Fig. 87. With it the threading tool is set so the axis of the hole and the center line of the threading tool point are at right angles.

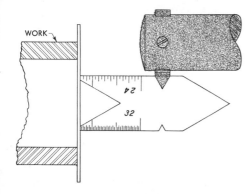

Fig. 87. A center gage can be used to set the internal threading tool square with the workpiece.

Suggestions on How To Cut Internal Threads. To feed the threading tool into the work, the operator turns the feed on the compound rest so as to move the compound rest toward him. To move the tool clear of the work—that is, to move the thread-cutting tool out of the cut being made—the operator carefully moves the compound rest away from him.

The thread stop should, if possible, serve a double purpose. It should operate in one direction to limit the size of the cut. In the opposite direction, it prevents the

tool from running too far back, in which case it would rub off the top of the thread at the back of the hole.

To prevent unnecessary spring, the shank of the tool should be as short and as large as possible.

The threading operation must be carefully done until the thread fits the screw plug gage precisely to the degree of looseness specified on the blueprint.

Measuring the Threaded Hole

A male thread gage, or plug gage, is used to check the pitch diameter of threads in the hole. Gages of various sizes are available for this purpose. The size is stamped on the handle. The double-end plug gage, shown in Fig. 88, has a *Go* member and a *No Go* member. The Go end is made to the lesser dimension specified on the blueprint for the threaded hole. The No Go end is made to the larger dimension specified.

The Go end of the gage should enter smoothly and freely into the threaded hole. The No Go, or higher diameter, will not enter if the hole is the correct size. It is not advisable to allow the No Go member to do more than attempt to enter the hole in order to check a threaded hole accurately. This tool is frequently used in inspection of mass production parts.

Fig. 88. Double-end plug gage. (Taft-Peirce Mfg. Co.)

Tapping Work in the Lathe

The operation of tapping is done in the lathe. Like other operations, it can give good results if the nature of the work is understood and the operation is carefully performed.

When a tap is used in the lathe to cut an internal thread, it is usually held by a tailstock chuck. It is fed into the hole by turning the tailstock handwheel. The work is slowly rotated in the holding device mounted at the headstock. A floating holder or a tap wrench may also be used to hold the tap. The center is kept on the tip of the dead center. When this is done, the tailstock spindle and handwheel control the rate of feed.

A relatively slow speed is recommended for rotating the workpiece. Unlike a single-point threading tool, a tap is a multiple-cutting-edge tool. It has several cutting edges which cut simultaneously. The cutting pressure of the combined edges subjects the shank to considerable stress. A heavy pressure also is concentrated on the cutting edge of the tap.

When a thread has been cut to its full depth, the tap should be backed out slowly. Another point to remember is that, once started, the tap should be kept cutting until the thread is completed. But, if the

lathe must be stopped, the tailstock should be backed off. The tap is backed out two or three threads to reduce the starting shock when the operation is resumed. The tap provides the means of threading a hole of such small diameter that it could not be threaded by the procedures previously given. Taps are naturally limited in size; therefore, the size of hole that can be threaded by this method is limited to the sizes of taps available.

Setup for Tapping

The workpiece should be held securely in the lathe chuck or mounted on the faceplate. The hole to be tapped and the tailstock center must align perfectly with the center line of the lathe. The piece should be drilled with the proper drill for the tap to be used. (See tap drill size charts in the Appendix.) Usually, the tapping follows immediately after the drilling operation is performed. The workpiece remains in position.

Use lard oil as a lubricant when tapping steel, wrought iron, and other metals. Cast iron is tapped dry. (Many of the suggestions applying to tapping given in Chapter 3 are equally applicable here.)

Lathe Speeds and Feeds

For best results it is necessary that the work in the lathe be revolved at the correct speed. Too slow a speed not only wastes time, but leaves a rough finish—too high a speed burns the tool. The speed and feed used will depend on the operation being performed, the kind of metal to be machined, the kind of cutting tool and the working condition and power of the lathe itself.

Cutting Speed

The *cutting speed* is the peripheral or surface speed of the work with respect to the tool. In turning operations, cutting speed is usually measured on the uncut surface of the work ahead of the tool.

The formula for determining surface speed is:

$$S = 3.1416 \, DN$$

where S is the cutting speed in feet per minute, D is the diameter of the work in feet, and N is the rpm of the work.

The *depth of cut* is the distance between the bottom of the cut and the uncut surface of the work, mea-sured at right angles to the machined surface of the work. This is the difference in height between the machined and the work surfaces. The *machined surface* is the surface left by the cutting tool. The *work surface* refers to the surface to be machined.

Handbooks and tables list the proper speeds for machining various materials. Table XIII in the Appendix also lists recommended cutting speeds. Use the closest spindle speed setting on the machine as well as the closest work diameter when setting your speeds.

Feed

The *feed* is the relative amount of motion of the cutting tool into the work per revolution, stroke, or unit of time. In turning, feed may be defined as inches per revolution, and is actuated by the lead screw on the engine lathe. Feed per revolution is listed on the table of the Quick Change Gear Box, and is normally given in increments of thousandths of an inch.

NOTE: Please see review questions at end of book.

<table>
<tr>
<td>

Chapter

8

</td>
<td>

Shapers

Construction, Types,
Setups, and Operations

</td>
</tr>
</table>

The shaper belongs to the family of reciprocating machine tools. The workpiece is machined by successive strokes of a reciprocating head which carries the cutting tool. On a similar machine tool, the planer, the relationship is reversed: the workpiece is fed with a reciprocating motion into a fixed cutting tool. While the machines differ in size and shape, the cutting action is basically the same. Thus, the principles which are discussed in the following pages apply equally to shapers and planers.

Shaper Construction and Types

Today's metal-working shaper is a useful machine. It offers speed and flexibility of setup in performing many kinds of work. The shaper can be found in the toolroom, in the die shop, and in small manufacturing operations. With it work can be done that is difficult or impractical to perform on other machines.

The shaper, Fig. 1, is a tool used to machine flat surfaces by performing successive reciprocating (alternating forward and backward) cuts over the workpiece. The horizontal shaper, Fig. 1, with ram movement in the horizontal plane, is the type most commonly used.

Shaper Parts

The parts of the horizontal shaper will be discussed as they might be brought together in assembling the machine. Main features and uses of the vertical shaper will be given in the discussion of shaper types.

Base. The base rests directly on the shop floor—or on the bench, if a bench-type machine. The base is

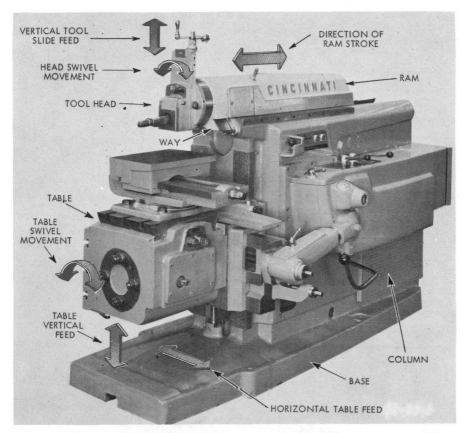

Fig. 1. The horizontal shaper is the most widely used type of shaper. (Cincinnati Shaper Co.)

a casting which serves as a foundation or platform for the machine.

Certain surfaces of the base are machined to fit parts that are fastened to the base. A rim on the base forms a trough to retain the excess oil that drips as machining operations are performed.

Column. The column, or frame, is mounted on the base. It is a hollow casting shaped like a box with openings at the top and bottom. It encloses the mechanism which drives the ram and houses the automatic feed. The ram *ways* at the top of the column form a guide for the ram. See Fig. 1.

The vertical face on the front of the column has been precision machined at right angles to the ram ways on the top of the column. The crossrail moves on this front face.

Crossrail. The crossrail, Fig. 2, is a long casting located across the front of the column. It allows horizontal movement of the table which

257

slides upon it. An elevating screw controls the up-and-down (vertical) movement of the table. A cross-feed screw, called a lead screw, is mounted horizontally in the cross-rail. The lead screw controls side-wise movement of the saddle, to which is attached the table.

Saddle. The saddle, or "apron," Fig. 2, is a flat casting located on the crossrail. This unit of the shaper supports the table.

Table and Table Support. The table is a boxlike casting, Fig. 2, with openings top and bottom. The rear face of the table is clamped to the front face of the saddle. The front face of the table, on many shapers, is used as a clamping surface for a table support. The top and two sides of the table are used to locate and hold the work directly, or to locate and hold a vise or fixture which in turn secures the work-

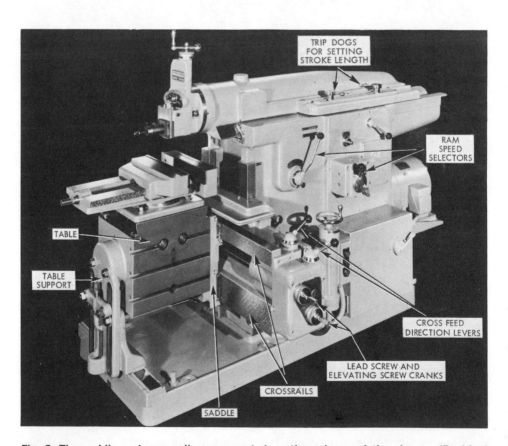

Fig. 2. The saddle and crossrails are mounted on the column of the shaper. (Rockford Machine Tool Co.)

piece. The surfaces on the top and two sides have T-slots to accommodate the bolts that clamp the work.

The table support, Fig. 2, if one is used, extends from the front face of the worktable to the base of the machine. It supports the outer end of the table, especially when a heavy cut is made.

There are two types of tables. The *plain*, or standard table, is capable of the horizontal and vertical movements previously described. The *universal* table, Fig. 3, can, in addition to the conventional table movements, swivel on a specially designed saddle on an axis parallel to the ram movement. This is a definite advantage when it is necessary to machine work surfaces to specified angles.

Ram. The ram, Fig. 2, another important part of the shaper, moves back and forth horizontally on the top of the column, carrying the tool with it. It slides on carefully designed and precision machined *ways* cut in the top of the column. The stroke of the ram can be adjusted to any length up to the maximum stroke for the particular machine. The ram is propelled by either mechanical or hydraulic power.

Tool Head. The tool head, Fig. 4A is clamped to the forward end of the ram. It consists of the parts which hold the cutting tool and those parts which guide the tool ver-

tically and adjust it for the desired cut. The head has a lead screw and a handle that permits feeding the clapper box and tool up and down by hand. An adjustable micrometer collar, graduated in thousandths of an inch, tells the distance the tool is raised or lowered when the downfeed screw is turned.

Clapper Box. The clapper box, or *tool block*, as it is sometimes called, Fig. 4A, is an important part of the tool head. When the ram is moving forward on the cutting stroke, the clapper box is forced back against the base of the tool head. It is thus properly supported. The clapper or tool block is hinged

Fig. 3. The universal shaper is named from its universal table which can be tilted for work that is to be done at specified angles. (Cincinnati Shaper Co.)

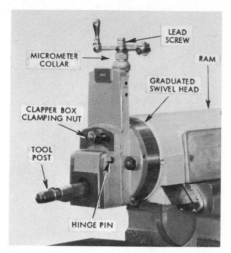

Fig. 4A. The tool head consists of the parts which hold the cutting tool and those parts which guide the tool and adjust it for making the desired cut. (Cincinnati Shaper Co.)

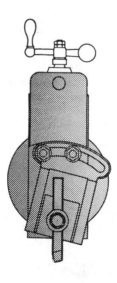

Fig. 4B. The clapper box may be swiveled through a small arc to allow the tool to swing out from the work on the return stroke.

to swing outward. This allows the tool to lift slightly and swing clear of the work on the return stroke. In this manner the cutting edge of the tool does not drag (except for its own weight) over the surface of the work on the return stroke of the ram.

The clapper box is attached to the tool slide with a pivot screw and clamping nut. The clapper box may be swiveled through a small arc in either direction, clockwise or counterclockwise, Fig. 4*B*, within the limits of the clamping nut slot. Proper adjustment of the clapper box will allow the tool to swing out from the work on the return stroke when machining vertical or angular surfaces. For horizontal cuts, the clapper box is usually set vertically.

The cutting tool, which is somewhat like a lathe tool in shape, is held in the tool post on the clapper box. On the forward or cutting stroke, the action of the cutting tool holds the clapper box securely against the base of the tool head.

Shaper Size or Capacity

The size of a shaper is given by the maximum length in inches of the stroke of the ram. The maximum length of stroke may range from 6 inches on a small bench-type shaper to 36 inches on a heavy-duty machine. A 16-inch shaper, for example, can be adjusted for any stroke from 0 to 16 inches in length.

The dimension for the length of the stroke indicates, in addition to the size of the machine, the dimension of a cube that can be held and planed in the shaper. A 16-inch shaper has a traverse table feed that can be used to plane a surface 16 inches wide. The vertical distance between the tool head and the worktable, in its extreme lower position, is great enough to surface a 16-inch cube resting on the table. Thus, a cube measuring 16 x 16 x 16 inches can be shaped on a 16-inch machine.

Types of Shapers

Horizontal Shapers. There are two types of horizontal shapers, the mechanical crank-type and the hydraulic type, which differ primarily in the method used to drive the ram. The parts and operating features described earlier are identical for both types. The drive mechanisms

for each will be briefly discussed.

Crank-Type Shaper. The crank-type shaper, Figs. 1 and 3, takes its name from the mechanism used to reciprocate the ram. Fig. 5 shows the crank drive. In construction, the crank-operated shaper employs a crank mechanism to change rotary motion to reciprocating (back and forth) motion. A large gear, called a "bull wheel," receives its rotary motion from the electric motor through belts and a speed-box drive shaft.

Hydraulic Shaper. The hydraulic shaper, Fig. 6, is similar in outer construction to the crank-operated shaper. The main difference consists in the method used to move the ram.

Hydraulic pressure, as a means to drive the ram, is of great practical value. The principle is based on Pascal's law, which states that a fluid enclosed in, say, a pipe, will transmit, when pressure is applied, equal

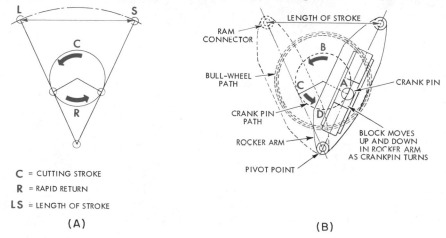

C = CUTTING STROKE
R = RAPID RETURN
LS = LENGTH OF STROKE

(A)

(B)

Fig. 5. The crank-operated shaper employs a crank mechanism to change rotary motion to back-and-forth motion.

261

pressure in all directions and to all surfaces it touches.

In the hydraulic shaper, the ram is moved back and forth by a piston moving in a cylinder under the ram. The flow of oil from a high-pressure pump acting against first one side of the piston and then the other moves the ram. This flow of oil gives

a positive drive to the ram. There is no chance of backlash, as there is with a gear drive. A wider range of cutting speeds and feeds is possible with the hydraulic type of shaper than with the mechanical shaper.

Vertical Shaper. The vertical shaper, or slotter, Fig. 7, has an operating mechanism similar in princi-

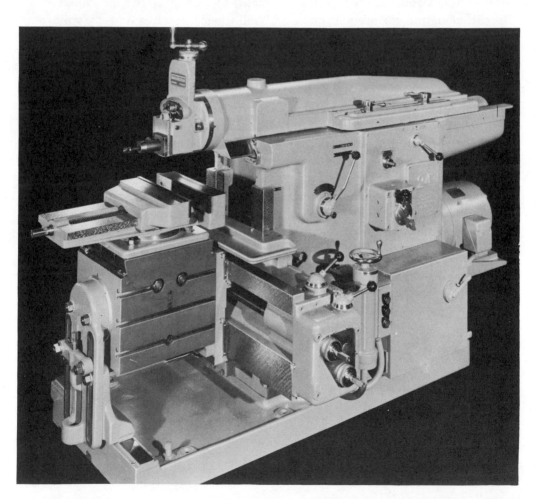

Fig. 6. Hydraulic pressure is used to operate the hydraulic shaper. (Rockford Machine Tool Co.)

ple to that used on crank shapers. The important difference between this machine and the horizontal shaper is the vertical ram. The construction of the table is quite different, too.

The vertical shaper is used in cutting internal slots and keyways of various shapes, and both external and internal gears. It is also used to cut intricate patterns in die work. The jobs that can be done on this machine are like those done on the standard shaper, but the vertical shaper performs them in a number of different ways. Vertical shapers are of three types: crank-driven, rack-driven, and screw-driven.

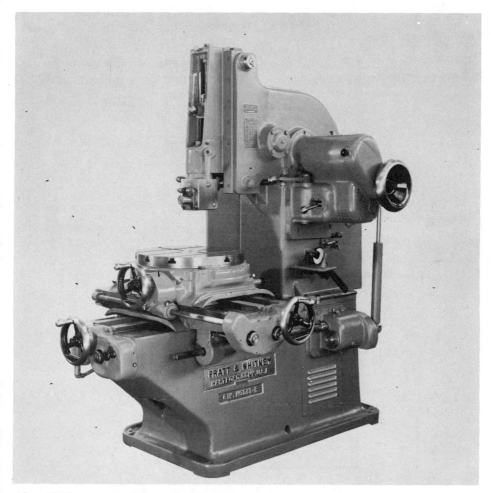

Fig. 7. The vertical shaper is used to cut internal slots and keyways of various shapes as well as internal and external gears. (Pratt & Whitney Div., Colt Industries, Inc.)

The rotary table on the vertical shaper can be fed in two directions as well as rotated. The flexibility of the rotary motion, together with the two-way horizontal feeding of the table, makes the vertical shaper a valuable tool for cutting keyways, slotting, and internal work. The ram may be tilted to give inside clearance for die work.

Cutting Tools Used with Shapers

Interchangeable bits ground for different cutting operations may be quickly fastened in the toolholder or removed. The tool bit may be swung at different positions depending on the nature of the cut.

Whether an interchangeable bit and toolholder or a solid tool is used, clamping the tool properly is important. It is a good policy to keep the slide up and the grip on the tool short, as shown in Fig. 8, so as to have all the rigidity possible.

Grinding of Shaper Tools. Shaper tools are ground either by

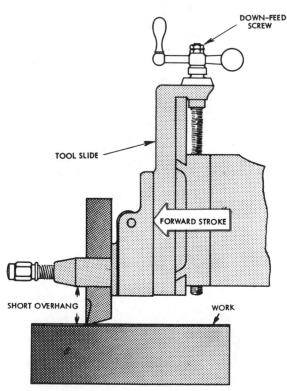

Fig. 8. The tool slide should be kept up and the grip on the tool should be as short as possible for the greatest possible tool rigidity.

hand or by special grinding machines. The shapes and clearance angles for shaper tools are very important. The type of toolholder will determine the clearance of the heel of the tool. Some holders hold the tool bit in a vertical position. Others hold the tool bit at an angle approximately 20 degrees to the vertical. If the tool bit is positioned exactly vertically, 3 degrees of heel clearance is sufficient.

Shaper tools should always have proper clearance on the side of the tool as well as at the end of the tool. A side clearance of 2 degrees is commonly used. The cutting edges should be stoned after grinding. Stoning makes the cutting edge last longer. Fig. 9 shows various shapes of tool bits used in machining work on a shaper.

Shaper Work-Holding Devices

The work must be held securely and solidly while it is machined in the shaper. It must not spring out of shape or "give" during the cut.

Vise. Usually the work is held in a vise bolted to the machine table with T-bolts. The vise consists of a fixed and a movable jaw. The standard vise, Fig. 10, has a graduated base on which the vise body can be swiveled or turned 360 degrees.

Standard vises are commonly the double-screw or single-screw type. The double-screw type can develop greater pressure and has a swivel jaw for taper work.

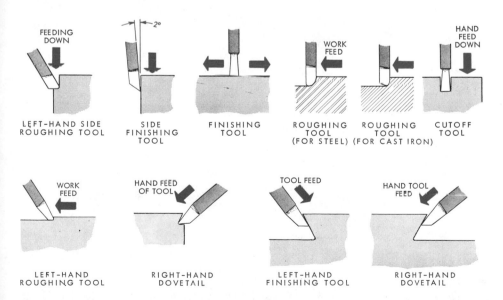

Fig. 9. Cutting tools for various machining operations performed on the shaper are shown.

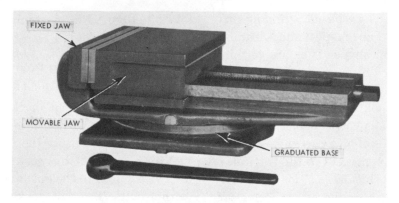

Fig. 10. The standard vise consists of a fixed jaw, a movable jaw, and a graduated base. The base can be turned 360°. (Cincinnati Shaper Co.)

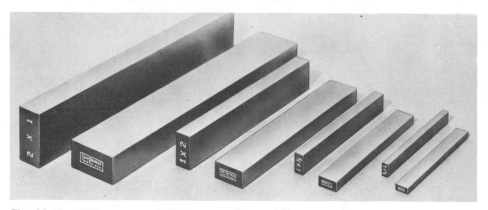

Fig. 11. Parallels are placed under workpieces to raise the piece to a suitable height. (Taft-Peirce Mfg. Co.)

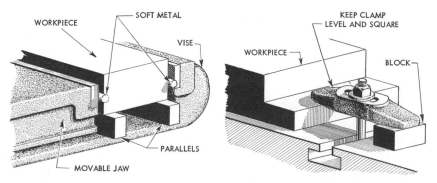

Fig. 12. *Left:* Work shown here is properly mounted in a vise. Note the soft metal rods used to keep the vise jaws from marring the workpiece. *Right:* The proper method of clamping a workpiece to the machine table.

266

Parallels. Parallels, Fig. 11, are square or rectangular bars of steel used in precision machining. They are placed beneath workpieces to provide a solid seat and to raise the work to a suitable height.

Bolts, Clamps, and Other Devices. Common methods of holding work are shown in Fig. 12. Work properly mounted in a vise is shown at left in the illustration. The proper way of clamping a workpiece directly to the table is shown at right.

Fig. 13 shows the use of V-blocks to hold a round piece of work to the table. Fig. 14 shows the use of angle plates bolted in position on the table and the use of C-clamps.

Shaper Setups and Operations

Up to this point in this chapter, the shaper parts, cutting tools, and work-holding devices have been discussed. Shaper operations will now be taken up one by one. Since it is always necessary to make a setup before the operation, the necessary adjustments in making setups will be detailed here.

Adjustments Necessary in Making Setups

The shaper is a relatively simple machine to operate but, like other machines, it must be adjusted properly or it will not give satisfactory results. The purpose of this section is to instruct the operator in the mechanical adjustments necessary

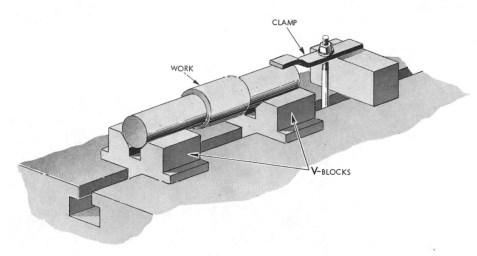

Fig. 13. V-blocks can be used to hold a round workpiece to the table. A clamp is used at both ends of the work, and a stop in front. The clamp and stop are omitted here to permit a clear view of the work in the V-blocks.

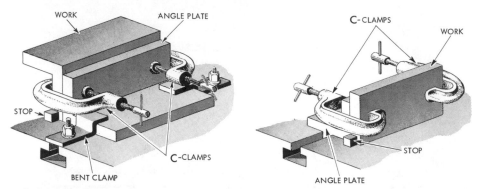

Fig. 14. C-clamps can be used to hold work on an angle plate. Note the use of the bolted bent clamp in the illustration on the left.

to get the shaper to function correctly during the various operations.

Ram Adjustment. The following two adjustments are provided for the ram. *First*, the stroke length is regulated by turning the stroke-adjusting shaft, Fig. 15, which is located at the operator's side of the

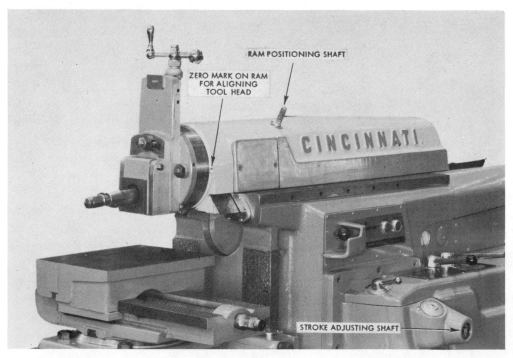

Fig. 15. Adjustment devices for the ram on the shaper are shown here. (Cincinnati Shaper Co.)

machine. Turning this shaft in one direction lengthens the stroke; turning it in the opposite direction shortens the stroke. *Second*, if, after the length of the stroke has been set correctly, the travel of the tool does not cover the work, the position of the ram is changed to make it correct. To reposition the ram the shaper must be stopped at the extreme end of the back stroke, then the self-locking ram-positioning shaft is adjusted, Fig. 15. The ram-positioning shaft is turned until the ram stroke allows the tool to move ¼ inch beyond the workpiece on the forward or cutting stroke, and ½ inch behind the workpiece on the return stroke, Fig. 16. Remember to make the adjustment for

length of stroke before beginning the adjustment for position over the workpiece.

Head Adjustment. Movement of the tool slide and the cutting tool is controlled by the handle at the top of the head. This handle is attached to the down-feed screw inside the slide, Fig. 4. Turning the handle clockwise lowers the slide. For convenience in making accurate adjustments of the slide and tool, the feed screw has a micrometer dial graduated in thousandths of an inch.

To make accurate adjustments of the tool, all "backlash" must be taken out of the down-feed screw before the dial is set to a definite figure. This backlash is the lost motion between the threads of the

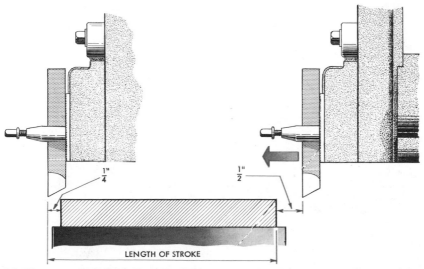

Fig. 16. The ram stroke should allow the tool to move ¼ inch beyond the workpiece on the forward stroke and ½ inch beyond the workpiece on the back stroke.

screw and the threads in the nut which the screw travels through.

To feed the head vertically, the head must first be squared by lining up the zero mark on the head with

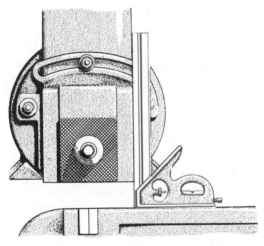

Fig. 17. The combination square can be used to square the head of the shaper.

Fig. 18. Angles may be shaped by setting the tool head at the desired angle and using the tool head feed. (Delta Power Tool Div., Rockwell Mfg. Co.)

the zero mark on the ram, Fig. 15. To check further on squareness, use a combination square, Fig. 17.

In the setup for shaping a dovetail and in certain other setups, it is necessary to position the tool head at an angle as shown in Fig. 18. Be sure not to run the ram back into the column with the slide set at an extreme angle, as the slide will strike the column when the ram moves back on the return stroke.

Adjusting the Clapper Box. For heavy cuts, the clapper box is set so the top slants slightly away from the cutting edge of the tool. This permits the tool to swing away from the work on the back stroke of the ram, protecting the cutting edge of the tool, Fig. 19.

Table Adjustment. The table is raised or lowered by the hand crank. The supporting clamp and saddle-clamping screws must first be loosened.

The table is moved back and forth (transversely) on the crossrails by the lead screw, which can be turned either by a hand crank or by mechanical power. The power drive is through a ratchet-feed mechanism.

Testing the workseat of a vise for parallelism to the stroke of the ram should precede any setup. This is done using a dial indicator clamped in the tool post. The workseat in the vise should be cleaned carefully, and accurately sized parallels should be used, Fig. 20.

Adjusting Vise to Ram. On most jobs it is necessary to adjust the solid jaw of the vise either parallel or at right angles to the stroke of the ram.

To test the vise jaws for being at right angles to the stroke of the ram, a dial indicator is clamped in the shaper tool post. The indicator button is brought to bear on the solid jaw of the vise as shown in Fig. 21. The table is moved at right angles to the ram, using the hand cross-

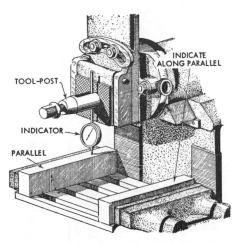

Fig. 20. A dial indicator can be mounted in the tool post. Used in conjunction with parallels, the workseat of the vise can be checked for trueness.

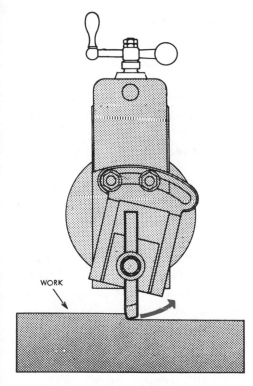

Fig. 19. The clapper box may be turned with the top away from the tool's cutting edge. The tool will swing away from the work at the end of the stroke.

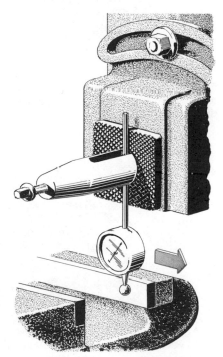

Fig. 21. Vise jaws can be checked for a perfect right angle position to the ram stroke using a dial indicator.

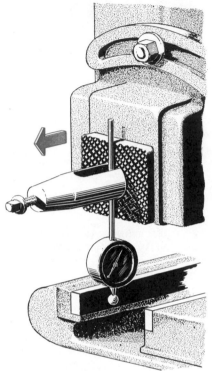

Fig. 22. A dial indicator can be used to check vise jaws for a position parallel to the ram stroke.

Fig. 23. In shaping, the workpieces can also be clamped to the side of the table. (Delta Power Tool Div., Rockwell Mfg. Co.)

feed. When the vise is set so that there is no movement of the needle on the indicator, this test is complete.

To test the vise jaws for being parallel to the ram stroke, the indicator button is brought to bear on the solid jaw of the vise as shown in Fig. 22. The ram is moved forward and back to determine any movement of the indicator needle. If there is none, it can be assumed that the vise jaw is parallel with the ram stroke.

Variety of Shaper Operations

The versatility of the shaper should again be emphasized, as there are a great many operations that can be done on it. Fig. 23 shows a setup used in shaping the end of a round shaft.

Another shaper operation and setup is shown in Fig. 24. An internal keyway is being cut, using an extension toolholder.

Fig. 25 shows a setup used to hold a workpiece when shaping splines in a shaft. The indexing centers make it possible to equally space the splines accurately around the cylindrical workpiece.

Other operations which can be readily performed on the shaper are serrating, dovetail cutting, and contour cutting.

Shaping Horizontal Surface

Much of the work done on the

shaper is shaping flat surfaces on pieces held in one or another of the holding devices. The horizontal or vertical surface produced is the result of a series of cuts made with a single-point cutting tool.

Fig. 26 shows a conventional shaper setup. This type of job probably accounts for 50 per cent of the work done on a shaper. The setup shows a piece of steel clamped in a single-screw vise. For horizontal cuts intended to remove excess metal regardless of finish, a roughing tool is used. The cutting tool should be clamped securely in the tool post in a vertical position square with the surface to be shaped.

Suggestions on How To Shape a Horizontal Surface. During the operation, the operator stands at the right and in front of the machine where the controls are within reach. The cut is started at the

EXTENSION TOOLHOLDER

KEYWAY

Fig. 24. Internal keyways can be shaped on the shaper. (Rockwell Mfg. Co.)

right end of the horizontal surface (nearest the operator). The first cut taken from a casting should be deep enough to get under the scale.

Fig. 25. Indexing centers are used to shape equally spaced splines about the circumference of a shaft. (Rockwell Mfg. Co.)

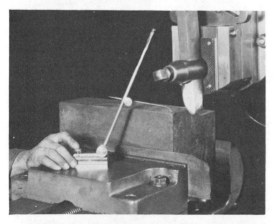

Fig. 26. A surface gage can be used to check the surface of work held in a shaper vise. (Lufkin Tool Co.)

Roughing cuts remove excess material. A finishing cut is then made to bring the work to size and produce a smooth finish. By setting the cross-feed about one-half the width of the cutting edge of the tool, each cut will overlap the last cut and a smooth surface will be obtained.

Shaping Vertical Surface

The work is mounted in the vise or directly to the table. The surface to be cut must be carefully lined up with the ram. An indicator can be fastened to the ram and the ram fed across the workpiece by hand to test trueness.

Either a straight or offset toolholder can be used for vertical cuts if the tool is properly ground. The toolholder must be so adjusted that it will not strike the vertical surface. The slide and toolholder, when in the lower cutting position, should not extend any more than absolutely necessary. The cutting edge of the tool should be set in an approximately horizontal plane.

For vertical cuts, the clapper box must be set at an angle from its vertical position, Fig. 27. This prevents the tool from dragging and scoring the planed surface during the return stroke. It should be emphasized again that the tool slide, the tool, and the clapper box must be carefully set for vertical cutting.

Suggestions on How To Shape a Vertical Surface. In making the cut, the tool is fed down carefully by hand about .010 inch at the end of each return stroke of the tool.

The finishing cut is made with a finishing tool properly ground for the material of the workpiece. It is adjusted in the tool post so that only the nose of the tool contacts the vertical surface, Fig. 27. A slight clearance is provided for the side of the tool. When the workpiece is cast iron, the edge of the casting is slightly beveled with a file to prevent the sand and scale on the casting from dulling the finishing tool. To prevent chipping the edge of the work at the end of the cutting stroke, it is wise to bevel the workpiece with a file to the depth of the cut.

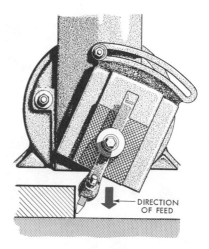

Fig. 27. When a vertical surface is to be shaped, the clapper box is set at an angle. The toolholder is adjusted so that it will not strike the surface being shaped.

Setup for Angular Shaping

When cuts are made at any angle other than a right angle to the horizontal or the vertical, they are called *angular cuts.*

Angular cuts are usually made on the shaper by swiveling the head to the desired angle either to the right or left of the vertical position. As the tool slide is lowered at an angle, the tool is fed along the surface to be cut, Fig. 28.

It is not always necessary, however, to swivel the head to machine one surface at an angle to another. The work may be set on tapered parallels in the vise in such a position that the layout line, representing the angular surface, is in a hori-

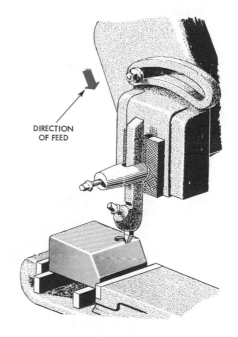

Fig. 28. Angular cuts are usually made by swiveling the head so that the feed is positioned for the angle desired.

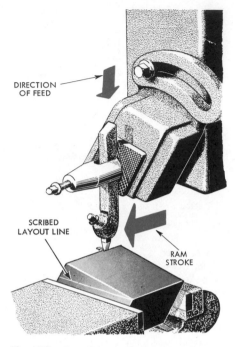

DIRECTION
OF FEED

SCRIBED
LAYOUT LINE

RAM
STROKE

Fig. 29. Angular surfaces may be machined by horizontal shaping. Note that the workpiece is set on tapered parallel bars so that the layout line is in a horizontal position.

zontal position, Fig. 29. The angular surface is then machined by horizontal shaping. If tapered parallels are not available or if the angle is an odd one, the work can be mounted in the vise with the aid of a surface gage, Fig. 30. If parallels are not used to support the work, care must be used that the work does not move under pressure of the tool, or the proper angle will not be cut.

Another method of shaping angular surfaces is to tip the table if the shaper is equipped with a universal table, Fig. 3. The work is placed horizontally in the vise. Then the table is set at the desired angle and a horizontal cut is taken as in horizontal shaping.

For cutting chamfers, the side cutting tool is appropriate, Fig. 9. Fig. 31 shows the machining of a 55-de-

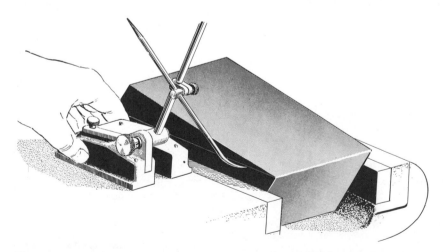

Fig. 30. Angular work can be mounted in a vise for horizontal shaping with the aid of a surface gage.

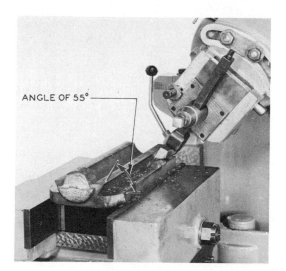

ANGLE OF 55°

Fig. 31. Machining a 55° dovetail. Note that the head is swiveled at an angle of 55°. (Rockford Machine Tool Co.)

gree dovetail on an iron casting. On jobs of this kind, the head is swiveled to the desired angle—in this case 55 degrees. Then the tool is fed downward either by hand or by power. The proper shape of tool is of course necessary to secure the desired corner and finish. The work is held in the vise on parallels. The tool used here is the replaceable bit.

Safety Precautions—Shaper

The two most common causes of injury around the shaper are (1) flying chips and (2) getting caught between the workpiece and the tool. Remember that it is dangerous to allow fingers near moving parts. If the tool is throwing off chips, the operator should keep his eyes out of the line of stroke of the ram. Injuries may happen if the operator tries to take measurements when the ram is in motion. On some machines, injury can result if removable handles are left in place while the machine is running.

Observe the following precautions:

1. When belts and pulleys are used to drive a shaper, they should be enclosed in standard guards of angle iron and wire mesh, or angle iron and expanded or sheet metal.

2. When the shaper is so located that the rear of the ram at the extreme limit of its possible travel comes within 18 inches of a wall, post, or other obstruction, or such travel projects into an aisle or walkway, the space between the end of the ram travel and the obstruction

or aisle or walkway should be enclosed by standard iron pipe railings or their equivalent.

3. Men operating shapers should be provided with and required to wear goggles unless the machine is provided with a chip guard.

4. Shaper operators should not wear gloves, loose or torn clothing.

5. When clamping the work in place on the machine table, the clamps should be so placed and blocked that they will have full purchase on the work and not spring out of shape.

6. When clamps are used, the bolts and nuts should be tightened only with wrenches which fit properly, as otherwise the wrench may slip off and the operator may be injured.

7. When a vise is used to hold the work, the operator should make sure that the bolts holding the vise in position and place are securely tightened.

8. Stop the machine before making adjustments to the machine or work, or before reaching across the table.

Courtesy of the National Safety Council—

NOTE: Please see review questions at end of book.

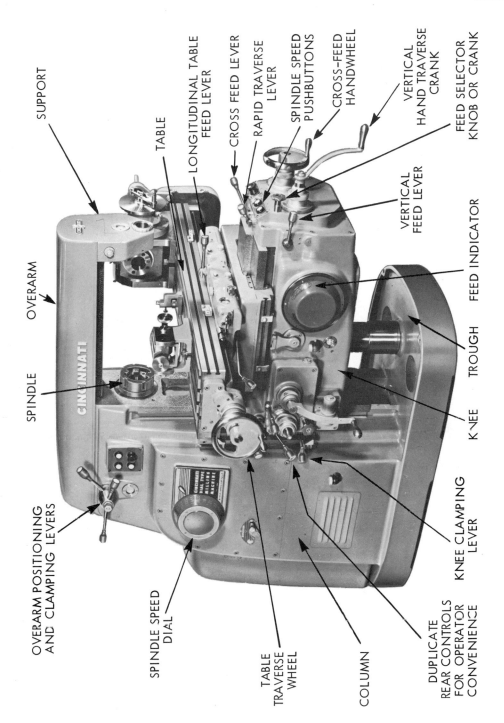

PROBLEM: PLAIN OR SLAB MILLING

COLUMN and BASE ☐ OVERARM

B

MOUNTING THE CUTTER

(A) Cutting Tool components — Arbor, Plain Spacers, Cutter with Key, Bearing Spacer, and Snug Nut. (1) A cleaned Arbor is inserted into a cleaned Spindle, and (2) secured by a Draw Bar which is thoroughly tightened with a wrench. (3) The Plain Spacers, Cutter with Key, Bearing Spacer, and Snug Nut are then positioned on the Arbor as shown at (B).

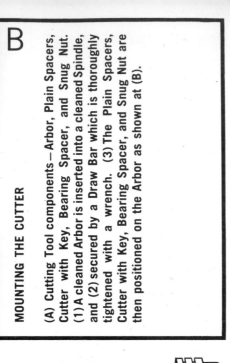

IMPORTANT MILLING MACHINE CONTROLS
AND SETTINGS

THE OVERARM LEVERS

The overarm levers which are located on the upper column are used to position and clamp the overarm. Securing the overarm will prevent vibrations of the cutter arbor when the cutter contacts the workpiece.

THE MASTER CONTROLS

The master control buttons, located on the right-hand side of the column, are the master start, stop, and coolant controls. The master start control must be activated before the machine can be operated.

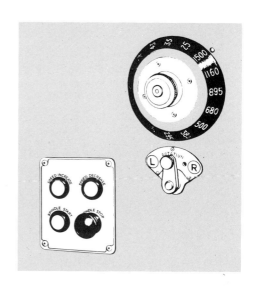

SPINDLE SPEED CONTROLS AND REVERSE LEVER

The spindle speed selector dial is located on the column. Speed settings range from 18 to 1800 RPM. On most machines the operator manually dials the desired spindle speed.

More automatic machines allow the operator to increase or decrease spindle speed by simply pushing the speed increase or speed decrease pushbutton. Speed changes are made automatically.

The reverse lever located beneath the spindle speed dial controls the direction of rotation of the spindle.

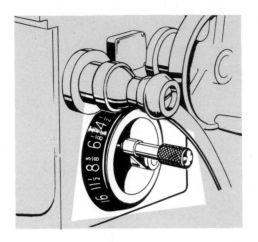

THE FEED SELECTOR

The feed selector is located at the front of the knee. Most milling machines have a dial crank-type feed selector (shown at left). More automatic types have a simple knob (Step 13, Page E), which is turned to obtain different feed settings.

The feed selector determines the rate of longitudinal, cross, or vertical feed, depending on which directional feed control is depressed. Feeds can range from 3/8 to 90 inches per minute.

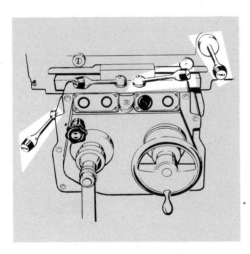

POWER FEED LEVERS

Longitudinal, cross, and vertical power feed levers are located at the front of the knee. To engage a power feed, set the feed selector to the desired rate of feed and then move the appropriate power feed lever to the position which will give the feed desired (right, left, up, etc.).

Manual controls with micrometer collars permit the operator to align the workpiece with the cutting tool before starting power operation.

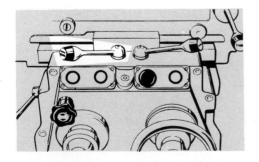

RAPID TRAVERSE LEVER

A powered rapid traverse lever at the front of the knee permits the operator to quickly change the position of the knee or table in whatever direction the feed is engaged. When the rapid traverse lever is released the original feed rate is resumed.

Milling Machines:

Types, Construction, Accessories, and Attachments

Chapter

9

Milling is the process of producing machined surfaces by progressively removing a predetermined amount of material from the workpiece which is fed to a rotating milling cutter. One of the characteristic features of the milling process is that each milling cutter tooth removes its share of stock in the form of individual chips. See Figs. 1, 3, and 4.

The quality of finish, or the physical character of a machined surface produced by the milling process, is determined by the method of milling. There are basically two methods of milling, *peripheral milling* and *face milling*.

Peripheral Milling

The milled surface in this method is produced by cutting teeth located on the periphery (outer edge) of the cutter body. The milled surface produced by peripheral milling corresponds to the contour of the milling cutter, which may vary from a flat surface to a form shape, Fig. 2. Milling cutters are discussed in greater detail in the following chapter.

Peripheral milling is performed by milling machines with horizontal spindles. The operation illustrated in the Milling Machine Trans-Vision is an example of peripheral milling. The milling cutters are mounted on arbors which are supported at the outer end by an overarm in order to increase the rigidity of the setup. The cutters may be located at a definite distance from the nose of the spindle through the use of spacing collars.

The quality of finish produced by peripheral milling is partly determined by two factors: (1) the geometry of the machining process, and (2) the plastic flow of the material being removed from the work-

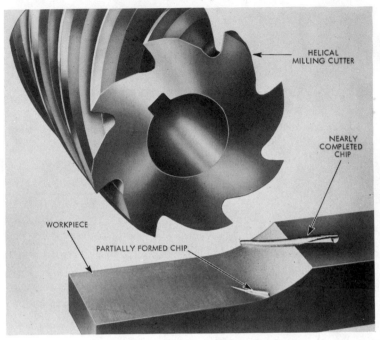

Fig. 1. Each tooth on the milling cutter removes its share of workpiece material in the form of individual chips. (Cincinnati Milling Machine Co.)

piece in the form of chips. Other factors which will influence the quality of the work are: correct selection of feeds and speeds, rigidity of the machine and setup, and the proper grinding of the cutting tool.

The geometry of the machining process depends roughly on the method of milling and the angle at which the cutting edge contacts the workpiece surface. The geometry of the peripheral milling process results in characteristic lines (tooth marks) on the workpiece approximately parallel to the axis of cutter rotation, and spaced at a distance equal to the feed per tooth.

Irregular marks may also result from the flow or resistance to flow of the type of metal being milled. The type of metal being machined, therefore, also determines the quality of the surface finish.

Feeding the Work. There are two basic methods of feeding the work into the milling cutter in peripheral milling. When the workpiece is fed in the direction of the cutter rotation, the operation is referred to as *down or climb milling*. See Fig. 3. When the workpiece is fed against or opposite the rotation of the cutter, the operation is referred to as *conventional or up milling*, Fig. 4. In

Fig. 2. In peripheral milling, teeth located on the outer edge of the cutter produce the milled surface. *Left:* Most peripheral cutters are flat in profile (called plain, slab mills) and produce a flat surface. *Right:* The cutter may also be shaped to produce a formed or contoured surface. (Cincinnati Milling Machine Co.)

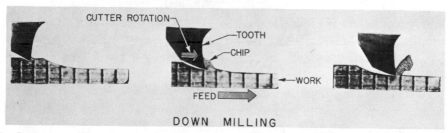

Fig. 3. *Down milling* occurs when the workpiece is fed in the direction of the cutter wheel's rotation. (Cincinnati Milling Machine Co.)

up milling the tooth marks are produced by the direct action of the cutting edge of each tooth as it engages the surface produced by the cutting edge of the previous tooth.

In down milling similar marks may result from the intersection of the plane of shear with the surface milled by the previous tooth.

The uniformity of tooth mark

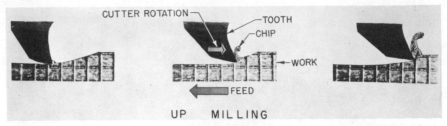

Fig. 4. When the workpiece is fed in the direction opposite to the cutter wheel's rotation, the operation is called *up milling*. (Cincinnati Milling Machine Co.)

Fig. 5. Variations in tooth spacing can be caused by *(from top to bottom)* one low tooth, one high tooth, or spindle run-out (wobble in the spindle). Cutters with high or low teeth should be replaced. When the run-out spindle condition occurs the spindle should be tightened or repaired. Note that the bottom two diagrams show the proper tooth spacing for a small diameter cutter and a large diameter cutter (Cincinnati Milling Machine Co.)

spacing and the character of the profile of the milled surface will depend on a number of factors:

1. *Variation in tooth spacing.* This is determined in the manufacture of the cutter.

2. *Variation in the distance of the cutting edge of the teeth from the axis of cutter rotation.* This usually results from inaccuracies in the cutter sharpening operation, which will produce high and low teeth in the cutter. See Fig. 5, A and B.

3. *Cutter run-out produced by springing the arbor when mounting it in the machine.* A sprung arbor is usually the result of tightening the arbor nut without having the arbor support in place. Cutter run-out may also be caused by nicks or dirt particles in the tapered mating surfaces of the spindle and arbor, and between the end faces of spacing

collars and the clamping nut. See Fig. 5C.

4. *Variation in the deflection of the arbor when under a load.* The load, or force, is caused by the cutting action of the cutter, which is keyed to the arbor.

5. *Diameter of the cutter.* See Fig. 5, D and E.

6. *Method of milling* (up or down milling).

7. *Feed per tooth.* See Fig. 5.

Face Milling

The milled surface in this method results from the combined action of cutting edges located on the face, or end, of the cutting tool, Fig. 6, as well as the cutting edges on the periphery. The milled surface produced by face milling is flat and there is no relation to the contour of the teeth, except in milling a

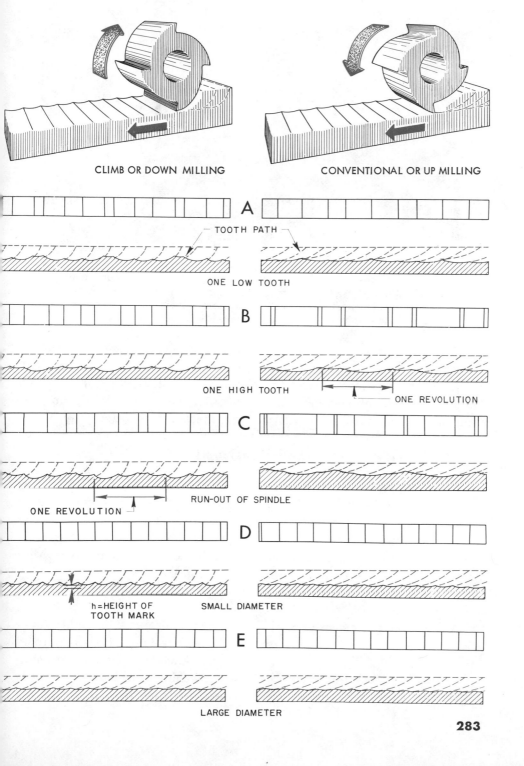

CLIMB OR DOWN MILLING

CONVENTIONAL OR UP MILLING

A

TOOTH PATH

ONE LOW TOOTH

B

ONE HIGH TOOTH

ONE REVOLUTION

C

RUN-OUT OF SPINDLE

ONE REVOLUTION

D

h=HEIGHT OF
TOOTH MARK

SMALL DIAMETER

E

LARGE DIAMETER

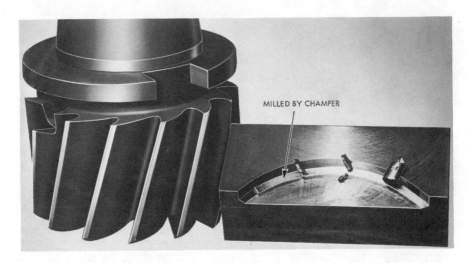

MILLED BY CHAMFER

Fig. 6. Face milling operations involve cutting by both the bottom surface as well as the peripheral cutting edges. The face mill shown above is mounted on a holder which in turn is mounted in the spindle. The chamfer on the peripheral edges of the cutter teeth provides most of the cutting action in face milling. (Cincinnati Milling Machine Co.)

shoulder. See Fig. 6.

Face milling operations are done on milling machines with both horizontal and vertical spindles. The face milling cutter is usually mounted on an adapter inserted in the spindle of the machine tool. Sometimes, however, the face mill may have a shank which is an integral part of the cutter body.

Milling Machine Types and Construction

The milling machine is one of the most versatile of the chip-producing machine tools. It is capable of accurately producing one or more machined surfaces on a piece of material. Its adaptability is especially valuable for quantity production, job shop operations, repair shop, tool room, and experimental work.

Types of Milling Machines

In order to meet the many requirements expected from the milling process, a number of milling machine types are available in a wide range of sizes. These machines may be classified as follows:

1. Knee-and-Column Type Milling Machines
2. Manufacturing Type Milling Machines
3. Special Type Milling Machines

If the basic operations performed by the knee-and-column type milling machine are basically understood, the skills developed will carry over to the manufacturing or special type milling machines. Therefore, the majority of the discussion will center around the knee-and-column type machine shown in the Milling Machine Trans-Vision.

Knee-and-Column Type Milling Machine. This is the standard milling machine, and its name is derived from the design of the two major elements: (1) the column-shaped main frame, and (2) the knee-shaped projection from the column, which supports the saddle and the worktable.

There are three basic styles of knee-and-column milling machines: (1) plain, (2) universal, and (3) vertical.

Plain Knee-and-Column Type. The plain, horizontal milling machine, Fig. 7, is a common type of knee-and-column machine. It has a horizontal spindle which projects at right angles from the column face. The spindle is hollow and tapered

Fig. 7. The plain knee-and-column milling machine has longitudinal, transverse (cross), and vertical table feed. (Cincinnati Milling Machine Co.)

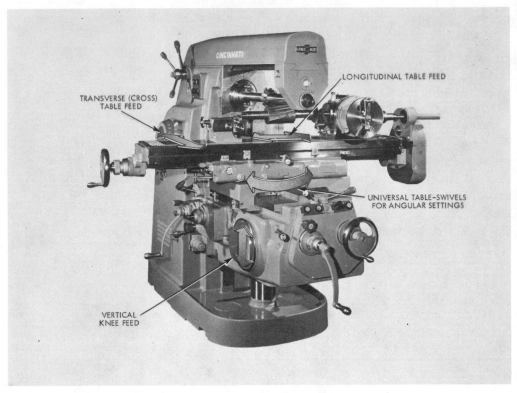

TRANSVERSE (CROSS) TABLE FEED

CINCINNATI

LONGITUDINAL TABLE FEED

UNIVERSAL TABLE-SWIVELS FOR ANGULAR SETTINGS

VERTICAL KNEE FEED

Fig. 8. The universal knee-and-column mill has all of the table feeds of the plain type. In addition, however, its table can be tilted at an angle to the column face. (Cincinnati Milling Machine Co.)

internally to receive cutter arbors which are mounted in it. The knee is supported by a vertical screw which permits the knee to be raised or lowered vertically on accurately machined ways on the column. The table can be moved manually or by power. It is mounted on precision machined ways on the saddle which rests on the knee. The three directions of table feed are longitudinal (parallel to the column face), crosswise (toward or away from the column face) and vertical. See Fig. 7.

Universal Knee-and-Column Milling Machine. The universal knee-and-column milling machine is shown in Fig. 8. The distinguishing feature of this machine is the table. The table has the same movements as the plain horizontal milling machine—longitudinal, crosswise, and vertical. In addition, however, the table can be set, or swiveled, at an angle to the column face. The type featured in the Trans-Vision is a universal knee-and-column milling machine.

Instead of being mounted directly on the saddle, the table is mounted on a swivel block which can be rotated about the center of the universal saddle. Numerous tool-room operations, including the milling of helical gears, drills, and fluted cutters, are economically performed on a universal machine.

Vertical Milling Machine. The vertical milling machine, Fig. 9, differs from the horizontal type. It has a vertical spindle, rather than a horizontal spindle. The spindle is mounted in a vertical head which can be fed up or down manually or by power.

The vertical-spindle milling machine is in many ways like the drilling machine. The spindle is located in about the same position relative to the table. The cutting tool is also located in the spindle. Therefore, an arbor is not used on a vertical mill. The table has the vertical, longitudinal, and cross-feed move-

Fig. 9. The vertical milling machine also has the conventional table feeds. In addition, this type of machine has a vertical head which can be raised or lowered. (Cincinnati Milling Machine Co.)

ments characteristic of the knee-and-column type machine.

Ram and Turret Mill. The ram type milling machine may also be classified as a knee-and-column type machine. The ram-type construction permits the movement of the milling head over the table, and it can be swiveled as desired for vertical or horizontal spindle arrangements. The ram and turret mill is one of the most useful mills for making tools and dies. See Fig. 10.

Manufacturing Milling Machine. The manufacturing or bed-type milling machine is a production machine. The spindle, supported in bearings, is located in an adjustable head that can be raised or lowered.

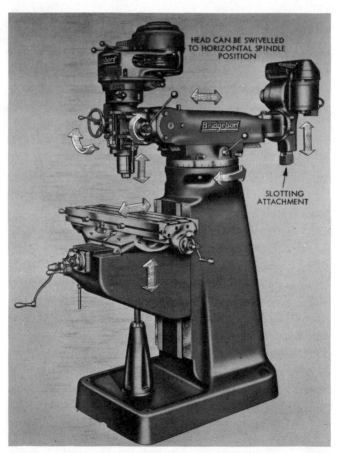

Fig. 10. The ram and turret milling machine has two heads: one for milling, and the other for slotting. The slotting head can be positioned over the table by rotating the ram 180° from the position shown. (Bridgeport Machine Tool Co.)

SPINDLE HEAD
ADJUSTABLE

LONGITUDIAL
TABLE FEED

Fig. 11. Manufacturing milling machines are heavy duty machines specially designed for continuous operation. The spindle can be raised or lowered, but the table is capable of longitudinal motion only. (Cincinnati Milling Machine Co.)

The table, however, *cannot* be raised or lowered. It moves in a horizontal position with longitudinal feed only. It does not have a cross-feed movement, Fig. 11.

These machines are heavy-duty fixed-bed types, specially designed for continuous operation on medium-to-large size parts, and they are capable of taking heavy conventional milling or climb milling cuts with either high-speed steel or sintered (heat treated) carbide cutters. These machines are available with or without tracer-controlled vertical traverse of the spindle carrier. They are built in plain (one spindle) and duplex (two independent spindles) styles.

Special Type Milling Machines. To obtain maximum efficiency and economy in the mass production of large, unusual workpieces, specially designed milling machines are often required. Milling machines may be custom designed and built for a specific operation. The use of such a machine is usually limited to the operation for which it was designed, or to very similar operations.

More frequently, standard machines may be modified to suit the requirements of the individual purchaser. The initial costs to the purchaser of modified or specially designed machines is recovered in reduced cost per unit over hundreds of thousands, possibly millions, of

289

workpiece units. Such machines permit the most efficient production rate and reduce the time required by the operator for adjustments and extra operations to a minimum.

Basic Considerations of Milling Machines

All milling machines are identified by four basic factors: size, horsepower, model, and type. These are given in sequence as follows.

Size or Capacity. The size of a milling machine is based on longitudinal table travel in inches. Longitudinal, cross, and vertical travel are closely related, but for size designation, only longitudinal table travel is used. At present, there are six standard sizes of knee-type milling machines with longitudinal table travel as follows:

Standard Size	Longitudinal Table Travel (inches)
No. 1	22
No. 2	28
No. 3	34
No. 4	42
No. 5	50
No. 6	60

In the future, it will be unnecessary to indicate the longitudinal table travel in inches for any knee-type machine if the size number is given. The size of a machine is indicated by the abbreviation *No.,* followed by the numeral; for example: 3hp *No.2.*

Horsepower. Horsepower is based on the rated horsepower of the spindle drive motor. These motors vary in size from 3 to 50 horsepower. The designation includes the rated horsepower of the spindle drive motor, given numerically. The number is followed by the horsepower abbreviation in lowercase letters without a period: 3hp, 50hp. Where a two-speed, spindle-drive motor is employed, the higher horsepower is given first: 20/10hp, 50/25hp.

Model. The model designation is determined by the manufacturer, as for example: 3hp No. 2HL.

Type. The types of milling machines are those which have been discussed previously.

Major Milling Machine Parts

The main parts of the knee-and-column type milling machine will be discussed as they would be brought together in assembling the machine. A knowledge of the design, location, and function of each of these parts is necessary for anyone who wants to become an expert operator of this many-purpose machine tool. The student is urged to review the Milling Machine Trans-Vision as often as necessary to familiarize himself with the elements and their functions. Familiarity with the parts and motions of the machine will enable the student to better visualize the setups and operations discussed in the following chapter.

Column and Base. The main casting of the milling machine is the base and column, identified by *russet* color in the Trans-Vision. This is usually cast in one piece and may include a tank for liquid coolant used on the cutter. The column is a hollow casting with thick walls. It is strongly braced, since it supports the other parts of the machine. The inner space houses the driving motor and the gear mechanism for transmitting power to the spindle and table.

On the face of the column is a wide slide, usually of a dovetailed design. See the Milling Machine Trans-Vision, page B. This slide is precision scraped to assure proper alignment of the sliding knee. The upper part of the column contains the revolving spindle. The spindle lies in a horizontal position and revolves in properly designed bearings.

The top of the column is designed to support the overarm, identified by the color *gray* in the Trans-Vision, Page B, which is another important part of the machine. The design of the overarm depends upon the make of the machine.

Knee. Attached to the column is a sliding knee or bracket, identified by the color *red* on Page C. It can be raised or lowered on the column for jobs of various sizes and to adjust the depth of cut. The knee

is a casting with two of its sides machined at right angles to each other. The vertical machined side of the knee slides on the smooth, accurate ways of the column face. The top of the knee is machined at right angles to the vertical surface which slides on the column. It supports the saddle, which slides on the knee toward or away from the column face. The knee is strong and rigid. A sturdy elevating screw raises or lowers the knee. Telescoping screws are often used.

Saddle. The saddle, identified by the color *orange* on Page D, is mounted on top of the knee. It supports the table. A precisely machined surface, usually dovetailed, on the underside of the saddle gives adequate bearing with the top of the knee. The bearing surfaces, or dovetails, on the top side hold the milling machine table.

Table. The table, identified by the color *yellow* on Page D, is mounted on the saddle. The *plain* milling machine has the following table movements: (1) vertical — raising the knee on the column; (2) crosswise or transverse—sliding the saddle on the knee; and (3) longitudinal—sliding the table on the saddle. These movements are made with the aid of micrometer dials graduated in thousandths, which permit the accurate setting or positioning of the table.

On the *universal* miller, which is

the type featured in the Trans-Vision, the table can also be *swiveled* horizontally on the saddle.

The table has T-slots running lengthwise on its top surface. These are used to fasten the work-piece or the work-holding device to the table with T-bolts. The vise mounted on the table in the Trans-Vision is held by T-bolts.

The table is designed to catch the coolant which flows over the cutter.

Overarm. The overarm is mounted in the top part of the column above the spindle and parallel to it. It is adjustable. It can be clamped at various distances out from the column as shown on Page C. The arbor support slides on the overarm.

Arbor Support. The arbor support is a casting which clamps onto the overarm. It extends down from the overarm and supports the milling arbor. The bearing in its lower end should be in perfect alignment with the spindle axis.

Two arbor supports are usually furnished with the machine. See Fig. 7. One is called the *inner support*

because it supports the arbor near the middle. It has a large hole in it. The bearing sleeve, that is placed between the collars on the arbor, fits into the hole. The *outer support* has a smaller hole to hold the ground end (pilot) of the arbor. The arbor support must be securely clamped to the overarm once it is in position, as shown on Page C (Step 5).

Some machines use overarm braces which are mounted to the knee and arbor supports. The braces tie the knee to the overarm. They support the arbor securely during heavy cuts.

Spindle. The spindle revolves in bearings in the upper part of the column. The spindle is hollow its entire length. The front end has a tapered hole to receive the standard tapered shanks of the milling arbors. On most milling machines, a draw-in bar is inserted from the rear end of the spindle. The bar pulls the arbor into place and holds it firmly. Some newer models eliminate the draw-in bar and permit the operator to mount and secure the arbor at the spindle.

Milling Machine Accessories

There are two main classes of accessories for the milling machine: (1) arbors, collet holders, and adapters to which the cutters are mounted,

and (2) devices to hold the work-piece for the setup and machining operation. The arbors and adapters will be taken up first.

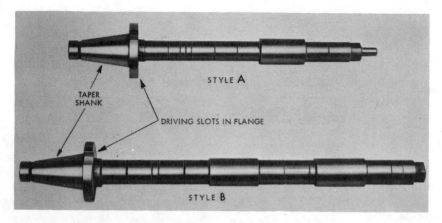

Fig. 12. *Top:* The style A arbor has a pilot at the outer end. *Bottom:* The style B arbor does not have a pilot. A bearing sleeve keyed to the arbor fits into the bearing in the arbor support. (Cincinnati Milling Machine Co.)

Milling Machine Arbors, Collet Holders, Adapters

Arbors, Fig. 12, are used to hold and drive cutters. Their precision and trueness greatly influence the accuracy and economy of milling operations. The arbor is a precision tool as much as any fine measuring instrument. It must be carefully handled in use and in storage.

In recent years, the manufacturers of milling machines and arbors have attempted to standardize spindles and arbors. A standard, tapered, spindle end having a No. 50 taper has been adopted. In addition, a uniform numbering system for national standard tapered arbors has been approved. The following specifications are listed in sequence: taper size, diameter, style, length from shoulder to nut, and size of bearing. For example, Arbor No. 51¼A 18-4 means an arbor with a No. 50 standard taper, 1¼ inches in diameter, style A, length 18 inches from shoulder to nut, and a No. 4 bearing.

The new arbors (with a steep taper shank to fit the national milling-machine standard spindle) are made in three styles: style A, style B, and style C. Each arbor has a particular application. The operator must determine which arbor to use for the job. For example, A and B arbors could not be used on a vertical mill.

Arbors are firmly held in the hollow milling-machine spindle, usually by means of an arbor draw-in bar. Driving contact is through two drive keys on the spindle nose which fits into corresponding slots on the arbor flanges. These give positive drive to the arbor.

Fig. 13. The style *A* arbor is mounted in the spindle so that the pilot fits into the bearing in the arbor support. (Cincinnati Milling Machine Co.)

Style A Arbor. Style A, Fig. 12, *top*, is an arbor with a pilot on the outer end. The pilot fits into the bearing in the arbor support suspended from the overarm, see Fig. 13. As the pilot turns in the bearing mounted in the support, the arbor is firmly supported.

Style B Arbor. Style B, Fig. 12, *bottom*, does not have a pilot. It is used wherever heavy cuts are made and the milling operation does not require maximum support clearance. A bearing sleeve fits over the arbor and is keyed to it. The sleeve revolves in the bearing in the intermediate-style arbor support which is used with the arbor, see Fig. 14. Proper support can thereby be placed along the arbor, close to the cutters for maximum rigidity. A style B arbor is shown being mounted on Pages B and C of the Trans-Vision.

Style C Arbor. Style C arbor is a short arbor requiring no arbor support, Fig. 15. It is used for holding the smaller sizes of *shell end mills* and *face milling cutters*, which

Fig. 14. The style *B* arbor has a sleeve which is carefully ground to fit into the bearing in the arbor support. This type of arbor is used for heavy cuts. (Cincinnati Milling Machine Co.)

are too small to be bolted directly to the spindle nose. The Style C arbor is sometimes called a *shell end mill arbor*. Solid lugs on the outer end of the arbor are used to drive the cutter.

Adapters. An adapter is used for holding face-milling cutters, which center on the outer face of the spindle.

Collet Holders. A collet holder is a form of sleeve bushing for reducing the size of the tapered hole in the machine spindle. With it cutters with smaller tapered shanks can be inserted directly into the spindle. A collet holder takes the place of an arbor by holding the cutter.

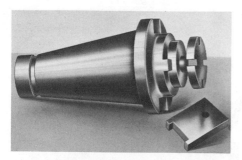

Fig. 15. The style *C* arbor is used for holding smaller sizes of cutters and is supported by the spindle without use of an arbor support. (Cincinnati Milling Machine Co.)

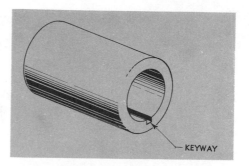

Fig. 17. A spacing collar is used to position a cutter on an arbor.

There are three types of collet holders. One type is inserted in the spindle and must be removed in order to change cutters. The extended type permits cutters to be changed without removing the collet holder from the spindle. A third kind, a spring collet holder, is used for holding straight-shank drills, reamers, and end mills. Collets of various diameters can be used.

Keys, Keyways. All cutters and saws used on milling machine arbors have keyways machined in them. By means of key stock, the cutter and the arbor revolve as one, Fig. 16. The square key fits into the

keyway of the arbor and the keyway of the cutter. Since the keyway of the arbor runs along its entire length, cutters can be mounted at any position along the arbor and keyed to it. A number of spacing collars, Fig. 17, are needed for each arbor. The illustration on Page B of the Trans-Vision shows how spacing collars are used to position the cutter a desired distance from the spindle.

Bearing Sleeve. The bearing sleeve, Fig. 18, is used on a style B arbor. It looks like a spacing collar,

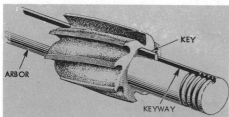

Fig. 16. Keys and Keyways prevent the cutters from turning on the arbor.

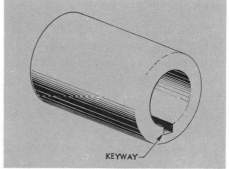

Fig. 18. An arbor bearing sleeve is used on style *B* arbors. The bearing sleeve has a greater diameter than a spacing collar.

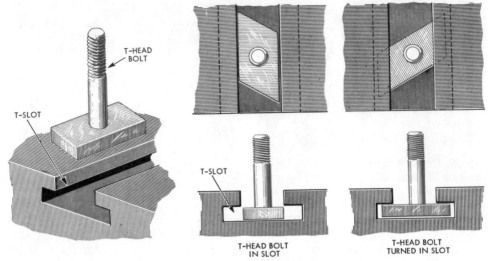

Fig. 19. T-head bolts are used for clamping.

but it differs in purpose and diameter. It is mounted on the arbor in the desired position when making the setup. Then the arbor and sleeve are inserted in the intermediate arbor support. The bearing sleeve has a keyway and may be keyed to the arbor. The outside diameter is carefully ground to fit the bushing of the arbor support.

Work-Holding Devices

To make the various setups for performing milling machine operations, numerous work-holding devices are used. The work must be securely fastened and well supported so it will not move during the milling operation.

T-Slots, Bolts, Straps, Clamps. The T-slots, which run the length of the top surface of the milling machine table, are used in every setup.

The T-slots should be kept free of chips, and the correct size bolts should be used. T-head bolts are used for clamping, see Fig. 19.

A variety of shapes and sizes of straps and clamps are needed in making the many kinds of setups.

Blocks and Jacks. Solid blocks and step blocks, Figs. 20 and 21, are used in making various setups. In addition to these, small jacks are sometimes used.

Stops. In making setups on the milling machine, stops are often required. They assure the operator that the workpiece will not shift out of place under the strain which the cutter puts on the work.

V-Blocks. V-blocks are used in certain milling operations to hold round work.

Angle Plates. An angle plate is an L-shaped piece of cast iron or

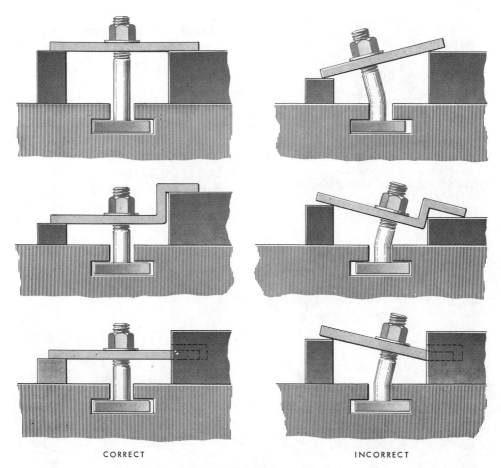

CORRECT   INCORRECT

Fig. 20. Correct and incorrect methods of clamping work for a milling operation are shown.

steel accurately machined to an angle of 90 degrees. For milling certain shapes of work, the workpiece can be mounted on the angle plate as a work-holding device.

Parallels. Parallels are square or rectangular bars of steel or cast iron used frequently in making setups on the milling machine. They are made in pairs of various sizes and lengths and are hardened and

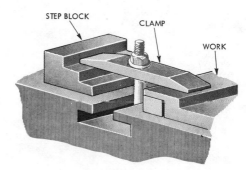

Fig. 21. A clamp and step-block can be used to hold work securely to the milling machine table.

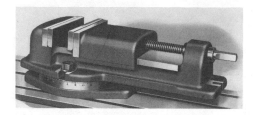

Fig. 22. The swivel vise has a graduated base which permits work to be placed in it at an angle to the axis of the milling cutter. (Cincinnati Milling Machine Co.)

ground to precision measurement.

Plain Vise.   The plain vise is designed so that when it is bolted to the table the jaws will be either exactly parallel or at right angles to the table T-slots. The plain vise is generally used more frequently than any other work-holding device. The Trans-Vision shows a plain vise mounted on the table on Page C.

Swivel Vise.   The swivel vise, Fig.

Fig. 23. The rotary table is used in performing flat, angular, and circular operations without changing setups. (Cincinnati Milling Machine Co.)

22, is a plain vise mounted on a base which is graduated in degrees on the full circle. This permits angular setting of the vise jaws in respect to the axis of the cutter.

Rotary Table. The rotary table, Fig. 23, permits a piece mounted on it to be revolved by hand or power through a circle. Note the degree ring at the base of the rotary table. The workpiece can be indexed at various stations through 360 degrees. It is used for performing flat, circular, and angular operations without change in setup.

Milling Machine Attachments

Vertical Milling Attachment

The vertical milling head is an easily operated unit that broadens still further the range of the already flexible horizontal milling machine. This attachment can be set at any angle from a vertical to a horizontal position. The angle is indicated by graduated scales. The spindle can be placed in a vertical position and the workpiece can be fed up into the cutter, as is shown in Fig. 24. The table can be moved in the usual directions with this attachment, which makes possible operations that are not ordinarily possible on the horizontal milling machine. Many jobs can best be milled by a face-milling cutter, Fig. 24, or an end mill in a vertical spindle.

Indexing

Indexing can be defined as the act of machining the circumference of a workpiece into any desired number of equal divisions. Indexing is required when milling gear teeth, splines, or drive shafts, spacing holes on a circle, and many other operations. It is used in toolrooms and shops where a large variety of work is handled.

Fig. 24. The vertical milling attachment can be set at any angle from vertical to horizontal and is used primarily for facing operations. (Cincinnati Milling Machine Co.)

Dividing Head

The indexing attachment used on the milling machine, Fig. 25, allows the workpiece to be rotated on its axis so numerous operations requiring indexing can be quickly and accurately performed. The indexing attachment, referred to as the *dividing head*, consists of the head, footstock, and steady rest. Work to be milled can be mounted on the centers of the head and footstock.

The following description of the use of the dividing head will help the student in preparing himself to use this important device. When setting up for indexing, a chart or handbook is usually referred to. Since reliance on a chart alone is not the best policy, the principles and the calculations required to set up without the chart are discussed.

Index Crank. In Fig. 25, the index crank may be seen bearing on the outside of the large index plate —the plate with the circles of holes on its surface. Revolving this crank by hand causes the dividing-head spindle to revolve for indexing purposes. Movement is transmitted through a worm and worm gear in

Fig. 25. A dividing head and footstock are shown mounted on the table of a universal milling machine. (Cincinnati Milling Machine Co.)

the head to the workpiece mounted between the centers.

Forty complete turns of the crank result in one complete turn of the work. This ratio is the manufacturers' standard. If, for instance, ten teeth are to be cut in a pinion, it requires $\frac{1}{10}$ of 40 turns or four full turns of the crank for each tooth or division. For 26 divisions, the number of crank turns for each division would be $\frac{1}{26}$ of 40. Simply dividing 40 by 26, the result is $1\frac{14}{26}$; by further reduction, this becomes $1\frac{7}{13}$ turns of the crank. This is where the hole circles come in, which allow fractional turns to be made accurately.

Index Plates. There are two types of index plates on the indexing head, each located in a different position and having different uses.

One plate, known as the *direct index plate* is located on the nose of the spindle where the work is mounted. This plate will be discussed in the section on direct indexing. The second type of plate is referred to as the *plain index plate* and will be discussed here.

The indexing attachment (dividing head) made by the Cincinnati Milling Machine Company, Fig. 25, has one standard index plate with a different series of circles (sets of holes) on each side. The Brown and Sharpe indexing head is equipped with three separate index plates.

The one standard plate used on Cincinnati milling machines is reversible. Circles of shallow holes are drilled on both sides. On the first side, the circles have 24, 25, 28, 30, 34, 37, 38, 39, 41, 42, and 43 holes respectively. The reverse side has circles made up of 46, 47, 49, 51, 53, 54, 57, 58, 59, 62, and 66 holes.

Referring again to our problem of figuring the proper spacing of 26 divisions or teeth for a pinion, it was necessary to turn the crank $1\frac{7}{13}$ turns. Here the plate helps us. The circle of holes divisible by 13 is the circle with 39 holes. Then $\frac{7}{13}$ of a turn would be $\frac{7}{13}$ of 39 or 21 holes. To index 26 divisions on a gear, the complete indexing movement for each of the 26 divisions becomes one full turn of the crank, plus a partial turn of 21 holes on the 39-hole circle. Fig. 26 shows the circle having 39 holes, all other circles having been left off the drawing.

With the three index plates for the Brown and Sharpe machine, each plate contains a series of holes arranged in concentric circles. This makes it possible to get a wide variety of fractional parts of a whole turn of the crank. The number of holes in each of the circles is:

Plate No. 1....15, 16, 17, 18, 19, 20
Plate No. 2....21, 23, 27, 29, 31, 33
Plate No. 3....37, 39, 41, 43, 47, 49

If the 26 divisions of this same problem above were to be accurately spaced, plate No. 3 with its 39 holes would be used.

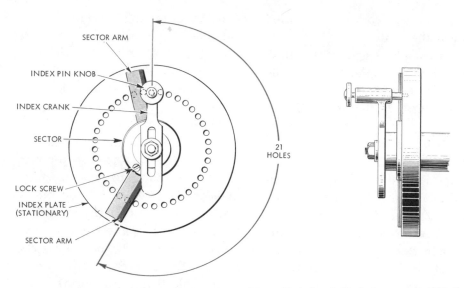

INDEX PIN KNOB

SECTOR ARM

INDEX CRANK

SECTOR

LOCK SCREW

INDEX PLATE
(STATIONARY)

SECTOR ARM

21
HOLES

Fig. 26. Front and side of an index plate with a 39 hole circle being used. (Cincinnati Milling Machine Co.)

Adjustable Sectors. When indexing, it is not necessary to count the holes each time. Instead, the adjustable sector with two arms are used, Fig. 26. The length of the index crank is first adjusted so that the index pin drops freely into a hole in the 39-hole circle, Fig. 26. The sector screws are next loosened. With one sector arm resting against the index pin, as in Fig. 26, the other is adjusted by moving it clockwise until the arms are 21 holes apart, not counting the one the pin is in. The adjustment is then locked by tightening the lock screw.

Upon completion of the first cut, and with the workpiece moved clear of the milling machine cutter, the workpiece is indexed for the next cut. The crank is disengaged by withdrawing the index pin and giving it one full turn in this case (26 teeth being cut on a pinion). Then the fractional turn is added by stopping at the forward sector arm. The sector is next moved in the same direction to bring it up with the crank, Fig. 27. Indexing is usually clockwise. This procedure is repeated for each new pass of the cutter.

The position of the crank is changed with each indexing movement. It is very important that the fractional turn, if any, is not omitted. The sector is also moved each time immediately after the crank has been turned. When using the crank, the pin must be pulled completely clear of the sector to avoid hitting it and losing the position.

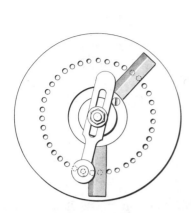

Fig. 27. After the first indexing movement of 1-7/13 turns, the new position of the crank and sector is as shown here. (Cincinnati Milling Machine Co.)

If the crank is turned too far, it should be reversed far enough to take up any lost motion (backlash) before advancing to the proper hole. A clamp is also provided for locking the spindle of the index head in the desired position during the milling operation.

Types of Indexing Heads

There are two types of indexing or dividing heads: plain and universal. The plain dividing head is limited to plain indexing and direct indexing. Direct indexing, sometimes called quick indexing, is limited to the factors of 24 only (on most dividing heads). In plain indexing, it is possible to obtain almost any number of divisions desired except prime numbers greater than 50 and certain odd numbers greater than 50.

The universal indexing or divid-

ing head may be used for direct (quick), plain, and differential indexing.

Classifications of Indexing

There are three methods of indexing: (1) rapid or quick indexing, usually called direct; (2) plain indexing; and (3) differential indexing. The latter requires a universal indexing head.

Direct Indexing. Direct indexing is a timesaving method of indexing which can be done on some kinds of milling jobs, such as: straddle milling a square nut, milling hexagons, fluting (grooving), milling taps, and similar work involving a small number of divisions.

Direct indexing is accomplished by means of the direct index plate located on the nose of the spindle. This plate contains 24 equally spaced holes that make it possible to index directly any number of divisions which is a factor of 24. These numbers of divisions are: 2, 3, 4, 6, 8, 12, and 24.

On the Cincinnati dividing head, the direct (rapid) index plate has 3 circles having 24, 30, and 36 holes. These circles make possible the following numbers of divisions by rapid indexing:

24-hole circle..2, 3, 4, 6, 8, 12, 24
30-hole circle..2, 3, 5, 6, 10, 15, 30
36-hole circle..2, 3, 4, 6, 9, 12, 18, 36

Plain Indexing. Plain or simple indexing requires the use of the side index plate, the sectors, and the index crank shown in Fig. 25. Following is a problem in simple or plain indexing, using the side dividing plate. The reduction ratio of the standard dividing head is 40 to 1. The formula to use in selecting the correct number of turns of the index crank and the correct hole circle required for the number of divisions in the problem is $\frac{40}{N}$, where N equals the number of divisions required. Our problem, in this discussion, is to machine 56 teeth in a spur gear blank. The first step, therefore, requires that the constant 40 be divided by the number of divisions required, which is 56.

This ratio, or fraction, $\frac{40}{56}$, can be reduced by dividing both the numerator (40) and the denominator (56) by 8 as follows:

$$\frac{40 \div 8}{56 \div 8} = \frac{5}{7}$$

It should be remembered that the numerator always represents the number of holes that should be indexed for each division required. The denominator always represents the hole circle to use. The smallest hole circle available in the Cincinnati standard index plate is 24. The fraction ⁵⁄₇ is obviously too small, so it must be raised. To do this, a number must be found which, when multiplied into the denominator (7) of the fraction, will equal a hole

circle which we have available on our standard Cincinnati index plate.

$$\frac{5 \times 4}{7 \times 4} = \frac{20}{28}$$

or 20 holes in the 28-hole circle

$$\frac{5 \times 6}{7 \times 6} = \frac{30}{42}$$

or 30 holes in the 42-hole circle

$$\frac{5 \times 7}{7 \times 7} = \frac{35}{49}$$

or 35 holes in the 49-hole circle

The above figures apply only to the use of the Cincinnati index plate. If the Brown and Sharpe standard plates are used, the problem is worked in the same manner, but the fraction ⁵⁄₇ must be multiplied by different figures to obtain a plate having the correct hole circle available.

Differential Indexing. Differential indexing enables a wide range of divisions to be indexed, which cannot be obtained by plain indexing. Change gears are used in connection with the index plates furnished with the headstock.

Angular Indexing

Angular indexing, or indexing by degrees, is easily accomplished with the dividing head of the milling machine. One complete turn of the crank, as outlined previously, is equivalent to $\frac{1}{40}$ of a turn of the work. Since there are 360 degrees in one complete turn of the crank, one turn equals 360 divided by 40, or 9 degrees of the work.

Suppose a number of slots are to

be milled 15 degrees apart. For 15 degrees, the turns required would then be 15 divided by 9 or 1⅔ turns of the crank. One turn and 16 holes on the 24-hole circle would do it. Fig. 28 illustrates angular indexing to obtain an index wheel with faces 60° apart.

Additional Parts for the Dividing Head

For the dividing head to function, several important parts are needed. One of these is the dividing-head driver. Several styles of drivers are used. One fits over the live center in the spindle. It is prevented from turning on the center by setscrews. The slot in the driver receives the tail of the milling-machine dog, which is locked in the slot. Another type of driver is secured to the face of the spindle, which has a standard milling-machine taper. A slot in this driver fits over the expanding or driving key attached to the spindle face. This acts as a positive drive and prevents any movement between the spindle and the driver. See Fig. 28.

The milling-machine dog is used when the work is held between the driving-head center and the tailstock center. The dog is slipped over the end of the work and held in position by the setscrew. The tail of the dog has two flat sides which are held firmly in the driver. The dog holds the work firmly, so it

does not move except through the dividing head.

The dividing-head chuck is used for those jobs that are not held between centers, but that do require a holding device like a chuck. The chuck is similar to the chuck used on the lathe. The work may be held entirely in the chuck or may be supported on one end by the tailstock center.

The tailstock (or footstock, as it is sometimes called), Figs. 25 and 28, supports the outer end of the work on a center. The base of the tailstock has tongues or blocks which fit into the slots in the table and line up the tailstock with the dividing-head spindle. A horizontal sliding bar is used to adjust the center in a horizontal position. A vertical sliding bar is used to raise or lower the center for adjustments. In addition, the unit containing the sliding bars may be swiveled in an angular position to a little above or below the horizontal position.

Besides the parts mentioned above, there are dividing-head centers, Fig. 25, similar in many ways to the lathe centers. There is also a steady rest. It is used with the dividing head whenever more support is necessary under the workpiece, due to pressure from the cutter. It holds steady such work as long, slender pieces or light material that would give way under the cutting pressure of the cutter.

Fig. 28. An angular indexing operation. The workpiece is rotated by the index a certain number of degrees. (Cincinnati Milling Machine Co.)

Safety Precautions—Milling Machine

1. The floor about the milling machine should be kept free of oil and material that might cause slipping or stumbling.

2. Bearings should be kept in good condition to prevent vibration.

3. The cutting compound should be changed periodically and the container should be cleaned out at regular intervals.

4. Whenever an operator observes an unsafe condition in his machine, he should report it without delay to his foreman or instructor.

5. Horseplay about a milling machine, or distracting the operator's attention from his work while the machine is in motion, is decidedly unsafe and should not be allowed.

6. All guards should be put in place over moving belts and pulleys before starting the machine.

Courtesy of the National Safety Council

NOTE: Please see review questions at end of book.

Milling Machines:

Milling Cutters, Setups, and Operations

A number of factors must be considered in setting up a milling-machine job for efficient operation. First of all, the best possible combination of milling machine and milling cutter must be selected. Then, provision must be made for locating the work properly with relation to the cutter, and for holding it rigidly throughout the cutting cycle. Finally, the best combination of cutter diameter, number of teeth in the cutter, tooth shape, cutting speed, and rate of feed must be selected. Usually, it must be decided at the same time whether a cutting fluid is to be used. The kind of fluid and method of delivering it to the cutter must be determined.

Steps in Making Setups

Setting up the milling machine for an operation or series of operations involves the following details.

1. Selecting the kind of cutter depending on the operation.
2. Choosing the cutter-holding device and mounting it in the spindle.
3. Selecting and mounting the holding device for the work.
4. Mounting work in holding device.
5. Mounting the cutter on the arbor or adapter.
6. Adjusting the table for height, travel, etc.
7. Selecting the correct cutting speed for the kind of metal in the workpiece. Setting machine speed.
8. Determining the correct rate of feed per minute and setting machine.
9. Bringing workpiece into proper position with cutter.
10. If a coolant is used, adjusting the flow.

11. Starting the machine and taking a trial cut. Making the necessary adjustments to get desired results.

This pattern of thinking and planning must be carried out, and the procedure, in whole or in part, carefully followed in the best order. These steps apply regardless of the operation to be performed.

Milling Cutters

The many operations performed on the milling machine are generally accomplished with a wide variety of milling cutters. Plain or slab cutters, side milling cutters, face mills, form cutters, end mills, and metal-slitting saws are some of the more common ones. One of the outstanding characteristics of the milling cutter is its number of teeth. Some cutters have few teeth, others have many. Milling cutters are often referred to as multi-tooth tools.

Cutter Materials

Cutting tool materials used in a milling cutter are of three general types: *high-speed steel, cast cutting alloys*, and *tungsten carbides*. The proper selection of the cutting tool material will depend on the material to be cut, the available equipment, the length of the run, the time element involved, and the overall cost.

High-Speed Steel. This is a general name given to high alloy steels. These tool materials were developed to retain keen cutting edges in service at elevated temperatures. Tools made of this material are heat treated to proper hardness after forming to shape.

Cast Cutting Alloys. This group of alloys is composed primarily of cobalt, chromium, and tungsten which must be cast and ground to size or shape. No heat treatment is necessary. The hardness of these materials is inherent in the metal itself and is retained at higher temperatures than is the case with high-speed steel.

Tungsten Carbides. Tungsten carbide and cobalt are combined to form these carbides. In some grades, tatalum and/or titanium carbides are added. The carbides are pressed or extruded into required shapes and sintered (heat-treated or fused at very high temperatures) into their final hard condition as cutting tool blanks. These tool materials have the ability to retain their keen cutting edges at higher temperatures than either high-speed steel or cast cutting alloys.

Milling Cutter Tooth Form

The common tooth form of milling cutters is shown in Fig. 1. Notice the positive radial rake angles at *A* and *B* in the illustration. Also notice the clearance angles falling away from the tip of each tooth,

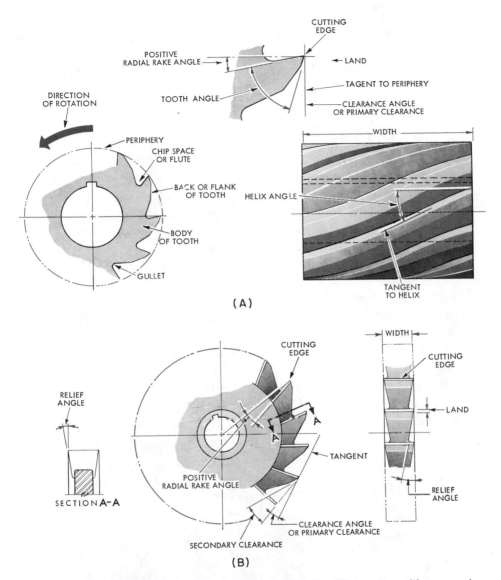

Fig. 1. At A: A diagrammatic view of a plain, helical slab milling cutter with nomenclature. At B: A diagrammatic view of a side milling cutter with nomenclature.

and the large chip space at the throat of each tooth.

Types of Milling Cutters

There are a number of different milling cutters, each one designed for a certain purpose. A given cutter will function well under proper conditions. It would be unwise, however, to use a cutter under conditions

309

for which it was not intended. The milling machine operator must know each cutter by name and the operations it will perform.

Plain Milling Cutters. The plain milling cutters are of plain cylindrical form. They have teeth only on the periphery. They are used to produce a flat surface parallel to the axis of rotation. They have an accurately ground bore for mounting on an arbor. The teeth may be straight, as shown in Fig. 2, *left*, or spiral (actually, *helical*) as shown in Fig. 2, *right*.

Most straight tooth plain milling cutters are narrow, usually less than ¾ inches wide. They generally have a large number of teeth, which is characteristic of cutters which are used for light milling operations.

The plain helical milling cutter is more commonly used in plain mill-

ing (slab milling) operations than the plain straight tooth cutter. Most helical milling cutters have fewer teeth and are used for heavy milling operations. However, helical milling cutters with many teeth and small helix angles are also used for light operations. The helix angle of a helical milling cutter can vary from 25° to 60°. Generally, the larger the helix angle, the heavier the cutting operation for which the cutter is designed.

Standard plain helical milling cutters are made for either right- or left-hand rotation, Fig. 3, with either a right- or left-hand helix. Viewing the cutter from its outer end, if the teeth point in a counter-clockwise direction, the cutter is a right-hand cutter. If the teeth point in a clockwise direction, the cutter is a left-hand cutter. If the flutes lead to the

Fig. 2. The standard plain milling cutter has teeth only on its periphery. The teeth may be straight (*left*) or helical (*right*). The helical mill is more commonly used. (The Cleveland Twist Drill Co.)

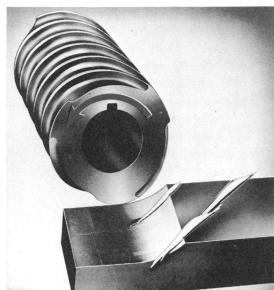

Fig. 3. *Right:* Shown is a right-hand cutter with a left-hand helix. *Left:* Shown is a left-hand cutter with a right-hand helix. (Cincinnati Milling Machine Co.)

right or left the cutter has a right- or left-hand helix respectively.

Side Milling Cutters. The side milling cutter, Fig. 4, is a plain milling cutter of cylindrical form. It is similar in appearance to the plain straight tooth cutter with the exception that the cutting edges extend down one or both sides of the cutter teeth as well as along the periphery. Such a cutter mills with its sides as well as its periphery.

Side milling cutters are used for *straddle mill work* where the opposite sides of the workpiece are milled simultaneously. Straddle milling will be explained in the section on Milling Operations.

The *staggered-tooth* side milling

Fig. 4. The side milling cutter can cut with one or both of its sides as well as with its periphery. (Pratt & Whitney Div., Colt Industries, Inc.)

311

cutter is designed for heavy cutting and deep slotting operations where the slot is relatively narrow. The teeth of this type of side milling cutter have alternating, opposite spirals and cutting edges, Fig. 5.

Interlocking side milling cutters, Fig. 6, are useful for extremely accurate work. This type of cutter is selected rather than the staggered-tooth type if the slot to be cut is relatively wide. The width of the cut can be accurately varied by inserting spacing materials between the interlocking cutters. The shearing action of the staggered-tooth and interlocking type cutters alternates

Fig. 5. Alternating side cutting edges on this staggered-tooth cutter eliminate the side thrust common to side milling cutters with teeth on one side only. (The Cleveland Twist Drill Co.)

from right to left. With side thrust eliminated, the cutting action of both types is smooth and rapid.

End Mills. End mills, Fig. 7, have teeth on their periphery and also on their *ends*. They are used for end milling, surface milling, slotting, and many other operations. Like reamers, end mills can be solid or the shell-type with replaceable heads. The two different styles are shown in Fig. 7.

Because end mills have cutting teeth on the end of the mill, they usually are held by their shank. This type has either a straight shank or a taper shank, to fit various collets and adapters.

End mills may be made for either right-hand or left-hand rotation. The hand (right or left) of the helix may be either the same or opposite to the hand of cut. End mills having the same hand of helix and cut (for example, right-hand cut and right-hand helix) generally are preferred. This arrangement gives a positive rake angle for the end teeth.

Face Milling Cutters. The face milling cutter, Fig. 8, is fastened to an adapter which is mounted in the spindle nose of the machine. Face milling cutters are larger than end mills. The cutting action of a face mill is identical to that for the end mill. Face mills are used for milling large flat areas. The cutting action of the face milling cutter is like that of a planer. The teeth on the peri-

Fig. 6. Interlocking half-side milling cutters (shown apart) are used for cutting relatively wide slots. (Goddard & Goddard Co.)

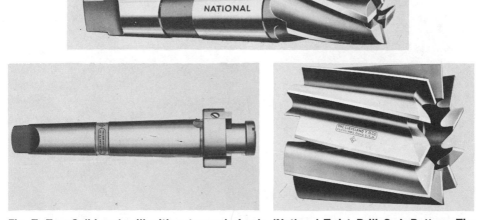

Fig. 7. *Top:* Solid end mill with a tapered shank. (National Twist Drill Co.) *Bottom:* The shell end mill must be mounted on the accompanying special taper-shank holder. (Cleveland Twist Drill Co.)

Fig. 8. The face mill shown has high-speed steel inserted teeth. (Cincinnati Milling Machine Co.)

phery do the principal cutting. The end teeth do a little finishing. These cutters are ground with considerable clearance back from the edge to reduce chatter.

Slitting Saws, Slotting Saws, Miscellaneous Cutters. *Slitting saws* are generally thin, straight-toothed, plain milling cutters, Fig. 9. They are used for slotting and parting. They may vary in thickness from a few thousandths of an inch to a quarter of an inch or more.

Alternate-tooth metal-slitting saws, Fig. 9, have the same general design as staggered-tooth side mills.

Screw-slotting cutters and metal-slitting saws are much alike in appearance and are often confused. Screw-slotting cutters make shallow, short slots like those in screw

Fig. 9. Slitting saws are used for parting and narrow slotting. *Left:* Standard plain metal-slitting saw. *Right:* Alternate tooth slitting saw. (Goddard & Goddard Co.)

Fig. 10. The T-slot milling cutter is used to mill the wide bottoms of T-slots after the narrow portion has been milled.

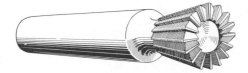

Fig. 11. The dove-tail milling cutter is an end mill used for milling dove-tail slots.

heads. They are made in 1¾-, 2¼- and 2¾-inch diameters in widths up to .020 inch only.

T-slot cutters, Fig. 10, are a special form of end mill for making T-slots. They are designed for milling the wide bottoms of T-slots after the narrow portion has been milled with a side or end mill.

Dovetail cutters, Fig. 11, are another form or variation of the end mill.

Woodruff keyseat cutters have profile teeth and are made in both

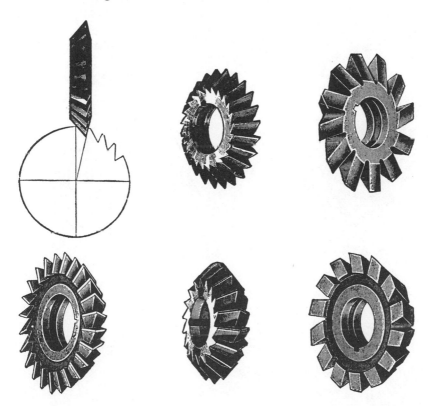

Fig. 12. Angle cutters are used for milling surfaces at various angles to the axis of rotation of the cutter.

shank and arbor types. They are used for cutting semicircular keyways in shafts.

Angle Milling Cutters. Angle cutters, Fig. 12, are used for milling surfaces at various angles to the axis of rotation. These cutters are often used when making other milling cutters. Generally, they have cutting surfaces on each side and different angles on each side, the customary angles in such cases being 40 degrees, 43 degrees, 45 degrees, or 48 degrees on one side—and 12 degrees

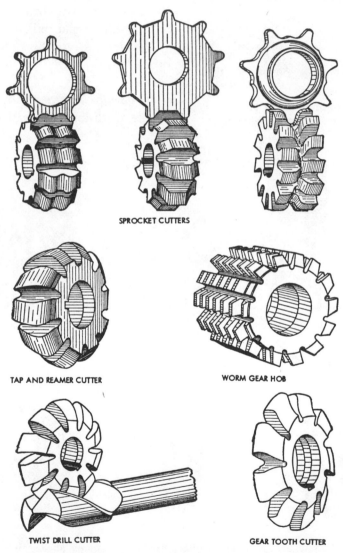

SPROCKET CUTTERS

TAP AND REAMER CUTTER

WORM GEAR HOB

TWIST DRILL CUTTER

GEAR TOOTH CUTTER

Fig. 13. Some standard shapes of form cutters are shown here.

on the other. The common single-angle cutters vary from 40 degrees to 80 degrees, either right- or left-hand. Double-angle cutters, as shown in the center of the lower row, Fig. 12, can be had with either 45-degree, 60-degree, or 90-degree included angle.

Form Cutters. The form cutter is a special milling cutter shaped to meet various specifications. Convex and concave solid or interlocking cutters, involute gear cutters, T-slot cutters, spline cutters, and thread cutters are the most common form cutters, Fig. 13. The increased production and the accuracy of the completed work more than compensate for the high cost of form cutters.

Setups

The successful milling of a workpiece depends to a large extent on the proper setup of tool and workpiece. Wrong selection of cutters, lack of rigidity in the setup, and improper choice of speeds and feeds will result in poorly machined surfaces of the workpiece. To avoid the above errors, suggestions are presented throughout the rest of the chapter which are intended to help the student "think through" the proper setups and operations for a variety of jobs.

The student is again urged to review the Milling Machine Trans-Vision. To derive the maximum information from the pages to follow, the student should have a clear idea of the parts of a milling machine, and how they function. The student is further advised to refer to the Trans-Vision whenever the setup or operation discussed by the text relates to the setup illustrated by the Trans-Vision.

Adjustments Necessary in Making Setups

Adjusting the table. There are three possible directions of table movement: (1) *vertical*—the table can be raised or lowered to position the workpiece on the cutter height by raising or lowering the knee, on which the saddle and table are mounted; (2) *traverse* (*cross*)—the table can be moved toward or away from the face of the column, thereby positioning the workpiece directly beneath the cutter; (3) *longitudinal* —the table can be moved longitudinally to the right or left, parallel with the face of the column, to feed the workpiece into or away from the cutter.

These movements are each controlled by both a power feed lever and a manual crank or handwheel. You will notice that power feed controls are located on the side of the knee near the column face as well as at the front of the knee. This is primarily for operating efficiency. Movement in any direction can also be speeded up by using the rapid traverse lever or levers. However, when setting the workpiece into position at the cutter prior to power operation, the manual controls are used since precise initial settings are required.

The vertical up or down movement of the table is accomplished when the elevating screw under the knee is caused to turn, thereby sliding the knee on the ways of the face of the column. Step 9, Trans-Vision Page D, shows the knee (and table) being raised into position beneath the cutter by means of the vertical hand crank at the front of the knee. The micrometer collar on the vertical hand crank permits the operator to set the machine to the exact height to mill the workpiece to within a thousandth of an inch of the desired dimension. See Steps 9, 10, and 11 of the Trans-Vision.

The table is aligned on the saddle (traverse movement) to position the workpiece in relation to the fixed position of the cutter. Traverse movement is accomplished by turning the saddle crank, or handwheel. Step 8, together with insert B, on Page D of the Trans-Vision, illustrates the function of the saddle crank in aligning the workpiece with the cutter. The saddle crank moves the table toward or away from the face of the column. A micrometer

collar on the saddle crank allows precise settings for depth of cut desired. For example, if a face mill were mounted in the spindle of the machine shown in the Trans-Vision instead of the slab mill setup shown, the saddle crank would be set to determine the size of the shoulder that could be cut in the workpiece.

The table can be moved lengthwise back and forth past the column face with the table handwheel located on the ends of the table. See Step 10 on Page D of the Trans-Vision. The handwheel also has a micrometer collar for setting up or adjusting the amount of metal to be cut in certain milling operations. In the Trans-Vision, this handwheel is turned to bring the workpiece from beneath the cutter so that the *vertical* hand crank can be set to the desired cut. Engaging the longitudinal power feed would then move the workpiece into the cutter for the desired depth of cut.

Because the manual feed cranks and handwheels will spin on some machines when the power feed is engaged, the handles can be removed or disengaged. Otherwise, they might turn and be a hazard to the operator. Many newer machines "throw out" or automatically disengage the manual controls when the power feeds are engaged.

Trip dogs mounted in T-slots at the front of the table disengage, or *trip*, the power feed. Trip dogs on more sophisticated machines can automatically move the table through a complete cycle of movements.

Adjusting Lead or Feed Screws. A micrometer collar on each manual feed control permits highly accurate table settings in three directions as described above. Each collar is graduated in thousandths of an inch. This gives micrometer adjustment for setting the work in a given position to the cutter. Since these micrometer adjustments will determine the amount of workpiece material removed, the operator should be familiar with making micrometer settings. A review of the chapter on precision measurements is recommended.

Compensating for Backlash. The operator of the machine must remember that backlash enters into the adjustment of the controls for the table. Backlash is the lack of motion or lost motion when the rotation of a screw in a nut is changed from clockwise to counterclockwise. In adjusting the position of the table, for example, the screw makes a partial turn (the backlash), during which the saddle does not move. To compensate for this loss of movement, easily measured on the graduated dials in thousandths, turn the handle about one-half turn in the direction opposite the proposed adjustment. Then turn the handle to bring the work to the point from which subsequent settings in the

same direction are to be made.

Adjusting Stops. Adjustable stops along the length of the table serve as automatic trips. These may be placed and then locked to limit the travel of the table in either direction. See Step 14 (Trans-Vision). These cause the table to stop and to reverse its direction of travel.

Determining Direction of Feed

In making the setup for horizontal milling, there is a choice in the direction of feed in relation to the direction of cutter rotation for horizontal feeds. The two choices are up or conventional milling and down or climb milling. The two methods were earlier described in the section on cutters.

Up or Conventional Milling. In up or conventional milling the cutter runs (rotates) against the direction of feed. That is, the work is fed against the cutter. See Fig. 14, *right*. The cut is made by the upward movement of the cutter teeth as the workpiece is forced into the revolving cutter. This form of feed and cutter rotation is called up or conventional milling. The feed forces the work against the revolving cutter and compensates for backlash in the table mechanism.

In this type of milling the cutter gets under the surface scale of the workpiece and pries it off. Each cutter tooth, since it is cutting up into the metal, starts its cut in clean

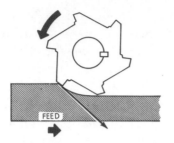 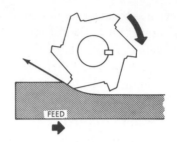

Fig. 14. *Left:* In climb milling the work is fed in the direction that the cutter rotates. *Right:* In conventional milling the work is fed in the opposite direction from the rotation of the cutter.

metal. It does not have to first break through surface scale because it cuts into the surface formed by the previous tooth. Cast iron and forgings, because of their scale, are usually milled by this method.

Down or Climb Milling. Down or climb milling differs from conventional milling in that the workpiece is fed in the same direction that the cutter tooth is rotating. Down milling requires a highly rigid setup and a machine with sufficiently high horsepower, since backlash must not occur. In newer machines, the backlash is eliminated by hydraulic feed or by a compensating nut on the feed screw.

Climb milling is shown by a straight arrow representing the feed in Fig. 14, *left*. The cutter tends to press the work down on the table.

Conventional milling, on the other hand, tends to lift the work, Fig. 14, *right*.

In milling thin, narrow, or irregular pieces, the downward pressure in climb milling is a distinct ad-

vantage. This pressure forces the work onto the table, the table onto the saddle, and so on. The whole setup is more rigid.

Climb milling gives a better quality of work, increased production, and longer cutter life. The cutter removes a piece of metal the thickness of the feed and makes a clean cut. However, climb milling should not be used where the work *surface* is hard or gritty or has hard scale. The cutter is continually pressing on this grit or scale before cutting into the clean metal. Only work free from rough surface scale is recommended for climb milling.

Selecting Cutters

One factor to consider in selecting a milling cutter for a given job is the number of teeth in the cutter. This cannot be determined by set rule. The best number of teeth must be determined from the thickness of the chip desired, the power available, the finish required, and the speed and feed.

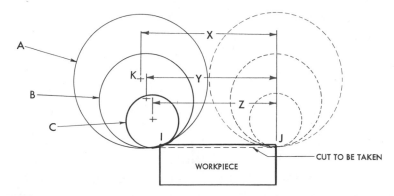

Fig. 15. The relationships between centers of cutters with different diameters is shown. Note that the cutter with the greatest diameter requires accordingly the greatest feed distance.

The usual practice is to use a coarse-tooth cutter with a large feed per tooth. These cutters require less power to remove a given amount of metal in a specified length of time than do fine-tooth cutters.

Determining Diameter of Cutter To Be Used. When selecting a cutter for a given operation, it is suggested that as small a diameter as possible be used for the sake of economy in cost, and to avoid unnecessarily large *run-in* allowances. The drawing in Fig. 15 shows the relative run-in required on large- and small-diameter cutters. Suppose a cut is to be taken across the top surface of the workpiece as shown by the dotted line. If a large face mill, A, is used, it will strike the piece at *I* when its center is at *K*. The line *X* shows how far the work must travel to have the large cutter reach point *J*. If the small cutter, *C* (for example, an end mill) is

used it will strike the work at *I* and leave the work at *J*, but will have traveled only distance *Z*. From the standpoint of machining time, we see that the end mill, *C*, with the smaller diameter, has less feeding distance than the larger cutter. In Fig. 15, circle *B* represents the smallest diameter of plain milling cutter that can be used on an arbor.

There are different methods of doing the same job. There is, however, a best method which usually more than repays for the little extra time spent in making the setup.

Effect of Machine Condition on Tool Selection

Rigidity. The best milling practice is to have ample rigidity in the machine tool no matter what tool material is to be used. The successful use of carbides requires a maximum of rigidity. Cast cutting alloys can be used under less ideal conditions. High speed steel is the most rugged cutting tool material and can be expected to perform reasonably well under less rigid conditions that would be ruinous to carbides.

Fixtures and Arbors. The use of rigid fixtures and arbors is a prime necessity for maximum economy in any milling operation, and is a must for successful use of carbide cutting tools. Arbors and their supports should be as rigid as possible. When using carbide cutters, arbors should be as large and well-supported as is practical, since every increment of increase in arbor diameter adds greatly to the rigidity of the setup with consequent improvement in tool life.

Horsepower. The speed and feed range for high speed steels and cast cutting alloys is fairly broad. Machines of limited horsepower can be set up and operated successfully with these cutting tool materials. Carbides, however, have more critical speed and feed requirements. Use of carbide cutting tools should not be specified or attempted unless the milling machine being used has adequate horsepower to be operated at the necessarily high speeds and feeds.

Speeds and Feeds for Milling

Speeds and feeds are important to milling operations. The speed dial, on machines so equipped, is used to set the machine for the proper speed of the cutter. A speed of 100 r.p.m. means that the cutter is revolving 100 revolutions per minute. Milling machines are provided with a large range of speeds, from 15 to 1800 r.p.m. After determining the correct speed at which the cutter is to revolve, the operator selects the speed on the face of the dial and turns the crank pin until the indicator points to the chosen speed.

Speed Selector

Milling machines are available with a wide range of control refinements. The more conventional model requires that the speed be set as described above. On most machines, the operator is required to reduce the spindle speed to an "idle," then to set the speed selector crank pin (on the speed dial at the side of the machine) to the new setting. The operator is required to "jog" the machine by alternately turning the spindle control on and off or, on some machines, to push the "jog" button, until the speed gears for the new setting are firmly meshed. Once the operator is certain that the gears are securely meshed, he can allow the spindle to come up to speed and resume operation. Page F of the Trans-Vision illustrates the manual speed selector which is representative of the type featured on most older models or less automatic models in use at present and the setting which most students will be required to make.

To change speed on the more automatic models like the one shown in the Trans-Vision, the operator simply pushes the *spindle stop pushbutton*. When the spindle has slowed, the operator then pushes the spindle *speed increase* or *speed decrease* pushbutton until the new speed setting he wants registers on the speed dial at the side of the machine. When the dial registers the desired speed, the operator releases the speed increase or decrease pushbutton. Speed gear changes are automatically and positively performed by the machine, eliminating the need for "jogging." Depressing the spindle *start* or *on* pushbutton resumes spindle rotation at the newly selected speed.

The speed settings for the machine illustrated in the Trans-Vision can be made with pushbuttons located both at the side of the column and on the front of the knee as described above.

Selection of Speed

There is no simple rule for determining the proper speed for the cutter. Experience is the best teacher. Some of the conditions that determine the proper choice of cut-

ter speed are: (1) the kind of material to be milled, (2) the kind of material the cutter is made of, (3) the depth of cut, and (4) the rigidity of the workpiece and cutter.

Regarding the first condition, the hardness of the material affects the choice of speed considerably. Cast iron, steel, brass, and aluminum, the more common kinds of workpiece materials, must be milled at different speeds. Cast iron and steel fall in a slow-speed classification, while brass and aluminum can be milled at high speed. Usually, the harder material requires a slower cutting speed. Tool steel requires a slower cutting speed than cold-rolled steel.

The second condition affecting cutter speed is the kind of material in the cutter. Carbon steel, high- and superhigh-speed steel, stellite, and tungsten carbide are common. Carbon tool steels cannot be operated at high speed. They lose their temper and cutting edge because of the heat generated in cutting. On the other hand, high- and superhigh-speed steel and stellite remain unusually hard at high speeds. The tungsten-carbide tipped tools, because of their exceptional hardness and comparative strength, are capable of the highest cutting speeds. Tungsten carbide is used to best advantage in the milling of cast and malleable iron, brass, aluminum, and bronze, besides some of the newer plastic materials. The high speeds permissible with tungsten carbide also produce finer finishes frequently required of milling operations.

Third is the depth of cut. To protect the cutter teeth in taking a heavy roughing cut, slower speeds are used. In finishing cuts, it is customary to increase the speed of the cutter. There is less metal to remove, and faster speeds give a better finish.

The fourth condition affecting the choice of speed is rigidity of the workpiece and cutter. The speeds which can be selected to mill various surfaces vary in direct proportion to the rigidity of the two items. Unless the workpiece is rigid and the cutter securely attached to the arbor, chatter will result, accuracy will be off, and the finish may not be satisfactory.

Speed Further Explained

A milling-machine speed dial is rated in r.p.m. of the spindle. The operator must learn to change this r.p.m. or speed rating into surface-feet-per-minute travel of the cutter. Surface feet per minute means the distance a single point on the circumference of the cutter travels, in lineal feet, during one minute of continuous rotation. Thus, the r.p.m. on a speed dial controls a cutter's surface-feet-per-minute rating. It is essential, then, that the operator know how to determine the proper speed, translate surface feet per minute to r.p.m., and set the machine accordingly. Some milling machines feature speed dials which can be read in both r.p.m. and surface feet per minute for any given setting, but the student should be familiar with the calculations for converting surface feet per minute to r.p.m.

Since the cutting edge is the effective part of the cutter, and is in contact with the workpiece, it is the speed of the cutting edge (its rate of travel in surface feet per minute) that designates the speed of cutting.

To illustrate this, assume a cutter 4 inches in diameter is revolving at 250 r.p.m. The cutting edge on this cutter will travel approximately 1 foot in making 1 revolution and, therefore, 250 feet in making 250 revolutions. Thus, the surface feet of travel is 250 feet per minute.

Determining Surface Feet

To determine the surface feet per minute traveled by the cutting edge on a cutter, we must first know the diameter. From that we can determine the circumference, which is then multiplied by the revolutions per minute of the spindle. Since the formula used is expressed in feet, while the cutter diameters are always given in inches, divide the product of the cutter diameter times 3.1416 times the r.p.m. by 12. See Table XIV, *Surface Feet per Min-*

ute to R.P.M. Conversion Table for Milling Machines, in the Appendix.

Surface Feet per Minute Tables. Milling-machine manufacturers and authorities in machine tool operation provide tables of recommended cutter speeds expressed in surface feet per minute. These tables are based on the kind of material in the cutter and the kind of material being milled. Table XV, *Recommended Cutting Speeds in S.F.M. for Milling Machines*, in the Appendix, is an example of such assembled information.

But even after determining the surface feet per minute or the travel of the cutting edge, the tables just referred to must be used with considerable judgment. In other words, a table of surface feet per minute is based solely on the kind of cutter material and the material in the workpiece. The rigidity of the workpiece, the depth of cut, the rate of feed, and other factors must also be taken into consideration. Experience in making the setups and performing operations is required before the operator can adequately make his choice of the correct speed. The tables are provided as a guide in gaining this experience.

Feed Selector

Just as milling machines are equipped with a rather wide range of speeds, they also have a broad selection of feeds. The term *feed*, as used in milling machine work, refers to the rate at which the workpiece is fed into the cutter expressed in inches per minute.

After the operator has determined the rate at which the workpiece is to be fed into the cutter, a feed setting is made on the feed selector. Most machines in use at present have a feed selector dial. Feed settings are changed by turning the dial crank on the feed selector dial either one half or one full revolution of the crank, depending on the manufacturer. Turning the crank pin one half or one full turn in either direction will change the feed to the next rate indicated on the dial for whatever direction the crank is turned.

The more automatic model shown in the Trans-Vision permits the operator to set and change feed rates by simply turning a small knob. The new feed setting registers in a window in the feed indicator which is positioned to the left of the feed selector (on the side of the knee). See Step 13, Page E of the Trans-Vision. Additional notes on feed setting are provided on Page G.

Whatever type of control is employed, the feed selector regulates the *powered* movements of the machine: the lengthwise or longitudinal rate of travel of the table, the crosswise movement of the saddle, and the vertical travel of the knee. Some milling machines are also geared to feed per revolution of the cutter.

Factors Affecting Feed Selection

One very important factor which determines the rate of feed for a milling operation is the degree of smoothness required on the finished surface. This finish is usually specified on the blueprint. The more common factors affecting feed selection are discussed below.

Knowing a milling machine's power capacity is important in the selection of feeds. Without such information, it is entirely possible for an operator to feed a workpiece into a cutter too fast. The machine may stall or the cutter break. The power required to remove metal is determined by the number of cubic inches of material removed each minute. The horsepower rating of any milling machine is calculated on this basis. The feed, therefore, must be selected with this factor in mind.

Another condition affecting the choice of feed is the rigidity of the machine and setup. The rigidity of the machine itself is determined largely by the care which it has received over a period of years. The adjustment of the gibs which determine the sliding fit between the various movable parts is important too. The importance of a rigid setup cannot be over-emphasized. Any unsupported part of a workpiece will result in chatter and inaccurate milling if the feed rate is not reduced accordingly.

The kind of material to be milled affects the choice of feed. Brass or aluminum, for example, are more easily machined than steel and cast iron, and can be fed faster.

The width and depth of cut are reflected in the power required to remove metal. When the dimensions of the cut have been determined from the specifications, the feed selected must not exceed the power of the machine. But even when the capacity of the machine is not exceeded, it is frequently advisable to select a lower feed rate. Sometimes, for instance, the rigidity of setup must be allowed for.

The type of finish desired, as explained earlier, governs the choice of a feed rate more than any other factor. Having selected a good combination feed and speed for a roughing operation, the finishing cut can be done by increasing the speed of the cutter and reducing the feed. This reduces the chip load per tooth and makes for a finer finish. Where finish is not important, the feed rate remains the same as for roughing and the speed rate is increased.

Where unusual accuracy is demanded by the specifications, a smaller amount of metal is removed per unit of time. This is regulated in part, at least, by changing the rate of feed. If the workpiece is fed into the cutter at a slower feed, obviously a smaller amount of metal will be removed. There is a direct relationship between the feed rate and the accuracy that can be machined into the finished part. Generally, the lower the feed rate, the higher the degree of accuracy possible in a finished part.

Chip Load Table. It is common to express the feed, or the inches per minute of table travel, in terms of *chip load*. Chip load refers to the amount of metal removed by each tooth during one revolution of the cutter. It is expressed in decimals of an inch of table travel for each tooth while it is making one revolution. Thus for a given cutter RPM, the faster the workpiece is fed into

the cutter, the greater will be the chip load. That is, each tooth on the cutter will remove more material while making one revolution.

Manufacturers and authorities in machine tool operation provide chip load tables similar to Table XVI in the Appendix, *Permissible Feed Per Tooth For Milling Cutters*, to assist operators in determining the proper feed rate. In such tables, the feed rate is expressed in thousandths of an inch per tooth per revolution. For efficient operation, it is important always to use the recommended rate.

Milling Operations

The following discussion on milling operations takes up each operation separately, the order of presentation having no particular significance. All the operations discussed can be performed on the plain knee-and-column milling machine. Some operations, of course, require special attachments. These same operations can also be performed on a universal milling machine as well as on other types of machines.

Plain or Slab Milling

Milling flat horizontal surfaces is sometimes called plain milling or surface milling. Plain milling is the machining of plain, flat, horizontal

surfaces with cylindrical mills whose lengths are usually much greater than their diameters, Fig. 16. Plain milling is also called slab milling. This term is used for the production of a flat surface parallel to the axis of the milling cutter. If a job calls for slabbing, a helical slab mill cutter is chosen. This reduces vibration and results in better milling. The setup in the Milling Machine Trans-Vision is for a slab milling operation.

Suggestions on How to Set Up and Mill Horizontal Surfaces. The first step in this setup is usually mounting the arbor in the spindle. Be sure the tapered hole in the spindle is clean before mounting the

Fig. 16. A plain milling cutter is shown in operation. (Cincinnati Milling Machine Co.)

arbor in the spindle. Likewise, the tapered shank of the arbor must be free from chips and dirt, and wiped dry. One of the essentials of good milling practice is cleanliness, especially in such important assemblies as the arbor and spindle. Arbors with damaged or dirty taper shanks should not be placed in the machine spindle. Run-out will result and the spindle taper may be damaged. The arbor is secured to the spindle with the draw bar as illustrated in the Milling Machine Trans-Vision, Page B.

Next, select the cutter of the right diameter and shape. Be sure it is sharp. Never use a dull cutter. Considerable time may be wasted if the cutter is larger in diameter than necessary.

Heavy-duty slab milling cutters are mounted on the arbor so that the end thrust caused by the helix

angle of the cutter teeth will be directed against the spindle bearings. The cutter should always be mounted on the arbor as close to the column of the machine as possible. Shims for cutter spacing must be in good mechanical condition. Cutters should slide on the arbor easily and should fit snugly. A loose cutter will cause inaccuracies and poor finish.

All cutters should be keyed to the arbor. Do not rely on friction to drive the cutter. In addition, the collars should be pulled up tightly by means of the snug or end nut. The snug nut on the end of the arbor should not be tightened or loosened until the arbor support is in place to prevent the springing of the arbor. Place the arbor support in position properly to support the arbor before tightening the snug nut with the open-end wrench.

Mounting Workpiece and Cutter. The workpiece can be mounted on the table before or after the cutter is positioned on the arbor. If the workpiece is mounted before the cutter, possible injury to the hand from the cutter may be prevented when the workpiece is being positioned. To mount the workpiece *after* the cutter is in position, run the table to the right or left with the longitudinal feed handwheel. The table should be moved so that its approximate center is far enough away from the cutter to permit safe and convenient mounting of the workpiece. See the Trans-Vision, Pages B and C. Both of the sequences for mounting workpiece and cutter described above are acceptable. Whichever is used, the operator must be careful. The workpiece should not be allowed to strike the cutter during the setup.

The workpiece is mounted securely in a vise bolted to the table. The vise is mounted at the center of the table. There should be no chips under the vise. Whether the vise has a swivel base or not, the vise should be set so that the jaws are either parallel or at right angles to the face of the milling machine column.

Place the vise in the position desired, bolt it to the table, and *indicate* it. The dial indicator may be fastened to the arbor, whether or not a cutter is mounted on the arbor, with the indicator button resting on the solid jaw of the vise. Fig. 17 shows a dial indicator being used to test the parallel alignment of the vise jaws with the column face. The hand feed is then used to adjust the table so that the vise jaw moves past the indicator. The vise is adjusted on the table until the dial needle shows little or no movement as the table is run back and forth by hand. When this condition is attained, the vise jaws can be assumed to be parallel to the machine column.

The work is now mounted on the

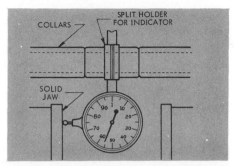

Fig. 17. A dial indicator can be used to align the solid jaw of the vise so that the jaws of the vise are exactly parallel or perpendicular to the face of the column.

vise. Parallels may be needed for certain jobs. If parallels are used, as in the setup shown in the Trans-Vision, the workpiece should be tapped with a soft hammer to make certain it is resting flat on the parallels. Then tighten the vise jaws.

Table Adjustments, Speed and Feed Settings. Adjust the table to proper height to give the cut desired. Adjust the trip dogs to give the proper movement of the table. Again, be careful not to have the work come in contact with the cutter when it is not revolving.

Calculate the revolutions per minute for the proper cutting speed for the given cutter and the given job. (See the preceding section, "Speeds and Feeds for Milling".)

Judge the amount of feed and the depth of cut. One cut may be enough. Again, however, one roughing cut and one finishing cut may be needed.

Proceed to make a trial cut. Consult your foreman or instructor immediately if necessary. CAUTION: Goggles should be worn to protect the eyes. Do not get the fingers near the revolving cutter.

Face Milling

Face milling, Fig. 18, means to produce a flat surface on a piece of work so that the resulting surface is at right angles to the axis of the cutter. The purpose of the operation may be to take a heavy cut or to take a very light finishing cut, leaving a very smooth surface.

The cutter used in face milling (described in the section on cutters) may be likened to a disk with teeth on the periphery and one face. It is fastened to the spindle nose, or to an adapter, so that all the face teeth are in contact when the work has passed fully under the cutter.

The teeth on the periphery do most of the cutting, and their action is like that of a planer tool, except that they move in a circular instead of a straight path. See the *insert*, Fig. 18. The face teeth, theoretically, do not cut at all, but actually they remove a small amount of stock as they sweep over the surface machined by the peripheral teeth.

Suggestions on Face Milling. Fig. 18 shows a typical face-milling

Fig. 18. Face milling operation. (Cincinnati Milling Machine Co.)

operation. In this setup, the workpiece is mounted securely to the table of the machine with T-bolts and straps. It is essential in making all setups that the workpiece be securely fastened to the table or work-holding device. The table is positioned up or down and toward or away from the cutter by means of the adjusting handles. Many of the suggestions on positioning the work and mounting the cutter given in the previous operation apply, as do the suggestions on feed and speed and safety.

End Milling

End milling is the operation of machining flat surfaces, either horizontal, vertical, or at an angle, using an end mill as a cutter. These cutters have teeth on the end face as well as on the periphery. End mills are made for use on a multitude of operations; the solid end mill is shown in use in Fig. 19. End milling is similar in many respects to face milling. However, end milling is

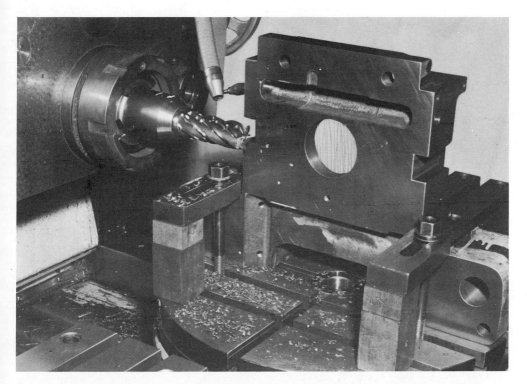

Fig. 19. The end mill shown has just finished a horizontal slot. The setup on the rotary table enables the operator to rotate the workpiece 180° to produce identical slots on each slide of the workpiece. (Cincinnati Milling Machine Co.)

limited to cutters of smaller diameter.

Suggestions on End Milling.
Fig. 19 shows a solid, spiral end mill held in a quick-change adapter in the spindle. The end of the cutter is being used to mill slots on each side of the workpiece, which is mounted on a rotary table.

Much of the work of end milling is done with the use of *shell end mills*. These cutters bridge the gap between the small, solid end mills and the larger face milling cutters. As can be seen from Fig. 20, the shell end mills have end and peripheral teeth. Most of the cutting action is accomplished by the peripheral teeth, just as in face milling. The end teeth do little cutting and

help to remove chips and to produce a finished surface. One of the advantages of using shell end mills is that these cutters are cheaper to replace when worn or broken than are the face mills.

In Fig. 20, a workpiece is shown mounted directly on the table and held in place with step blocks and end clamps. The shell end mill produces a finished horizontal and vertical surface on the workpiece at the same time. In Fig. 21, the workpiece is shown mounted in a swivel vise so that the shell end mill produces a vertical face and a shoulder at an angle across one corner of the workpiece. In both cases the shell end mills are mounted on style "C" arbors, which in turn mount in the spindle.

Fig. 20. The shell end mill, like the face mill, can produce two finished perpendicular surfaces on the workpiece at the same time. (Brown & Sharpe Mfg. Co.)

Fig. 21. A shell end mill is shown cutting an angular face and shoulder on the workpiece. The swivel vise is set to the desired angle of cut. (Brown & Sharpe Mfg. Co.)

Side Milling

The side milling operation consists of machining a vertical surface on the side of a workpiece, using a side-milling cutter, Fig. 22. This operation would be performed on either the outer surfaces of a workpiece, such as squaring one end of the work, or on the inner surfaces, such as shown in Figs. 22 and 23.

Plain side-milling cutters have teeth on each side as well as on the periphery. Half side-milling cutters have teeth on one side only and the periphery, and are sometimes used in pairs to mill parallel sides of the work. This kind of milling is covered later in the discussion of the straddle-milling operation.

Interlocking cutters with interlocking teeth make it possible to use two cutters side by side to mill slots of standard width. Washers or shims placed between the two staggered-tooth cutters regulate the width of cut. In this way, they can be used for slotting.

Suggestions on Side Milling. Fig. 22 shows the workpiece securely mounted in a vise that has been fastened to the table. Notice that the low part of the work has been positioned just out of the vise, while the upper part of the work will just clear under the arbor. The proper diameter of cutter is important in a job of this kind.

The staggered-tooth side milling cutter shown in this illustration is

Fig. 22. A staggered-tooth side milling cutter is shown being used to produce a square corner milled to precise depth. (Brown & Sharpe Mfg. Co.)

remarkably efficient on work of this kind because of the free cutting action that permits removal of a large amount of metal without destructive vibration and chatter, while giving a good finish.

On the work shown in Fig. 22, the cutter has been set to the surface already machined, and will remove enough stock from the high portion of the forging to bring it to the required dimensions, leaving a sharp corner where the surfaces form a right angle.

Fig. 23. The center of a solid steel workpiece is being milled to specified depth and width with an inserted-tooth side mill. (Brown & Sharpe Mfg. Co.)

Fig. 24. A square head bolt is shown being straddle milled using an indexing attachment. (Brown & Sharpe Mfg. Co.)

Fig. 23 shows an inserted-tooth cutter at work machining the inside surface of a block of steel that is to be milled out to form a yoke. A large-diameter cutter has been chosen so the workpiece will not strike the arbor when the cutter reaches the bottom of the slot. The work is held in a standard vise.

Straddle Milling

Straddle milling requires two half-side milling cutters. They are spaced on the arbor a definite distance apart using spacing collars and shims. This makes it possible to finish two or more parallel surfaces at the same time by using two or more side mills. Half side mills have teeth on one side as well as on the periphery and are used for work of this kind. See Fig. 24.

There are many different applications of straddle milling, each with its own special setup. Whenever any two surfaces need to be machined accurately in relation to one another, the operation is usually performed by straddle milling. One of the common applications of straddle milling is milling square or hexagonal surfaces.

Fig. 24 shows the milling of a square with the workpiece held in the dividing head turned to a vertical position. By using the rapid indexing feature of the dividing head, a square head on a bolt can be milled.

Straddle milling is sometimes combined with slab milling or with form milling.

Suggestions on Straddle Milling. Fig. 25 shows another setup for the straddle-milling operation. Here we see the half-side milling cutters actually straddling the work to machine four vertical surfaces. The following suggestions can be used in making such a setup.

Mount the arbor in the milling-machine spindle. Select four high-speed steel, half-side milling cutters of proper diameter, width, and hole size. One pair of cutters should be right-hand and one left-hand. The cutters are arranged on the arbor so that they are as close in to the column as the workpiece and its holder will permit. The necessary spacing collars and shims are used between the cutters to provide a micrometer distance across the finished surfaces on the workpiece. The intermediate arbor support can be placed in position after the needed collars and bearing sleeve are placed on the arbor. Additional collars are also needed to fill up the space between the bearing sleeve in the intermediate arbor and the arbor end or snug nut. After the arbor nut has been positioned, the outer arbor support can be mounted in place if the setup permits.

The setup can be made still more rigid if supporting rods are bolted to the arbor support as shown in

Fig. 25. Pairs of half-side mills are used here to mill four surfaces simultaneously. Note the support bars bolted to the arbor support for extra rigidity. (Brown & Sharpe Mfg. Co.)

Figs. 25 and 26. Fig. 26 illustrates a variation in the straddle-milling setup, where the intermediate arbor support is placed *between* the milling cutters.

The work is mounted in the vise and moved up to the cutter, with care being taken not to nick the cutter. The work is adjusted between the cutters and the saddle locked. The knee may need to be raised or lowered to position the work to give the desired depth of cut. A short trial cut is taken, the

Fig. 26. This straddle-milling setup places the intermediate arbor support *between* the half-side mills. The part being machined is the upper receiver for the M-16 E-1 automatic rifle. (Cincinnati Milling Machine Co.)

machine stopped, and the work moved clear of the cutter. The work can now be measured and necessary adjustments made.

Slitting and Cutting Off

Metal-slitting saws are used for milling narrow slots and for cutting off stock. The saws used are narrow saws that are slightly thicker at the outer edge than near the center to provide clearance behind the cutter teeth. Some saws have teeth only on the periphery while others have alternating side teeth with side-chip clearance. See Fig. 9. This lat-

ter type gives a much smoother finish on the sides of the slots than would be possible at the same depth of cut, using the regular plain-sided type.

Stock can be sawed off in a milling machine as shown by the example in Fig. 27. Here a metal-slitting saw is mounted on the arbor. Because of the stock being so thin, it was necessary, in this case, to block up under the stock with parallels. If the thin stock had been mounted in the bottom of the vise, the arbor and arbor support would have run into the vise jaws.

Form Milling

Form milling, Fig. 28, is the machining of irregular contours by using form cutters. There is an endless number of different forms and shapes that cutters can be made to in order to produce desired contours. The more standard cutters are the convex and concave cutters. These cutters were shown and discussed previously under the section on cutters.

Gang Milling

When more than two cutters are mounted on the arbor to machine surfaces of a workpiece, the operation is referred to as gang milling, Fig. 29. One of the more common setups for gang milling is that in which several helical cutters are mounted on the arbor to machine a large surface. Helical cutters with opposed spirals are used in this setup. Each helical cutter counteracts the thrust caused by the shearing action of the teeth of the other cutter.

Where helical cutters are used merely to produce a flat surface, it is important that the cutters be matched as to diameter. By using interlocking cutters, there is no chance of a slight rib being left on

Fig. 27. Thin stock can be sawed to length using plain or alternate-tooth metal-slitting saws. (Brown & Sharpe Mfg. Co.)

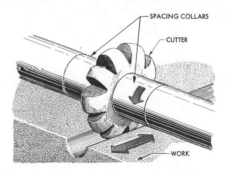

Fig. 28. Formed cutters are available to mill whatever contour is desired.

the work. There would be if this same operation were performed using plain slab mills mounted side by side on the arbor.

Gearing

Gears are vital factors in machinery. One of the first mechanisms invented using gears was the clock. In fact, a clock is little more than

Fig. 29. Gang milling operations will produce an endless variety of surfaces. However, gang milling requires extremely rigid and well-supported setups like the one shown here. (Cincinnati Milling Machine Co.)

a train of gears. Considerable study and research have been made on gears in recent years because of their wide use under exacting conditions. They have to transmit heavier loads and run at higher speeds than ever before. The engineer, the machinist, in fact, all of industry consider gearing the prime element in nearly all classes of machinery.

Gear Selection

Gearing is a means of transmitting power or motion from one shaft to another. The fundamental factors governing the selection of the proper gears for any drive are as follows:

1. Relation between the speeds of two shafts carrying the meshing gears.
2. Direction of rotation of the driving and driven shafts.
3. Amount of power or load to be transmitted.
4. Kind of motion desired.
5. Method of mounting the gears.
6. Efficiency required in the power transmission.

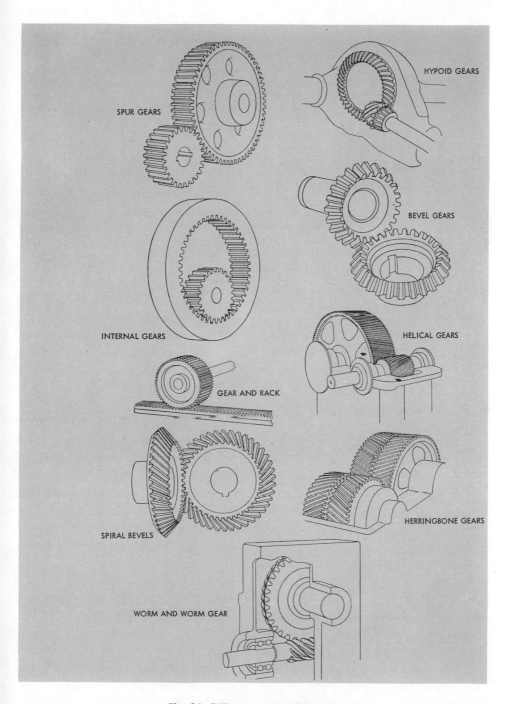

Fig. 30. Different types of gears.

Types of Gears

Gears are classified according to the position of the shaft on which the gears are mounted, Fig. 30. Representative gears belonging to each of three classes are as follows:

1. Gears with shafts parallel— spur, sprocket and chain, helical, herringbone, and internal.
2. Gears with shafts intersecting at any angle in the same plane —miter and bevel.
3. Gears with shafts that do not intersect—helical, worm and worm gear, and spiral.

Uses of Various Gears

The *involute spur gear* is used to a large extent in power transmission. The term *involute* refers to a particular shape of gear tooth and is explained a few paragraphs hence. This tooth gives a maximum ratio of reduction between two spur gears, ranging from 1 to 1 to 8 to 1. In order to have a higher ratio of reduction, gears are sometimes compounded. *Compound gearing* is used (for example, in the lathe) because with a given number of change gears a much larger range of combinations is possible than can be obtained with a simple gear train. With compounded gears, however, there are complications—lost motion, strains, and space required.

Helical gears are stronger, quieter, and smoother in operation. The teeth on helical gears slide one into another. They have more teeth in contact at a given time. However, they have one disadvantage: an end thrust caused by the angle of the teeth. This thrust is proportional to the angle of the teeth.

The herringbone gears, Fig. 30, eliminate this end thrust and provide a quiet and efficient drive.

Sprockets and chain are limited to a ratio of reduction from 1 to 1 to about 6 to 1. They are bulky. They are used for transmitting positive motion between two shafts at some distance apart.

Internal gears allow a larger ratio of reduction in comparatively small space. They provide more tooth contact, greater strength, and smoother and quieter rolling action.

The bevel gear is a common gear for connecting shafts which are not parallel to each other—that is, they intersect. When two meshing bevel gears have the same number of teeth, they are called *miter* gears. Where smooth running is required to avoid noise at high speeds, the straight bevel gear is impractical. The spiral bevel gear is better for high speeds.

The worm and worm gear are used to secure large ratios of reduction. They are smooth, quiet, and powerful. Worms are made with both single and multiple threads. Worms and worm gears are widely used as speed reducers with a corresponding increase in power.

Development of Gear Teeth

In toothed gearing, the surfaces of two cylinders or cones come in contact. They roll together in opposite directions.

Two cylinders, *A* and *B*, are shown in Fig. 31. The axes of *A* and *B* are in the same plane and parallel. The peripheries of the cylinders are in contact. If the cylinder *A* is rotated in the direction of the arrow, frictional contact will cause cylinder *B* to rotate in the opposite direction. It will also be evident that if the cylinders in Fig. 31 are of equal diameter, and consequently of equal circumference, a complete revolution of *A* will produce a complete revolution of *B*. If *A* is one-half the diameter of *B*, the latter will make but half a revolution to one complete revolution of *A*. Likewise, if cylinder *B* is one-half the diameter of *A*, it will make two complete revolutions to one of cylinder *A*.

This relation allows for no slipping of the cylinders on each other. To avoid slipping, positive drives instead of friction drives must be provided.

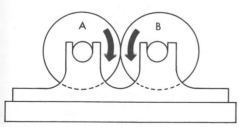

Fig. 31. Two cylinders in contact will revolve in opposite directions.

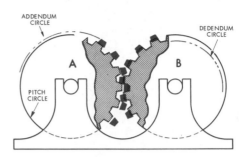

Fig. 32. Teeth on each cylinder will prevent slipping between the two cylinders in contact.

For transmitting power, the faces of these cylinders are provided with teeth, Fig. 32. The teeth are cut parallel to the axis of each cylinder and interlock with those of the other cylinder. Thus, they effectually prevent any slipping. By this means, we can produce a pair of involute spur gears which are used for power transmission.

Involute Gear

The term *involute* refers to the shape of the gear tooth curve. It is the curve traced by a point on a cord as the cord is unwound from a cylinder known as the base circle. The shape of the curve changes with the diameter of the base circle.

An involute curve must have a base circle from which the curve is generated. When two involutes are brought into contact on the line of centers, a pitch point is established. This point determines the diameter of the *pitch circle*. A fundamental

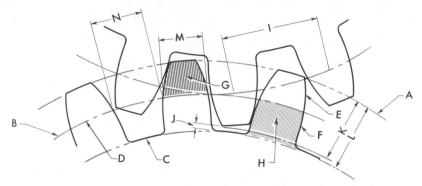

Fig. 33. Tooth parts. The letters in this illustration correspond to the parts explanation in the text.

characteristic of the pitch circle is that its diameter is a flexible dimension. This condition applies only to the involute form of tooth. Thus, involute gears can operate successfully at slightly varying center distances.

Tooth Parts

The various tooth parts are indicated by letters in Fig. 33. The student must be familiar with the technical terms for describing the teeth of gears, and he must know the methods of calculating, designing, and drawing their various parts. These terms are important to the student who is learning to mill gear blanks.

Names and Definitions. The tooth parts are named, defined, and explained in the following paragraphs:

A. The *addendum circle* has the same diameter as the outside diameter of the gear.

B. The *pitch circle* is the line of contact of two cylinders which would have the same speed ratios as the gears.

C. The *dedendum circle* is the circle at the bottom of the tooth space.

D. *Pitch diameter* is the diameter of the pitch circle.

E. The *tooth face* is that portion of the curved surface of the tooth between the pitch circle and the addendum circle.

F. The *flank* of the tooth is the portion of the curved surface of the tooth between the pitch circle and dedendum circle.

G. The *addendum* is the line-shaded portion of the tooth between the pitch circle and the addendum circle.

H. The *dedendum* is the dot-shaded portion of the tooth between the pitch circle and the dedendum circle.

I. *Circular pitch* is the distance

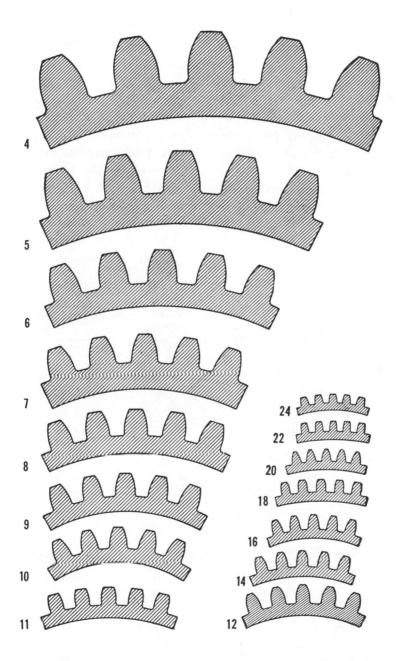

Fig. 34. Proportions of standard involute gear teeth of different diametral pitch.

from center to center of the teeth when measured on an arc of the pitch circle.

J. Clearance is the space between the bottom of a tooth space and the tip of a tooth fully meshed into that tooth space.

K. Working depth is the distance a tooth extends into the tooth space when in full mesh and having correct *clearance*.

L. Whole depth is the total height of the tooth from the dedendum circle to the addendum circle.

M. Chordal tooth thickness is the thickness of the tooth as measured

on the chord of an arc of the pitch circle.

N. The *tooth space* is the distance between adjacent teeth, measured as the chord of an arc on the pitch circle.

Fig. 34 shows the proportions of standard involute gear teeth of different diametral pitches with corresponding circular pitches.

Formulas for Figuring the Various Parts of a Spur Gear. Various formulas used in calculating some principal characteristics of spur gears are provided in Table I, *Spur Gear Formulas*.

TABLE I. FORMULAS FOR DETERMINING PARTS OF A SPUR GEAR

1. To find *Diametral Pitch* (*DP*):

$$DP = \frac{\text{number of teeth} + 2}{\text{outside diameter}} \quad \text{or} \quad \frac{N + 2}{OD}$$

$$DP = \frac{\text{number of teeth}}{\text{pitch diameter}} \quad \text{or} \quad \frac{N}{PD}$$

$$DP = \frac{3.1416}{\text{circular pitch}} \quad \text{or} \quad \frac{3.1416}{CP}$$

2. To find *Pitch Diameter* (*PD*):

$$PD = \frac{\text{outside diameter} \times \text{number of teeth}}{\text{number of teeth} + 2} \quad \text{or} \quad \frac{OD \times N}{N + 2}$$

$$PD = \frac{\text{Number of teeth}}{\text{diametral pitch}} \quad \text{or} \quad \frac{N}{DP}$$

$$PD = \text{outside diameter} - \frac{2}{\text{diametral pitch}} \quad \text{or} \quad OD - \frac{2}{DP}$$

3. To find *Number of Teeth* (*N*):

$$N = \frac{\text{pitch diameter} \times 3.1416}{\text{circular pitch}} \quad \text{or} \quad \frac{PD \times 3.1416}{CP}$$

$$N = \text{pitch diameter} \times \text{diametral pitch} \quad \text{or} \quad PD \times DP$$

$$N = \text{outside diameter} \times \text{diametral pitch} - 2 \quad \text{or} \quad OD \times DP - 2$$

4. To find *Outside Diameter* (*OD*):

$$OD = \text{pitch diameter} + \frac{2}{\text{diametral pitch}} \quad \text{or} \quad PD + \frac{2}{DP}$$

$$OD = \frac{\text{number of teeth} + 2}{\text{diametral pitch}} \quad \text{or} \quad \frac{N + 2}{DP}$$

$$OD = \text{number of teeth} + 2 \times \text{addendum} \quad \text{or} \quad N + 2 \times A$$

5. To find *Tooth Thickness* (*T*):

$$t = \frac{\text{circular pitch}}{2} \quad \text{or} \quad \frac{CP}{2}$$

$$t = \frac{1.5708}{\text{diametral pitch}} \quad \text{or} \quad \frac{1.5708}{DP}$$

6. To find *Circular Pitch* (*CP*):

$$CP = \frac{3.1416}{\text{diametral pitch}} \quad \text{or} \quad \frac{3.1416}{DP}$$

7. To find *Clearance* (*C*):

$$C = \frac{.157}{\text{diametral pitch}} \quad \text{or} \quad \frac{.157}{DP}$$

8. To find *Working Depth* (*WD*):

$$WD = \frac{2}{\text{diametral pitch}} \quad \text{or} \quad \frac{2}{DP}$$

9. To find *Whole Depth* (*WD*):

$$WD = \frac{2.157}{\text{diametral pitch}} \quad \text{or} \quad \frac{2.157}{DP}$$

10. To find *Addendum* (*A*):

$$A = \frac{1}{\text{diametral pitch}} \quad \text{or} \quad \frac{1}{DP}$$

$$A = \frac{\text{circular pitch}}{3.1416} \quad \text{or} \quad \frac{CP}{3.1416}$$

Machining Gear Teeth

Two general processes are used for cutting the teeth of gears: *form milling* and *generating*. Either of these two processes may be used for cutting the teeth of spur gears, internal gears, and racks.

Form Milling a Gear. The first process is milling with a properly formed revolving cutter, Fig. 35, as in ordinary milling machine work. It is applicable also to cutting helical gears, bevel gears, and a portion of the work on worm gears. The cutter must be shaped exactly to the form of the space between the teeth of the gear to be cut. It must revolve at a speed suitable to the metal, and it must be mounted on a suitable arbor to insure rigidity and to eliminate vibration. The gear blank must be carefully mounted and fed to the cutter so as to prevent vibration.

The rotation of the gear blank to give proper spacing of the teeth is taken care of by the dividing head of the milling machine. For complete information on how to index with the dividing head, see the indexing section in the preceding chapter.

Form Cutters for Involute Gear. Cutters properly formed to cut involute teeth will cut gears of considerable variation in number of teeth. This greatly reduces the number of cutters of each pitch required for cutting a complete range of work from pinions of 12 teeth to a rack. Where 8 cutters are required, the range of work is as follows:

No. 1 will cut from 135 teeth to a rack.
No. 2 will cut from 55 teeth to 134 teeth.
No. 3 will cut from 35 to 54 teeth.
No. 4 will cut from 26 to 34 teeth.
No. 5 will cut from 21 to 25 teeth.
No. 6 will cut from 17 to 20 teeth.
No. 7 will cut from 14 to 16 teeth.
No. 8 will cut from 12 to 13 teeth.

Fig. 35. A rotating cutter like the one shown here is used for form-milling involute teeth.

Fig. 35 shows the usual form of rotating cutter for milling the involute form of gear tooth. The teeth of these cutters are relieved or backed off. Their form is not changed when ground upon the front faces after they have become dulled by use.

Setup for Milling Spur Gear

The milling of a spur gear from a gear blank can be one of the most, if not the most, fascinating and profitable experiences of all milling operations. Like all other operations, however, it requires a very accurately made setup. Because the indexing attachment (dividing head, as it is sometimes called) is involved, a certain amount of careful planning and performance is required of the operator.

Speed of Cutter. In milling gears, the speed of the cutter should be slightly less than that of ordinary milling cutters. The cutting surface is not only over the points of the cutting teeth, but on both sides also. The speed at which the cutter should be rotated will depend upon the material being milled, as well as the depth of cut being taken.

Feed. The proper feed for gear cutting will be the same as for most milling-machine work where the same kind of material is involved. In some cases, the feed should be slightly less. For cast iron, it is usually about 1/16 inch per revolution.

Suggestions on How To Cut a Spur Gear. Clean the milling-machine table thoroughly of chips. Now mount the dividing head and tailstock units on the milling-machine table. Bolt them tightly to the table. Check the alignment of headstock and tailstock centers by measuring with a rule and square head from the column face of the machine to each center point. Press the gear blank on a proper size mandrel. Place a milling-machine dog on the large end of the mandrel, Fig. 36. Mount the mandrel between the dividing centers with the large end of the mandrel toward the headstock of the dividing head. The tail of the dog should be placed in the slot of the driving plate, and the setscrew should be tightened against the flat of the tail. The proper side index plate is now at-

tached to the side of the dividing head. The plate to use can be determined by the calculations explained in the previous chapter under the discussion of the indexing attachment (dividing head).

Centering Cutter over Gear Blank. The cutter is centered accurately over the blank so that the center line of the gear teeth will be radial to the axis of the gear. This can be done as follows: With the cutter mounted on the arbor and before the gear blank and mandrel are mounted on the centers, move the table to bring the center point of the tailstock center up to the cutter. Be careful not to nick the cutter. Adjust the table, using the cross-feed screw, to bring the point of the tailstock center on the center line of the cutter.

Setting Cutter to Proper Depth. Setting the cutter to proper depth is done as follows:

1. Adjust the table upward until the cutter just touches a piece of paper held between the gear blank and cutter.

2. Set the dial of the vertical adjusting screws at zero.

3. Move the table of the machine horizontally until the blank clears the cutter.

4. Raise the table vertically to the proper depth of the tooth using the dial of the vertical adjusting screws. It is important that the gear blank was previously turned (ma-

Fig. 36. Milling a spur gear with a form-cutter using the indexing attachment. (Cincinnati Milling Machine Co.)

chined) to the exact diameter.

5. The number of cuts to be taken is governed by the whole depth of the cut in the tooth. Be sure that all the necessary locking clamps on the machine and the dividing head are properly locked.

Taking the Trial Cut. The cut is started at the tailstock side of the gear blank. In this way, the gear blank is forced onto the mandrel, since the gear blank is being forced toward the large end of the mandrel.

When the form cutter has progressed through the gear blank, forming one space between two un-

finished teeth, the table is returned to the original starting position, the index crank is adjusted to turn the gear blank another space, and a new cut is taken.

Helical Milling (Spiral Milling)

Helical milling is the operation performed in milling helical cutters, reamers, drills, helical gears, worms, etc.

A common form of helical milling is that of milling a helical gear, as shown in Fig. 37. Helical gears have been called by several names in the past, "spiral" being the most common. The American Society of Gear

Fig. 37. When milling a helical gear, the universal milling machine must be used, since the table must be set at an angle corresponding to the helix of the gear to be milled. (Cincinnati Milling Machine Co.)

Standards is making an effort to call them by their proper name, which is *helical*, since the teeth wind around the outside diameter in the form of a helix.

The simplest form of helical gears is shown in Fig. 38. The shafts are parallel and the teeth of both gears are cut at an angle of 30 degrees to their axis. Helical gears operate like ordinary spur gears, except that the engagement of the teeth is continuous; this tends to make for noiseless action.

In cutting helical gears on the milling machine, it is necessary that a universal milling machine be used since the table must be turned at an angle. It is also necessary that an indexing head be used that can be *geared to the table* for rotating the gear blank at the same time that the table is fed toward the cutter.

The Helix. Before making the setup on the machine, it is necessary that a little study be made of the helix. If a line is drawn around a cylinder so that it advances a uniform distance along the cylinder as the cylinder revolves, the curve formed is called a *helix*. The line

349

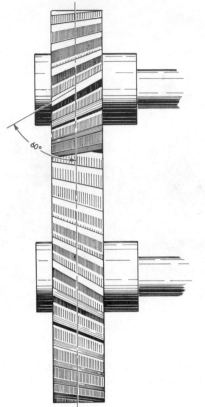

Fig. 38. The simplest form of helical gear is shown here.

drawn and the curve formed can best be illustrated by a piece of paper laid out flat to form a triangle as shown in Fig. 39. The area bounded by the three lines forming

the triangle represents a piece of paper cut and wound around a cylinder. The side *X*, or the hypotenuse, of the triangle represents the curve called a *helix*.

Lead of the Helix. The *lead* of the helix is the distance which the curve advances along the cylinder, parallel to the cylinder axis for each revolution of the cylinder. The lead is shown as the side *Y* of the triangle, Fig. 39.

After the lead of the helix has been determined for a helical gear to be cut, the angular setting of the milling-machine table must be calculated. The lead of the helix is always given on the blueprint or drawing in inches. The table then advances and rotates the workpiece as the cutter is machining the gear.

Helix Angle. In all cases of helical gear cutting, the universal milling-machine table must be set to the angle of the helix to be machined. Since the graduations on the milling-machine table are in degrees, it is necessary to determine the helix angle in degrees and fractions of a degree.

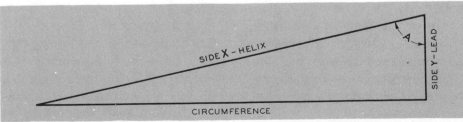

Fig. 39. This diagram can be used to find the angle of the helix, angle A, to which the universal table must be set.

The helix angle, A in Fig. 39, is the angle included between the hypotenuse, side X, and the lead, side Y. The helix angle varies in size with the circumference of the cylinder and the lead of the helix.

The usual practice is to obtain the value of the angle after working out a simple problem in plane trigonometry and then consulting a table of tangents. To do this, divide the circumference of the work by the lead of the helix, as in the following formula:

$$\text{tangent of helix } \angle A = \frac{\text{circumference of work}}{\text{lead of helix}}$$

To find the circumference of the workpiece, multiply the diameter by π ($= 3.1416$).

The Dividing Head. Helical milling is performed on a milling machine using a universal dividing head. The dividing head is so designed that gears can be connected from the dividing head to the table lead screw. Set up in this manner, the rotary movement of the dividing head is controlled by the longitudinal movement of the table. In other words, the workpiece can be given a definite rotary motion in relation to the longitudinal travel of the table.

Fig. 37 shows the milling machine setup for helical milling. Gears enclosed inside the cover rotate the workpiece on its axis in one direction for a right-hand helix, or in the opposite direction for a left-hand helix.

It is necessary to know the correct combination of gears for the dividing head and the table in order to turn the work at the correct speed to generate a helix. Standard change gears are furnished by the manufacturer with instructions.

Helical Milling Setup and Operation

Fig. 40 shows two combinations of milling-machine change gears: one for cutting a right-hand helical gear, and one for cutting a left-hand helical gear. A left-hand helical gear of the same angle as a right-hand gear may be cut by setting the table to the other side of zero from that used to cut the right-hand gear, and then inserting an idler change gear as shown in Fig. 40.

The gear blank is mounted on a mandrel, which in turn is mounted between the dividing-head centers. The gear blank revolves slowly as the table advances at an angle to the axis of the cutter.

The center of the cutter should be directly above the point of intersection of the axis of the mandrel carrying the blank and the arbor upon which the cutter is mounted. The cutter should be centered directly above the axis of the mandrel before the table is swiveled to the angle of the helix to be machined. This can be done using the method

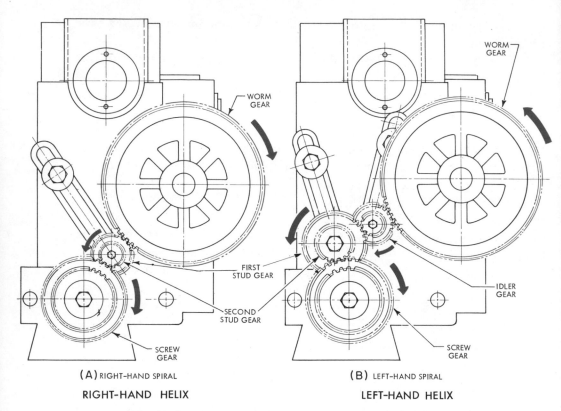

(A) RIGHT-HAND SPIRAL

RIGHT-HAND HELIX

(B) LEFT-HAND SPIRAL

LEFT-HAND HELIX

Fig. 40. Standard change gears for a right-hand and a left-hand helix.

explained previously in this chapter under the heading "Centering Cutter over Gear Blank."

Suggestions on Figuring and Milling a Helical Gear. The following problem will serve as a review of the preceding material on the calculation and machining of a helical gear.

PROBLEM: Assume that it is required to mill a helix on a blank 1½ inches in diameter with a lead of 21 inches.

SOLUTION: To determine the angle to which the table must be swiveled, use the formula:

$$\text{Tangent of the angle} = \frac{C}{L}$$

where

$C =$ circumference of work
$L =$ lead of helix

Substituting

$$\text{Tangent of the angle} = \frac{4.712}{21}$$
$$= .2244$$

Reference to a table of tangents shows that the angle whose tangent is .2244 is 12 degrees, 39 minutes. Set the universal table at this angle.

The next step is to select the proper gears for connecting the dividing head to the machine-table lead screw so that the workpiece will rotate as the machine table is moved horizontally. Since the lead of the helix to be machined is 21 inches, and the lead of the machine is 10 (true of all milling machines), the ratio is then 10/21. If a 10-tooth gear and a 21-tooth gear were large enough in diameters and available, it would be a simple matter to connect them and mill the helix. This cannot be done, however, as the two gears mentioned would not be large enough to span the distance from the dividing head to the lead screw of the machine and mesh properly.

The set of gears usually furnished with a machine for performing a helix-milling operation has the following number of teeth: 24, 28, 32, 40, 44, 48, 56, 64, 72, 86, 100. It is always necessary to compound the gearing in helical milling. In other words, it will be necessary to select four gears from the change-gear set furnished with the machine. When properly placed, they will produce the same results as the original two-gear ratio of 10/21.

To compound the gearing, it is necessary to factor the original ratio 10/21. To factor any number, two numbers are taken, such that, when multiplied together, they will equal the original number. For example, factors of 8 are 4 × 2. The factors of 10 and 21 are:

$$\frac{10}{21} = \frac{2 \times 5}{3 \times 7}$$

Now it is necessary to take one of the factors of 10, which is 2, and one of the factors of 21, which is 3, and find a number which, when multiplied by 2 and 3, will produce 2 gears available in the set of gears provided with the machine. Thus:

$$\frac{2 \times 24}{3 \times 24} = \frac{48}{72}$$

The same is done with the remaining factors of 10 and 21. Thus:

$$\frac{5 \times 8}{7 \times 8} = \frac{40}{56}$$

The problem is now complete. The four gears selected, which are available in the set of gears provided with the machine, are 48, 72, 40, 56.

It is important to know where to place the gears or the correct ratio will not be obtained. The following rule should be observed when placing the gears in position:

$$\frac{\text{Lead of machine (10)}}{\text{Lead of helix (21)}} = \frac{\text{driving gears}}{\text{driven gears}}$$

From the above, it is apparent that 48 and 40 are the driving gears and 72 and 56 are the driven gears. They are placed as follows:

Worm gear 72
First gear on stud 40
Second gear on stud 56
Screw gear 48

Consult Fig. 40 for the distinction

between the gears. To generate a left-hand helix, an idler gear would have to be introduced into the gear train to reverse the direction of rotation. The screw is the initial driving gear. The operator must be careful to place the driving and driven gears in the proper place or the spiral will not be cut as desired.

The number of the cutter to be used in milling a helical gear may be found by dividing the number of teeth in the gear by the cube of the cosine of the helix angle.

The Rack

The involute rack and pinion are shown in Fig. 41. Since the pitch line of the rack is a straight line, the base circle (which is necessarily parallel to it) becomes a straight line. The involute of the base circle is a straight line perpendicular to the line of action or obliquity. Hence, involute rack teeth have straight sides from the bottom of the tooth to the pitch line. Where involute racks are to mesh with pinions which have a relatively small number of teeth, the upper part of each tooth face of the rack is rounded off, through a distance equal to one-half the addendum, by an arc whose radius is equal to 2.10 inches divided by the diametral pitch. This is done to avoid interference of teeth. The motion of a rack must be reciprocating, with the direction of the pinion changing periodically.

Milling Rack Teeth. The rack may be compared to a spur gear that has been straightened out and fas-

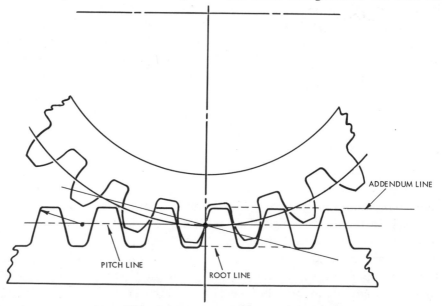

ADDENDUM LINE

PITCH LINE

ROOT LINE

Fig. 41. Involute rack and pinion gear in mesh.

tened to a flat surface. The center-to-center distance of the teeth must equal the circular pitch of a mating pinion.

When selecting the cutter, it is common practice to use a No. 1 spur gear cutter of the required diametral pitch. This number is intended for spur gears varying from 135 teeth up to a rack.

The depth of the tooth is calculated the same as for the depth of spur gear teeth.

The teeth may be indexed by means of the graduated dial on the crossfeed screw. The dial is set at zero for the first tooth, and the rack is fed into the cutter with the cross-feed screw. Fig. 42 illustrates the rack-cutting attachment used to cut rack teeth. This attachment is connected to the table lead screw which automatically indexes the successive cuts.

Measuring Gear Teeth for Accuracy

When milling a gear on the milling machine, it is necessary to test the teeth of the gear to see if they are being cut properly. The gear-tooth vernier, shown in Fig. 43, is a special form of caliper for measuring the thickness of a gear tooth at the pitch line or circle. This is referred to as the *chordal thick-*

Fig. 42. Special gear-cutting attachment. (Cincinnati Milling Machine Co.)

ness. The gear-tooth vernier will also measure the distance from the top of the tooth to the pitch line. This is called the *addendum*. Since it is provided with vernier scales to measure in both directions, it can be accurately adjusted to the required dimensions as found in tables for this purpose.

Table II shows various dimensions of the parts of gear teeth calculated for involute teeth designed upon the diametral-pitch system. This table shows dimensions for measuring the teeth of racks only. It is useful in comparing the different dimensions of the same pitch with one another, and in comparing similar dimensions used in the same pitch. It will enable the milling-machine operator to avoid tedious calculations.

The circular pitch, as defined previously, is the distance from one tooth to the next, measured around the pitch circle. The thickness of the tooth is one-half of the circular pitch, measured in the same manner. On a rack, the pitch line is straight. It can be measured easily with a vernier caliper like that in Fig. 43. With this instrument, measurements can be taken to the nearest thousandth of an inch.

Using the Gear-Tooth Vernier Caliper To Measure a Gear. The procedure for using the gear-tooth vernier caliper to measure the teeth of a gear is to first set the tongue for the proper addendum distance.

TABLE II. INVOLUTE GEAR TOOTH PARTS

Diametral Pitch	Circular Pitch	Thickness of Tooth	Addendum	Working Depth	Whole Depth
1	3.1416	1.5708	1.0000	2.0000	2.1571
1½	2.0944	1.0472	.6666	1.3333	1.4381
2	1.5708	.7854	.5000	1.0000	1.0785
2½	1.2566	.6283	.4000	.8000	.8628
3	1.0472	.5236	.3333	.6666	.7190
4	.7854	.3927	.2500	.5000	.5393
5	.6283	.3142	.2000	.4000	.4314
6	.5236	.2618	.1666	.3333	.3595
7	.4488	.2244	.1429	.2857	.3081
8	.3927	.1963	.1250	.2500	.2696
9	.3491	.1745	.1111	.2222	.2397
10	.3142	.1571	.1000	.2000	.2157
12	.2618	.1309	.0833	.1666	.1798
14	.2244	.1122	.0714	.1429	.1541
16	.1963	.0982	.0625	.1250	.1348
18	.1745	.0873	.0555	.1111	.1198
20	.1571	.0785	.0500	.1000	.1079
24	.1309	.0654	.0417	.0833	.0898

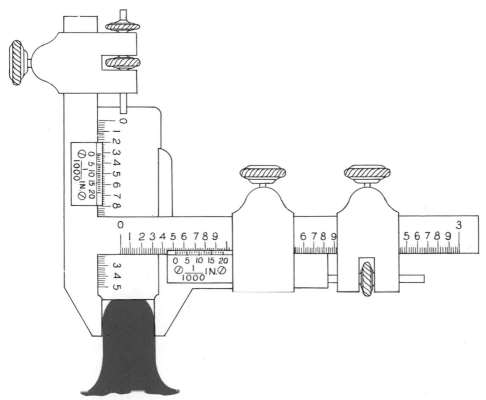

Fig. 43. The gear-tooth vernier is a special form of caliper designed to measure the chordal thickness of a gear tooth.

Next, adjust the horizontal screw to the proper width as given in tables in a handbook, see Table III. The vernier caliper should fit the gear tooth as shown in Fig. 44, *right*.

The addendum of a gear tooth is the distance from the pitch circle to the outside diameter of a gear. It has a definite value for each diametral pitch and number of teeth. When measuring gear teeth, the true addendum cannot be used; rather a "corrected addendum," as it is called, must be figured. If the cali-

pers are adjusted to the true addendum dimension, the measurement will be taken on the tooth sides on the line *AA*, shown in Fig. 44, *left*. However, the dimension actually required is at the tooth sides where the pitch circle intersects the tooth outline, as shown in Fig. 44, *right*. Notice that the points of the caliper jaws are slightly below the line *AA*.

Table III gives the corrected addendum and chordal thickness for various numbers of gear teeth. The

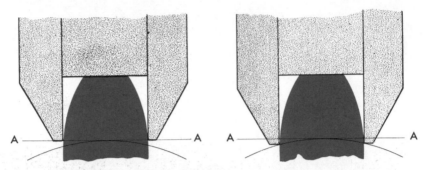

Fig. 44. Measuring the gear tooth.

values in this table are for gears of one diametral pitch. For any other diametral pitch, divide the given value by the required pitch.

The corrected addendum is the setting for the vertical scale on the gear-tooth caliper so that the jaws of the caliper will rest upon the gear tooth exactly at the pitch line. The

horizontal scale on the vernier is set for the thickness of the tooth as given in Table III. This is called the chordal thickness. Once the vertical scale is set for the corrected addendum dimension, it can be determined whether the chordal thickness is oversize or undersize from the reading of the horizontal scale.

TABLE III. SPUR GEARS—CORRECTED ADDENDUM AND CHORDAL
THICKNESS FOR 1 DIAMETRAL PITCH[1]

Teeth	Corrected Adden.	Corrected Thick. of Tooth	Teeth	Corrected Adden.	Corrected Thick. of Tooth	Teeth	Corrected Adden.	Corrected Thick. of Tooth
N	a	t	N	a	t	N	a	t
6	1.1022	1.5529	65	1.0094	1.5706	124	1.0049	1.5707
7	1.0877	1.5576	66	1.0093	1.5706	125	1.0049	1.5707
8	1.0768	1.5607	67	1.0092	1.5706	126	1.0049	1.5707
9	1.0683	1.5628	68	1.0090	1.5706	127	1.0048	1.5707
10	1.0615	1.5643	69	1.0089	1.5706	128	1.0048	1.5707
11	1.0559	1.5653	70	1.0088	1.5706	129	1.0047	1.5707
12	1.0513	1.5663	71	1.0086	1.5706	130	1.0047	1.5707
13	1.0473	1.5669	72	1.0085	1.5706	131	1.0047	1.5707
14	1.0440	1.5675	73	1.0084	1.5706	132	1.0046	1.5707
15	1.0410	1.5679	74	1.0083	1.5706	133	1.0046	1.5707
16	1.0385	1.5682	75	1.0082	1.5706	134	1.0046	1.5707
17	1.0362	1.5685	76	1.0081	1.5706	135	1.0045	1.5707
18	1.0342	1.5688	77	1.0079	1.5706	136	1.0045	1.5707
19	1.0324	1.5689	78	1.0078	1.5706	137	1.0045	1.5707
20	1.0308	1.5691	79	1.0078	1.5706	138	1.0044	1.5707
21	1.0293	1.5693	80	1.0077	1.5707	139	1.0044	1.5707
22	1.0280	1.5694	81	1.0076	1.5707	140	1.0044	1.5707

[1] Source: Gould and Eberhardt.

TABLE III SPUR GEARS—CORRECTED ADDENDUM AND CHORDAL THICKNESS FOR 1 DIAMETRAL PITCH[1] (*Continued*)

Teeth	Corrected Adden.	Corrected Thick. of Tooth	Teeth	Corrected Adden.	Corrected Thick. of Tooth	Teeth	Corrected Adden.	Corrected Thick. of Tooth
N	a	t	N	a	t	N	a	t
23	1.0268	1.5695	82	1.0075	1.5707	141	1.0043	1.5707
24	1.0256	1.5696	83	1.0074	1.5707	142	1.0043	1.5707
25	1.0246	1.5697	84	1.0073	1.5707	143	1.0043	1.5707
26	1.0237	1.5698	85	1.0072	1.5707	144	1.0042	1.5707
27	1.0223	1.5699	86	1.0071	1.5707	145	1.0042	1.5707
28	1.0219	1.5699	87	1.0070	1.5707	146	1.0042	1.5707
29	1.0212	1.5700	88	1.0070	1.5707	147	1.0041	1.5707
30	1.0205	1.5700	89	1.0069	1.5707	148	1.0041	1.5707
31	1.0199	1.5701	90	1.0068	1.5707	149	1.0041	1.5707
32	1.0192	1.5701	91	1.0067	1.5707	150	1.0041	1.5707
33	1.0186	1.5701	92	1.0067	1.5707	151	1.0040	1.5707
34	1.0181	1.5702	93	1.0066	1.5707	152	1.0040	1.5707
35	1.0176	1.5702	94	1.0065	1.5707	153	1.0040	1.5707
36	1.0171	1.5702	95	1.0065	1.5707	154	1.0040	1.5707
37	1.0166	1.5703	96	1.0064	1.5707	155	1.0040	1.5707
38	1.0162	1.5703	97	1.0063	1.5707	156	1.0039	1.5707
39	1.0158	1.5703	98	1.0063	1.5707	157	1.0039	1.5707
40	1.0154	1.5703	99	1.0062	1.5707	158	1.0039	1.5707
41	1.0150	1.5704	100	1.0061	1.5707	159	1.0038	1.5707
42	1.0147	1.5704	101	1.0061	1.5707	160	1.0038	1.5707
43	1.0143	1.5704	102	1.0060	1.5707	161	1.0038	1.5707
44	1.0140	1.5704	103	1.0059	1.5707	162	1.0038	1.5707
45	1.0137	1.5704	104	1.0059	1.5707	163	1.0037	1.5707
46	1.0133	1.5705	105	1.0058	1.5707	164	1.0037	1.5707
47	1.0131	1.5705	106	1.0058	1.5707	165	1.0037	1.5707
48	1.0128	1.5705	107	1.0057	1.5707	166	1.0037	1.5707
49	1.0125	1.5705	108	1.0057	1.5707	167	1.0036	1.5707
50	1.0123	1.5705	109	1.0056	1.5707	168	1.0036	1.5707
51	1.0120	1.5705	110	1.0056	1.5707	169	1.0036	1.5707
52	1.0118	1.5705	111	1.0055	1.5707	170	1.0036	1.5707
53	1.0116	1.5705	112	1.0055	1.5707	171	1.0035	1.5707
54	1.0114	1.5705	113	1.0054	1.5707	172	1.0035	1.5707
55	1.0112	1.5705	114	1.0054	1.5707	173	1.0035	1.5707
56	1.0110	1.5705	115	1.0053	1.5707	174	1.0035	1.5707
57	1.0108	1.5706	116	1.0053	1.5707	175	1.0035	1.5707
58	1.0106	1.5706	117	1.0052	1.5707	176	1.0034	1.5707
59	1.0104	1.5706	118	1.0052	1.5707	177	1.0034	1.5707
60	1.0102	1.5706	119	1.0051	1.5707	178	1.0034	1.5707
61	1.0101	1.5706	120	1.0051	1.5707	179	1.0034	1.5707
62	1.0099	1.5706	121	1.0051	1.5707	180	1.0034	1.5707
63	1.0097	1.5706	122	1.0050	1.5707	181	1.0034	1.5707
64	1.0096	1.5706	123	1.0050	1.5707	182	1.0033	1.5707

[1] Source: Gould and Eberhardt.

359

Safety Precautions—Milling Machine

The milling machine, like any other machine shop power tool, is not dangerous to operate if certain safeguards are installed and if the operator uses safe practice.

The principal hazard to the milling machine operator is that of injury to the arms, hands, or fingers by contact with the cutter. Hands are frequently cut when they are used to brush away chips from various parts of the machine. Eye injury can result from not using goggles.

Observe the following precautions:

1. When clamping the work in place, only wrenches which properly fit the nut and bolt heads should be used, and care should be taken not to spring the work out of shape.

2. The table should be kept free of all tools or other material which might fall and injure the operator.

3. Chips should not be removed from the table by hand; nor should they be blown off with compressed air. A brush should be used for that purpose.

4. Before inserting the arbor or adapters into the spindle, be sure both arbor and spindle hole are clean and free from nicks.

5. Sprung arbors should not be used. If an arbor is sprung, it should be straightened or a straight arbor obtained before the cutter and spacing collars are put in place.

6. To avoid striking the hands on the cutter while setting up, the setup should be done as far away from the cutter as possible.

7. The operator should not attempt to clean or oil the machine or make any adjustments to the work while the machine is in motion.

8. When an operator has finished an operation, or before he leaves his machine for any reason, he should shut off the power and make sure the machine has stopped running.

Courtesy of the National Safety Council

NOTE: Please see review questions at end of book.

TURRET LATHE A

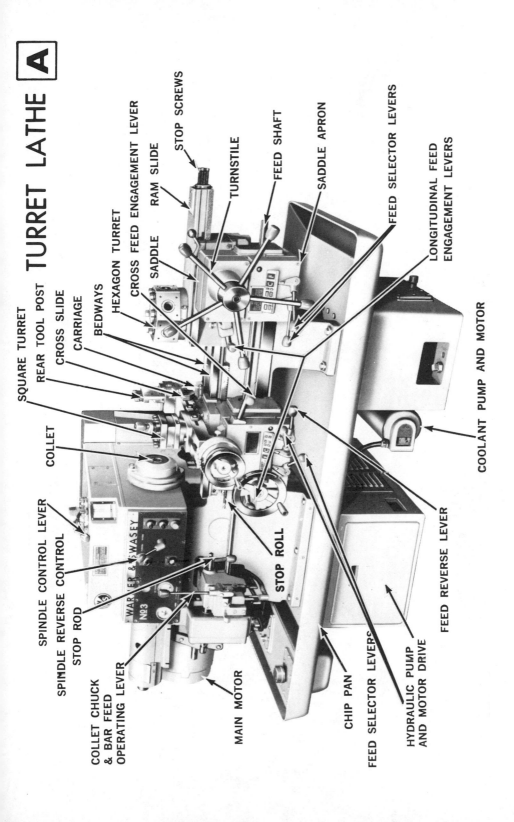

STOP SCREWS

CROSS FEED ENGAGEMENT LEVER

RAM SLIDE

SADDLE

TURNSTILE

HEXAGON TURRET

FEED SHAFT

CROSS FEED ENGAGEMENT LEVER

SADDLE APRON

BEDWAYS

FEED SELECTOR LEVERS

CARRIAGE

CROSS SLIDE

REAR TOOL POST

LONGITUDINAL FEED ENGAGEMENT LEVERS

SQUARE TURRET

COLLET

COOLANT PUMP AND MOTOR

SPINDLE CONTROL LEVER

SPINDLE REVERSE CONTROL

STOP ROD

WARNER & SWASEY

Nº3

COLLET CHUCK & BAR FEED OPERATING LEVER

STOP ROLL

FEED REVERSE LEVER

MAIN MOTOR

CHIP PAN

FEED SELECTOR LEVERS

HYDRAULIC PUMP AND MOTOR DRIVE

FEED REVERSE LEVER

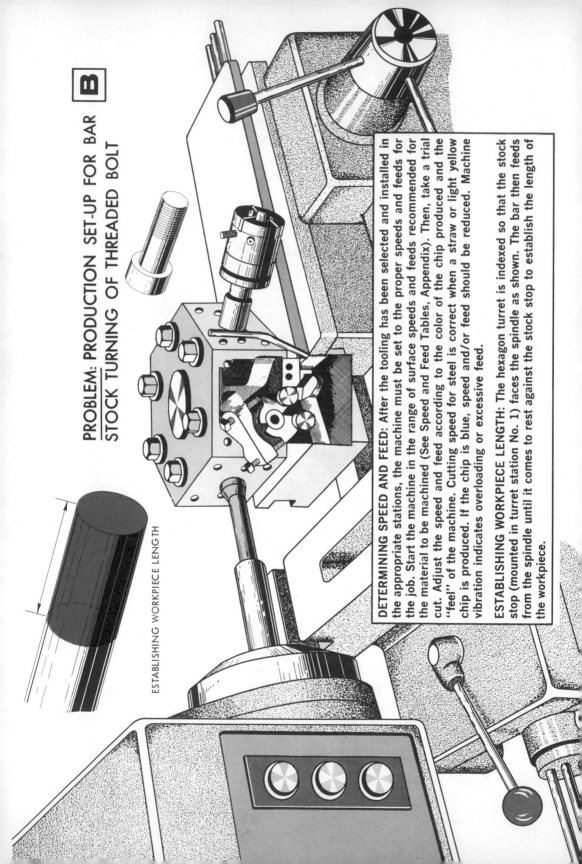

PROBLEM: PRODUCTION SET-UP FOR BAR STOCK TURNING OF THREADED BOLT

ESTABLISHING WORKPIECE LENGTH

DETERMINING SPEED AND FEED: After the tooling has been selected and installed in the appropriate stations, the machine must be set to the proper speeds and feeds for the job. Start the machine in the range of surface speeds and feeds recommended for the material to be machined (See Speed and Feed Tables, Appendix). Then, take a trial cut. Adjust the speed and feed according to the color of the chip produced and the "feel" of the machine. Cutting speed for steel is correct when a straw or light yellow chip is produced. If the chip is blue, speed and/or feed should be reduced. Machine vibration indicates overloading or excessive feed.

ESTABLISHING WORKPIECE LENGTH: The hexagon turret is indexed so that the stock stop (mounted in turret station No. 1) faces the spindle as shown. The bar then feeds from the spindle until it comes to rest against the stock stop to establish the length of the workpiece.

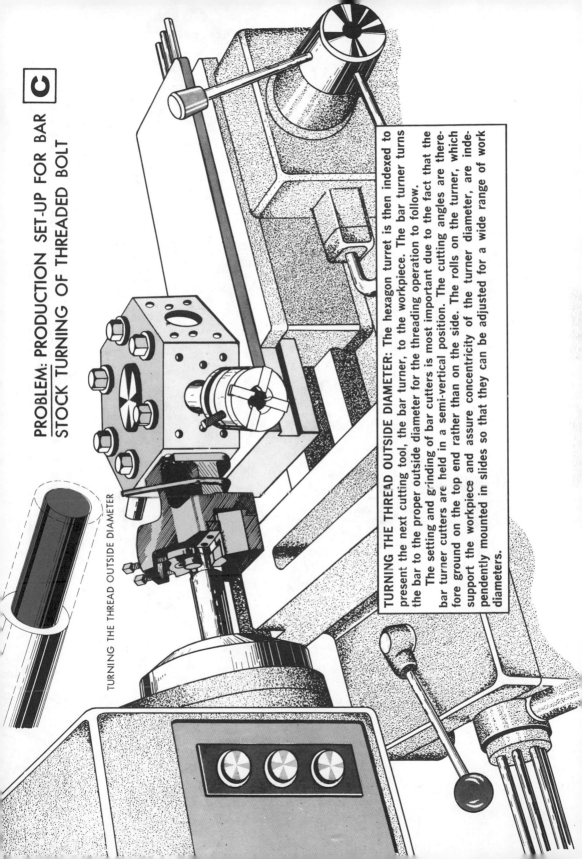

PROBLEM: PRODUCTION SET-UP FOR BAR STOCK TURNING OF THREADED BOLT

TURNING THE THREAD OUTSIDE DIAMETER

TURNING THE THREAD OUTSIDE DIAMETER: The hexagon turret is then indexed to present the next cutting tool, the bar turner, to the workpiece. The bar turner turns the bar to the proper outside diameter for the threading operation to follow.

The setting and grinding of bar cutters is most important due to the fact that the bar turner cutters are held in a semi-vertical position. The cutting angles are therefore ground on the top end rather than on the side. The rolls on the turner, which support the workpiece and assure concentricity of the turner diameter, are independently mounted in slides so that they can be adjusted for a wide range of work diameters.

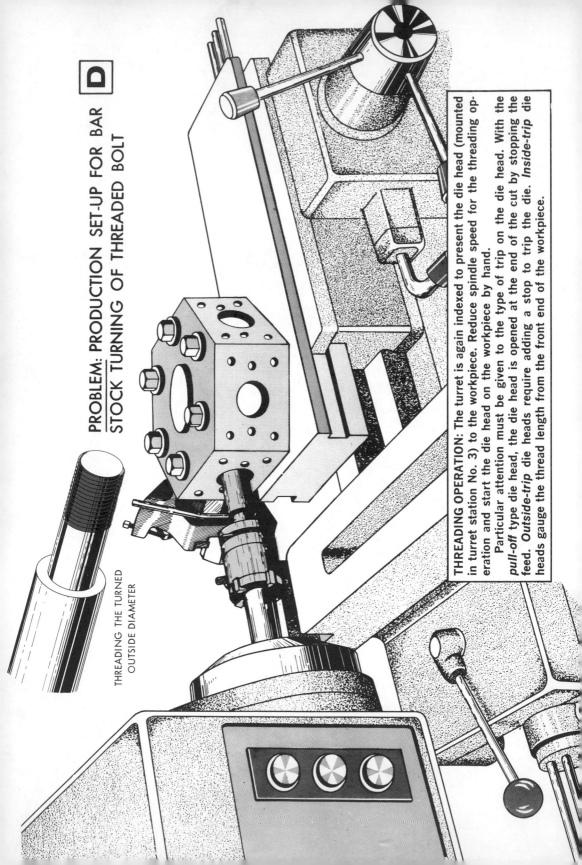

PROBLEM: PRODUCTION SET-UP FOR BAR STOCK TURNING OF THREADED BOLT

D

THREADING THE TURNED OUTSIDE DIAMETER

THREADING OPERATION: The turret is again indexed to present the die head (mounted in turret station No. 3) to the workpiece. Reduce spindle speed for the threading operation and start the die head on the workpiece by hand.

Particular attention must be given to the type of trip on the die head. With the *pull-off* type die head, the die head is opened at the end of the cut by stopping the feed. *Outside-trip* die heads require adding a stop to trip the die. *Inside-trip* die heads gauge the thread length from the front end of the workpiece.

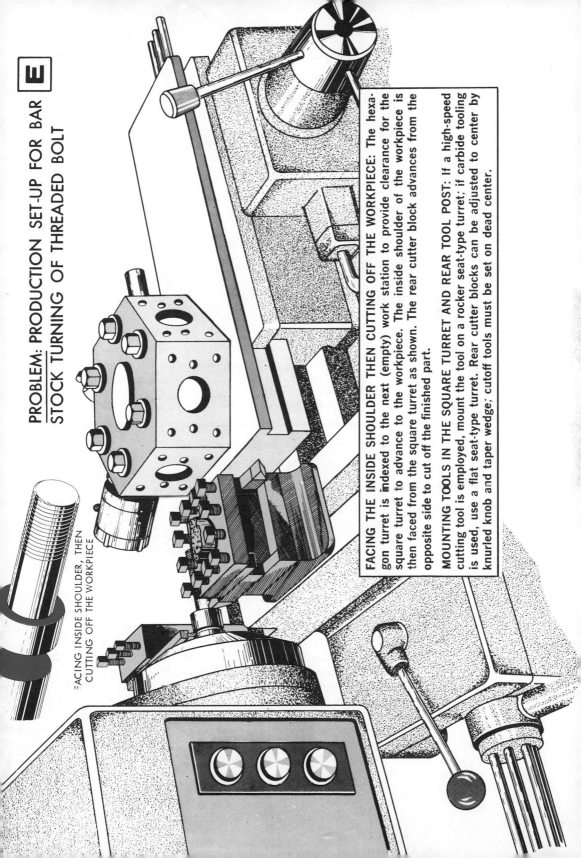

PROBLEM: PRODUCTION SET-UP FOR BAR STOCK TURNING OF THREADED BOLT

E

FACING INSIDE SHOULDER, THEN CUTTING OFF THE WORKPIECE

FACING THE INSIDE SHOULDER THEN CUTTING OFF THE WORKPIECE: The hexagon turret is indexed to the next (empty) work station to provide clearance for the square turret to advance to the workpiece. The inside shoulder of the workpiece is then faced from the square turret as shown. The rear cutter block advances from the opposite side to cut off the finished part.

MOUNTING TOOLS IN THE SQUARE TURRET AND REAR TOOL POST: If a high-speed cutting tool is employed, mount the tool on a rocker seat-type turret; if carbide tooling is used, use a flat seat-type turret. Rear cutter blocks can be adjusted to center by knurled knob and taper wedge; cutoff tools must be set on dead center.

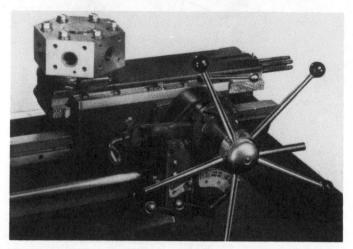

THE HEXAGON TURRET. Internal machining operations are usually performed with tools mounted in the hexagon turret. The hexagon turret moves back and forth longitudinally to pre-set as many as six cutting tools in sequence for the machining operations selected. On ram type turret lathes the hexagon turret is mounted on a ram slide which moves back and forth in a stationary saddle. See photo.

STOP SCREWS

SETTING THE HEXAGON TURRET STOPS. On ram type turret lathes the ram travel, or the distance through which the ram moves to present each of the cutting tools mounted in the hexagon turret to the workpiece, is controlled by six turret stop screws (photo) mounted in a stop roll at the end of the turret slide. The stop roll rotates automatically with the turret so that at all times the screw in the lowest position of the stop roll controls the travel of the face of the turret that is in the working position.

To set a hexagon turret stop, first make a trial cut to the desired length. Then stop the spindle, engage the feed lever, and clamp the turret slide. Then thread the stop screw in until the feed lever disengages, a signal that you are nearing the dead stop position. Finally, continue to turn the screw until it contacts the dead stop. The stop is now adjusted to disengage the tool after the desired length of cut has been made. Repeat the procedure for each tool mounted in the hexagon turret.

THE SQUARE TURRET. External machining operations such as turning, facing and forming are frequently performed from the square turret. The square turret is mounted on a cross slide which moves back and forth across the bed to present as many as four cutting tools in sequence to the workpiece. Major elements of the cross slide unit, identified by number in the illustration alongside, are (1) the cross slide, (2) the square turret, (3) the apron and (4) the gearbox.

SETTING SQUARE TURRET CROSS FEED STOPS. The square turret cross slide has feed trips or positive stops which control the cross feed of tools mounted in the square turret. Stops are set for the power cross feed by positioning the lugs along the slotted opening to the desired travel distance, then tightening them down (see photo). To accurately size a workpiece after the power feed has disengaged, the cross slide is fed by hand a few thousandths to the final reading marked by an adjustable clip on the feed dial.

SETTING SQUARE TURRET LONGITUDINAL FEED STOPS. Longitudinal travel of the universal cross slide carriage is controlled by stop screws, similar to those which control hexagon turret travel described above, and a stop rod. See illustration alongside. The stop screws are carried in a six-station stop roll carried near the base of the carriage. The notched, adjustable stop rod extends from the headstock bracket. The stop rod has a spring lockbolt plunger for endwise location. This gives the operator a quick, positive means of positioning the rod and keeps the overhang of the stop screws at a minimum. In addition to the individual adjustments provided by the six stop screws, the master adjusting screw located in the end of the stop rod permits changing the entire setup without changing the individual stop screw settings.

To accurately locate the cross slide along the bed, bring the carriage stop roll

firmly against the stop rod using the hand wheel. Then release the hand wheel—which allows the carriage to assume its natural position—and clamp the cross slide carriage to the bed.

Production Turning

Turret Lathes and Automatic Screw Machines

Turret Lathes

The turret lathe is a metal turning machine used to produce a number of identical parts in succession. The heart of this machine tool is its two multiple-tool holders, the hexagon turret and the square turret. Various cutting tools to perform different operations are mounted in the faces, or *stations*, of these turrets. Then, by rotating or *indexing* the turrets, and moving them back and forth, you can bring the various cutting tools against a rotating workpiece in a predetermined sequence. And by repeating this sequence over and over, you can produce any number of identical pieces.

Because workpieces vary widely in size and shape, turret lathes, too, vary in type and size. There are hand-operated turret lathes, auto-matic turret lathes, vertical and horizontal types, and others.

This chapter deals primarily with hand-operated horizontal turret lathes. Fig. 1 illustrates the basic elements common to machines of this type. The workpiece is held in a workholding device attached to the end of the machine's spindle. Power to turn the spindle comes from the main motor through the gears and speed selecting clutches in the headstock. The cutting tools required to perform the necessary operations on the workpiece are inserted and clamped in the turret lathe's two indexing tool turrets, the hexagon end-working turret and the square turret. After each cut, the turret is indexed to bring the next required cutting tool into working position.

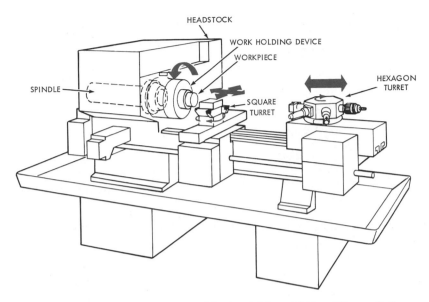

Fig. 1. Basic elements of the hand-operated horizontal turret lathe. (Warner & Swasey Co.)

Type of Parts Machined on Turret Lathes

Basically, all parts produced on turret lathes can be classified as one of two types, *bar parts* or *chucking parts*.

Bar Parts. These workpieces are produced from a length of solid bar or tube stock. The stock passes through the hollow of the turret lathe spindle and is gripped in a collet chuck mounted on the end of the spindle, Fig. 2. A portion of the bar — usually just enough for one workpiece — extends through the collet. As this portion is machined and then cut off, the collet is opened and enough bar for another workpiece is fed forward. Although most bar parts are machined from round

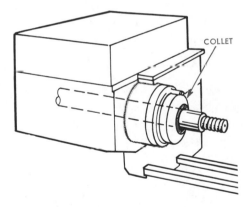

Fig. 2. On the bar machine, bar stock is fed through the hollow spindle and gripped in the collet chuck. (Warner & Swasey Co.)

bar stock, bar parts can also be made from square, hexagonal, or even specially-shaped extruded stock.

Chucking Parts. Chucking parts

369

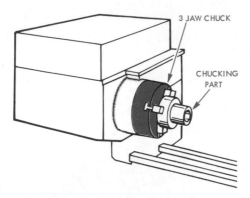

Fig. 3. Chucking part is shown held in a standard 3-jaw chuck. (Warner & Swasey Co.)

are workpieces turned from individual pieces of metal. These workpieces may start out as rough castings or forgings or as pre-cut lengths of bar stock called "slugs."

Chucking parts are gripped in one of several types of standard or specially designed workholding fixtures. The 3-jaw chuck, Fig. 3, is the most commonly used workholding device for chucking parts.

Bar and chucking parts vary in general size and shape. Most turret

lathes are equipped with tooling and workholding devices for only one of these two types of work. A turret lathe is thus referred to as either a bar machine or a chucking machine depending on how it is equipped.

Types of Horizontal Turret Lathes

Two distinctly different types of horizontal turret lathes are required to efficiently produce today's wide variety of workpieces: *Ram Type* turret lathes and *Saddle Type* turret lathes.

Ram Type Turret Lathes. Ram Type turret lathes of the type illustrated on Page A of the special introduction to this chapter are fast, easy handling machines designed for small, compact workpieces. Generally, Ram Type machines can produce bar parts up to 3″ dia. or chucking work up to approximately 20″ dia.

The hexagon end-working turret on Ram-Type machines mounts on

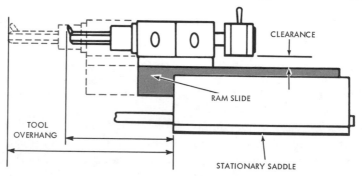

Fig. 4. On the Ram Type turret lathe, the ram slide moves along a stationary saddle. (Warner & Swasey Co.)

a ram slide which moves longitudinally in the saddle. When a job is set up, the saddle is clamped in a convenient working position along the bedways. The saddle remains fixed during machining; only the ram slide moves, as shown in Fig. 4. The length of the ram slide stroke governs the length of cut of the cutting tool mounted on the hexagon turret, and varies with the size of the machine from 4″ to 13″.

The hexagon turret of the Ram Type machine is designed for fast handling. It indexes automatically as the ram slide is moved away from the work at the end of a cutting stroke.

The square turret of the Ram Type machine travels on a cross slide at a right angle to the bedways. Used mainly for external machining of the workpiece, as many as four tools can be mounted in the square turret. In addition to the square turret, a cutter-holding tool post is located at the rear of the cross slide. Procedures for controlling the movement of both the hexagon and square turrets of the Ram Type turret lathe are explained in the special introduction to this chapter.

Saddle Type Turret Lathes. Saddle Type machines, Fig. 5, are basically larger machines suited to heavy duty jobs—particularly those jobs requiring long, accurate cuts. Standard Saddle Type turret lathes can handle bar work up to 12″ dia

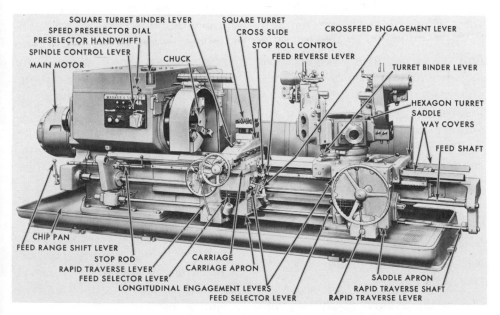

Fig. 5. Basic elements of the Saddle Type turret lathe. (Warner & Swasey Co.)

and chucking jobs up to about 36″ dia.

Aside from size, the major difference between the Ram and Saddle Type machines is in the construc-

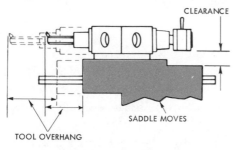

Fig. 6. On the Saddle Type machine, the turret and saddle move as a unit. (Warner & Swasey Co.)

tion of the saddle which carries the end-working hexagon turret. On Saddle Type machines, the turret mounts directly on the saddle instead of on a ram slide and, when taking cuts, the entire saddle moves as a unit along the bedways as shown in Fig. 6.

Mounting the turret directly on the saddle provides maximum rigidity regardless of the length of cut. Lengths of cuts taken from the hexagon turret on Saddle Type machines are therefore limited only by the length of the machine's bedways.

Turret Lathe Machining Operations

Nearly all turret lathe jobs involve a series of both external and internal machining operations. Before you can set up a job, you must first determine the types of cuts required—which requires that you be familiar with various machining operations—and the sequence in which they must be performed. This is not as difficult as it sounds however since even the most difficult workpieces are produced by combinations of the simple basic cuts to be described in this section.

Basic External Machining Operations

Basically, external machining performed on the horizontal turret lathe involves a sequence of the unit operations described in *Chapter 7, Engine Lathe Setups and Operations.* The workpiece shown in Fig. 7, for example, requires four basic external machining operations to generate the required shape: turning, taper turning, threading and grooving. These and other basic turret lathe external machining operations are described in the following pages.

Overhead Turning. Turning cuts can be made from the hexagon turret with an overhead turning setup, Fig. 8, which assures rigid tool support for heavy metal removal. Turning cutter holders can be mounted in the Multiple Turning Head shown, which lets you mount

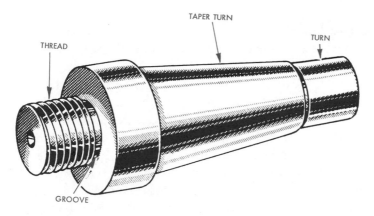

Fig. 7. Typical turret lathe external machining operations. (Warner & Swasey Co.)

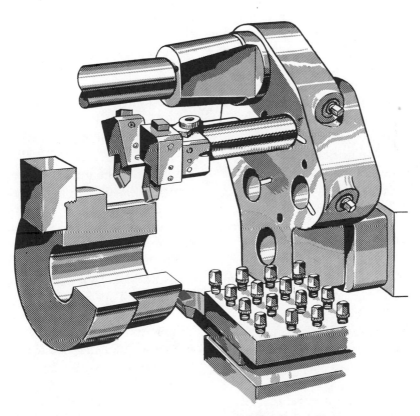

Fig. 8. Overhead turning from the hexagon turret while side turning from the square turret. (Warner & Swasey Co.)

several cutter holders in the same turret station, or in a Single Turning Head.

Use Reversible Plain Cutter Holders for rough turning cuts. Reversible Adjustable Cutter Holders are preferred for finishing cuts because they have a built-in micrometer adjustment which makes it easy to set cutters to size.

Side Turning. Turning cuts are also commonly taken from the square turret. When combining machining operations, turning cuts from the square turret, Fig. 8, can be performed at the same time you are drilling, boring—or even turning —from the hexagon turret.

To duplicate sizes accurately, always make final adjustments by advancing the square turret toward the workpiece to remove all backlash between the cross slide screw and nut. And always index the turret in the same direction.

Bar Turning. When turning a long bar from the hexagon turret a Bar Turner should be used, as shown in Fig. 9. The Bar Turner has rollers which center and support the workpiece, permitting heavier cuts and higher feed rates.

For a smooth surface finish, set the Bar Turner rolls *behind* the tip of the turning cutter to burnish the turned surface as shown in Fig. 9. When turning concentric diameters, mount the rolls *ahead of* the turning cutter on the previously turned diameter, as shown in Fig. 10.

Bar Turners have a built-in cutter relieving feature. At the end of a turning cut, use the Cutter Relieving Lever, Figs. 9 and 10, to move the cutter away from the workpiece. With the cutter retracted, you can back the Bar Turner off the workpiece without leaving cutter withdrawal marks on the finished surface. Push the lever forward to reset the cutter for the next workpiece.

Taper Turning. A Taper Attachment makes it possible to machine external or internal tapers from the square turret on Ram Type or fixed-center Saddle Type turret lathes. Taper Attachments are available for cross sliding hexagon turret machines for either the cross slide or the cross sliding turret.

The Taper Attachment's guide plate, set to the desired taper angle, controls the travel of the cutter to generate the taper on the workpiece.

Combined Feed Taper Turning. Tapered surfaces can be roughed out by combining square turret cross feeds and hexagon turret longitudinal feeds. Taper angles are varied by increasing or decreasing the cross or longitudinal feeds.

Facing. Facing produces flat surfaces at right angles to turned diameters. Most facing cuts are taken with single point cutters mounted in the square turret, Fig. 11, and are often made simultaneously with

Fig. 9. Bar turning from the hexagon turret. The cutter relieving lever disengages the tool at the end of the cut. (Warner & Swasey Co.)

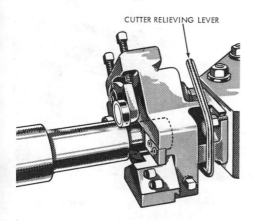

Fig. 10. To assure concentric diameters, set the rolls of the bar turner ahead of the cutting tool on the larger diameter. (Warner & Swasey Co.)

Fig. 11. Facing the workpiece from the square turret. (Warner & Swasey Co.)

drilling or boring cuts from the hexagon turret. Workpiece length is established by positioning the cross slide carriage along the bedways. Make sure the cross slide carriage is firmly clamped to the bedways during the facing cuts.

When square turret stations are at a premium, short facing cuts can be taken from the hexagon turret by using a Quick Acting Slide Tool. The cutter is fed at right angles to the spindle with the Slide Tool's hand operating lever.

On cross sliding turret machines, facing from either turret is commonplace. For cuts from the hexagon turret, hold the cutter in a 3-Slot Cutter Holder, Boring Bar, or in a Tool Base Cutter Block.

End Facing. Face the ends of longer bar parts with an End Former or with a Combination Turner and End Former. Both of these tools have rollers similar to those of the Bar Turner shown in Fig. 9, which support the workpiece and keep it from springing away from the cut-

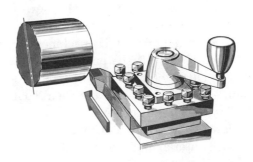

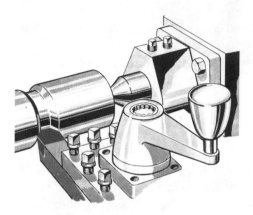

Fig. 12. Forming cut taken from the square turret. (Warner & Swasey Co.)

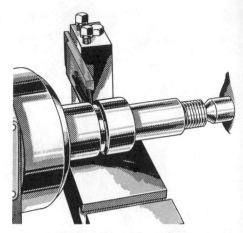

Fig. 13. The cutoff tool is usually mounted on the rear tool post opposite the square turret. (Warner & Swasey Co.)

ter. Be sure to set the rolls ahead of the cutter and feed the tool by hand. To assure a smooth finish, use slow cutting speeds.

Forming and Cutoff. Forming is a fast method of producing finished diameters and shapes. Most single chamfering, necking and grooving cuts are taken from the square turret, Fig. 12. The square turret holds four tools and, if needed, another cutter can be mounted on the Rear Tool Post opposite the square turret on the cross slide, Fig. 13.

Although not too common, you can also groove or chamfer from the hexagon turret, using the setup shown in Fig. 14.

After a bar part is machined, it must be cut off the end of the bar. The cutter blade is usually mounted in a Cutoff Holder on the Rear Tool Post, with the cutting edge upside down, Fig. 13. A Cutoff Holder with interchangeable carbide cutoff

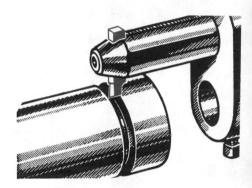

Fig. 14. Grooving from the hexagon turret. (Warner & Swasey Co.)

blades gives maximum rigidity for fast, smooth cutoff. Cutoff feed rates are very important when using carbide tools.

Threading. Although the choice of threading methods depends on the size, type and accuracy of the thread, small-diameter, standard-size external threads are generally cut with Self-Opening Die Heads, Fig. 15. The Die Heads spring open

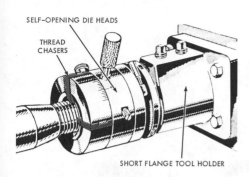

Fig. 15. Threading from the hexagon turret using an automatic-releasing die head (Warner & Swasey Co.)

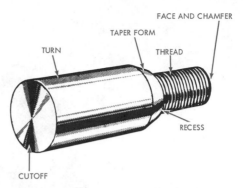

Fig. 16. The basic external machining operations required to generate the threaded shaft are identified (Warner & Swasey Co.)

and release automatically when the desired thread length has been machined.

When threading, you must start the Die Head on the work by hand and provide an even pressure on the turnstile to keep the hexagon turret moving with the Die Head.

Leading a Die Head by hand is satisfactory for most threads, however long or accurate threads require a more positive means for feeding the Die Head over the work. On Ram Type machines use a hexagon turret Leading-On Attachment. Or, if your machine is equipped with a cross slide Thread Chasing Attachment, you can provide positive lead by linking the cross slide to the hexagon turret

External Cuts Required to Machine a Threaded Shaft

Although not always as critical as the sequence of internal cuts, some external cuts must also be made in a set sequence. For example, before

you can thread you must first prepare the thread diameter by turning it to size. Although other external cuts may not be dependent on a preceding operation, nevertheless a small amount of time spent planning the best sequence of cuts can save time and effort on each piece you produce.

Let's look now at a typical external machining job. The threaded shaft shown in Fig. 16 requires a number of typical external cuts. The turret lathe setup required to machine the shaft is illustrated in Fig. 17. Here is the sequence of cuts required to machine the threaded shaft:

1. Feed out the stock to proper length against the Revolving Stock Stop mounted in Station 1.
2. Index the turret to the next station and turn thread diameter D using the Bar Turner mounted in Station 2 of the hexagon turret.

377

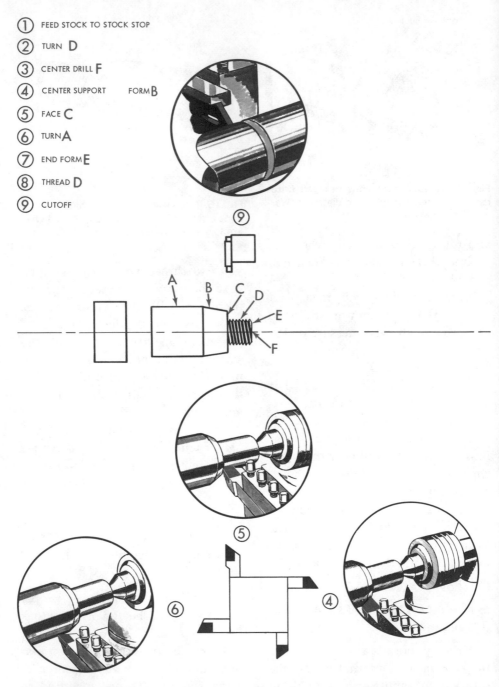

① FEED STOCK TO STOCK STOP

② TURN D

③ CENTER DRILL F

④ CENTER SUPPORT   FORM B

⑤ FACE C

⑥ TURN A

⑦ END FORM E

⑧ THREAD D

⑨ CUTOFF

Fig. 17. Turret lathe setup and sequence of operations required to produce the threaded shaft shown in Fig. 16. (Warner & Swasey Co.)

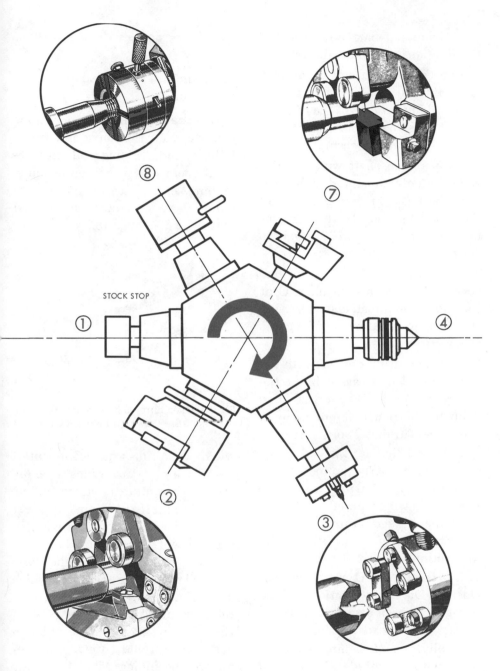

STOCK STOP

The chamfered bar end helps give the Bar Turner a gradual start.

3. Center drill the end of the bar (F) with the Center Drilling Tool mounted in Station 3 of the hexagon turret. This tool's rolls support the work and assure an accurate center.

4. Support the shaft with the Revolving Center mounted in Station 4 of the hexagon turret, and form the tapered section B with a Forming Tool mounted in the square turret.

5. Without withdrawing the Revolving Center, face shoulder C at the end of the thread diameter using the Facing Tool mounted in the square turret.

6. Still supporting the part with the Revolving Center, again index the square turret and turn the large diameter A from the square turret using the Turning Tool shown.

7. Face and chamfer the end of the shaft E with an End Former, or the Combination Turner and End Former shown mounted in Station 5 of the hexagon turret. Both of these tools have rolls which hold the work firmly against the cutter.

8. Cut the thread D using the Self-Opening Die Head (Station 6 of the hexagon turret) arranged with the proper size thread cutting chasers.

9. Finally, cut off the finished workpiece with a carbide Cutoff Tool mounted in the Rear Tool Post. Most cutoff tools simultaneously

chamfer or de-burr the bar stock as well.

Basic Internal Machining Operations

Internal machining operations performed on the horizontal turret lathe include drilling and such related operations as boring, reaming, tapping, etc. These operations are required to generate the finished part shown in Fig. 18.

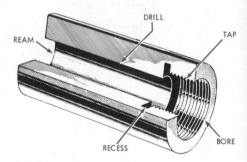

Fig. 18. Basic turret lathe internal machining operations are shown. (Warner & Swasey Co.)

The order, or sequence, of internal machining operations is more critical than for external machining. Thus, in Fig. 18, the main bore must be drilled before it can be reamed; the concentric ID must be bored before it is tapped; etc. The basic internal machining operations which are performed on the horizontal turret lathe are therefore described in the following pages in the order in which they would normally be performed, beginning with the family of drilling operations.

Start Drilling. The accuracy of

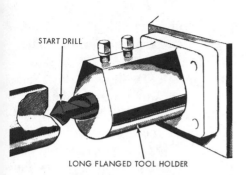

START DRILL

LONG FLANGED TOOL HOLDER

Fig. 19. Start-drilling from the hexagon turret. (Warner & Swasey Co.)

a drilled hole depends upon its start. Rough and uneven surfaces on bar stock, castings or forgings may cause longer drills to weave at the beginning of a drilling operation. But using a short, rigid Start Drill, Fig. 19, to spot a true cone in the workpiece will prevent the twist drill from leading off center.

When hexagon turret stations are at a premium, you can use a Combination Stock Stop and Start Drill to simultaneously position the bar stock to length and start drill.

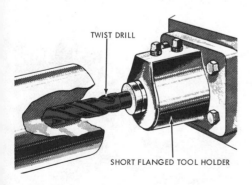

TWIST DRILL

SHORT FLANGED TOOL HOLDER

Fig. 20. Twist drilling from the hexagon turret. (Warner & Swasey Co.)

Twist Drilling. Twist Drills, Fig. 20, are essentially end cutting tools having one or more cutting edges. For efficient cutting, keep the drill sharp and correctly ground for the material you are machining.

It is good practice to use two drills on extremely large diameter drilling work: a smaller drill to pierce the hole, and a larger diameter drill to enlarge the hole to desired size. Twist drills are available with either straight or tapered shanks.

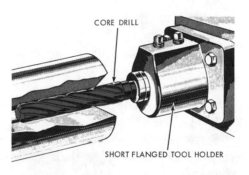

CORE DRILL

SHORT FLANGED TOOL HOLDER

Fig. 21. Core drilling from the hexagon turret. (Warner & Swasey Co.)

Core Drilling. Core drills, Fig. 21, are 3- and 4-fluted cutters used to enlarge previously drilled, cored or pierced holes. Since cored or pierced holes seldom run true, it is advisable to provide an accurate guide for the core drill by chamfering the work first with a start drill or a chamfering cutter mounted in a boring bar.

When core drilling long holes, use a short boring cut to provide a true

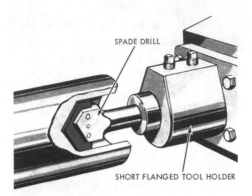

SPADE DRILL

SHORT FLANGED TOOL HOLDER

Fig. 22. Large diameter holes are drilled using a spade drill mounted in the hexagon turret. (Warner & Swasey Co.)

start for the core drill. Bore the hole ⅜″ to ½″ deep and to the same diameter as the core drill.

Spade Drilling. A spade drill, Fig. 22, is a flat cutter firmly clamped in a bar-type holder. Like twist drills, spade drills have two cutting edges.

Spade drills are commonly used to produce holes larger than 2″ in diameter. Twist drills of this size are generally too long and overhang the turret excessively, whereas the

spade drill, with its short shank, can be snugged-up for minimum overhang.

Deep Hole Drilling. Any hole longer than four times its diameter is considered a deep hole. Chip removal is a most important factor when drilling holes of this type. Many shops use oil-hole drills for deep drilling, Fig. 23, which supply coolant under pressure to the point of the drill, washing out chips and keeping the cutting edges cool.

To avoid drill breakage when drilling deep holes with smaller-diameter drills, withdraw the drill frequently to allow the chips to escape.

Center Drilling. Center drilling provides a means of supporting the workpiece during later operations on the turret lathe or on other machines. For accurate centers and longer center drill life, use a Center Drilling Tool with self-centering rollers which support the end of the part, Fig. 24.

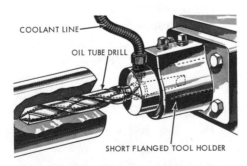

COOLANT LINE

OIL TUBE DRILL

SHORT FLANGED TOOL HOLDER

Fig. 23. Drills with coolant passages are recommended for extra-deep holes. (Warner & Swasey Co.)

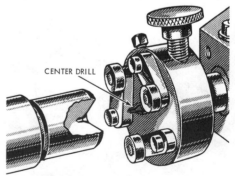

CENTER DRILL

Fig. 24. Hexagon turret setup for center drilling. (Warner & Swasey Co.)

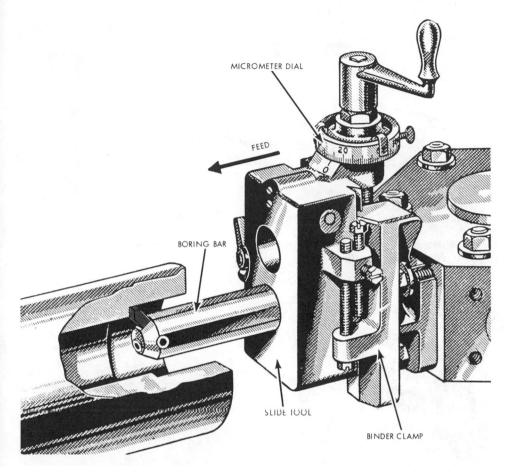

MICROMETER DIAL

FEED

BORING BAR

SLIDE TOOL

BINDER CLAMP

Fig. 25. Hexagon turret setup for boring. The slide tool has a micrometer adjustment for setting depth of cut. (Warner & Swasey Co.)

Boring. Boring cuts true up holes and produce accurate sizes. A boring bar held in a Slide Tool mounted on the hexagon turret provides the rigidity required for heavy, accurate boring cuts, Fig. 25. You can adjust to the desired bore size quickly and accurately using the Slide Tool's micrometer dial. Once the cutter is set, lock the Slide with the binder clamp.

Reaming. Reamers, Fig. 26, are finishing cutters designed to remove .003″ to .015″ of metal from a previously drilled or bored hole. Fluted reamers, found in most shops, can be either solid or expandable. *Solid* reamers are ground to one hole size and cannot be adjusted. *Expandable* reamers have blades which can be adjusted to compensate for wear or re-grinding.

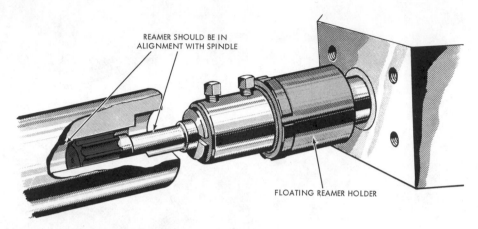

Fig. 26. Hexagon turret setup for reaming. A floating reamer holder is used with a solid reamer. (Warner & Swasey Co.)

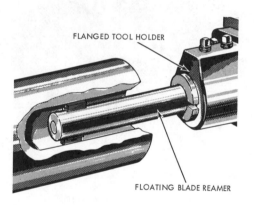

Fig. 27. Hexagon turret setup for reaming using a floating blade reamer. (Warner & Swasey Co.)

To produce straight, properly-sized holes, mount the reamer in a Floating Reamer Holder. The Holder permits the reamer to align itself with the existing hole. Guide the reamer by hand into the drilled or bored hole. Use coarse feeds and slow speeds.

Adjustable Floating Blade Reamers, Fig. 27, are recommended for large holes. Mount these reamers solidly in an Adjustable Tool Holder as shown. The reamer's two cutting blades provide the floating action necessary for proper alignment. For smooth finish and accurate size, keep the reamer blades sharp. Use coolants when machining steel.

Recessing, Facing and Chamfering. Internal grinding reliefs, thread clearances, chamfers, etc. require a longitudinal tool slide motion to position the cutting tool at the proper point inside the workpiece, and either a cross or vertical feed stroke to make the actual cut.

On Ram Type or fixed-center Saddle Type machines, a Quick Acting Slide Tool mounted on the hexagon turret provides the vertical motion required for these internal cuts, as shown in Figs. 28, 29 and 30. A recessing cutter, ground to shape and held in a boring bar, gives maximum rigidity for heavier cuts. Depth of

cut is accurately determined by stop screws on the Slide Tool.

On cross sliding Saddle Type machines, mount the boring bar directly in the turret or in a Flanged Tool Holder and use the cross feeding motion of the turret to take the cut.

You can also recess, face, or chamfer internally using a cutter held in the square turret. Since rigidity is

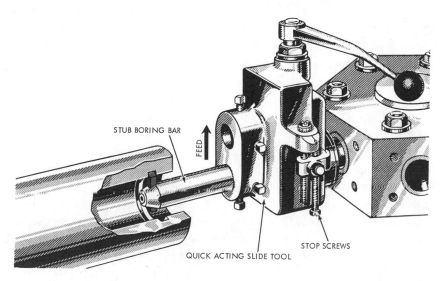

Fig. 28. Hexagon turret setup for internal grooving. Stop screws on the slide tool are set to the desired depth of cut. (Warner & Swasey Co.)

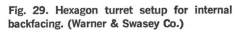

Fig. 29. Hexagon turret setup for internal backfacing. (Warner & Swasey Co.)

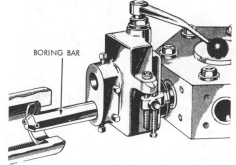

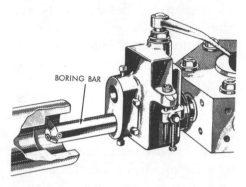

BORING BAR

Fig. 30. Hexagon turret setup for internal facing. (Warner & Swasey Co.)

limited by the cross section of the cutter and its overhang from the turret, lighter cuts using hand feed are recommended when performing these operations from the square turret.

Internal Threading. Small diameter internal threads are most commonly produced by using a solid tap held in a Releasing Tap Holder, Fig. 31. To tap a hole, you must advance the hexagon turret by hand and provide sufficient pressure to start the tap in the hole. Once the tap starts cutting, its own helix angle and the rotation of the workpiece combine to pull it into the

workpiece at the proper rate.

As the tap feeds into the workpiece, carefully apply pressure to the handwheel to keep the turret and tap holder moving at the same rate as the tap. Never force the tap faster than it normally moves into the work. And never allow the tap to drag and pull the turret along as it feeds in. Either practice—forcing or dragging the tap—results in a defective thread.

Set a turret stop to end the forward motion of the turret approximately $\frac{1}{8}''$ before the tap reaches the required length of thread, as explained in the special introduc-

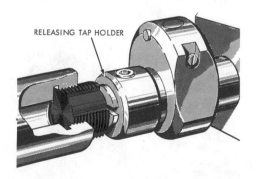

RELEASING TAP HOLDER

Fig. 31. Hexagon turret setup for internal threading using a releasing tap holder. (Warner & Swasey Co.)

tion to this chapter. When the forward progress of the turret and Tap Holder are stopped, the Tap Holder's automatic release will allow the tap to revolve with the workpiece.

To withdraw the tap, reverse the direction of spindle rotation. As the tap follows the thread back out of the hole, you must again keep the turret and Tap Holder moving at the same rate as the tap.

Holes of 1½″ dia and larger often are threaded using collapsing taps. These taps are designed so that the thread cutting edges pull away from the workpiece at the end of the forward tapping stroke. It is not necessary to reverse the spindle to withdraw a collapsing tap from the threaded hole. Collapsing taps can be held directly in the turret or in a Flanged Tool Holder.

Internal Cuts Required To Machine a Threaded Adapter

Internal cuts must be set up in the proper sequence; the performance of each successive machining operation depends upon the form and accuracy produced by the preceding cut.

For example, to generate the threaded adapter shown in Fig. 18 requires start drilling, drilling, boring, reaming, internal recessing and tapping—in that order. Fig. 32 illustrates the proper turret lathe set-up for the sequence of internal cuts

required to produce the part, as follows:

1. Start-drill the bar at A to provide a true, clean start for the twist drill in the next station. Use either a Start Drill or, if part length is critical, a Combination Stock Stop and Start Drill as the tooling for Station 1.

2. Index the turret to the next station (No. 2) and drill the hole B through the part. Use a Drill Socket to adapt the drill to a Flanged Tool Holder.

3. Index the turret to Station 3 and single-point-bore the previously drilled hole B to the thread diameter of proper size. For quick adjustment to size, mount the Boring Bar in a Slide Tool.

4. Index the turret to Station 4 and ream just-bored hole B. To assure proper alignment with the hole, a fluted reamer, mounted in a Floating Reamer Holder, should be used.

5. Index the turret to Station 5 and recess the thread clearance groove C. For this operation use a boring bar with a recessing cutter mounted in a Quick Acting Slide Tool.

6. Index the hexagon turret to Station 6 and thread the reamed hole B using a tap held in a Releasing Tap Holder. If this were an odd size thread it would be necessary to single-point the thread from the square turret or hexagon turret.

① START DRILL A
② DRILL B
③ BORE B
④ REAM B
⑤ RECESS C
⑥ TAP B

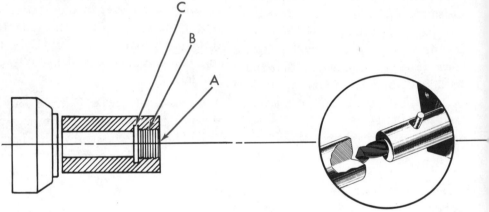

Fig. 32. Turret lathe setup required to produce the threaded adapter shown in Fig. 18. (Warner & Swasey Co.)

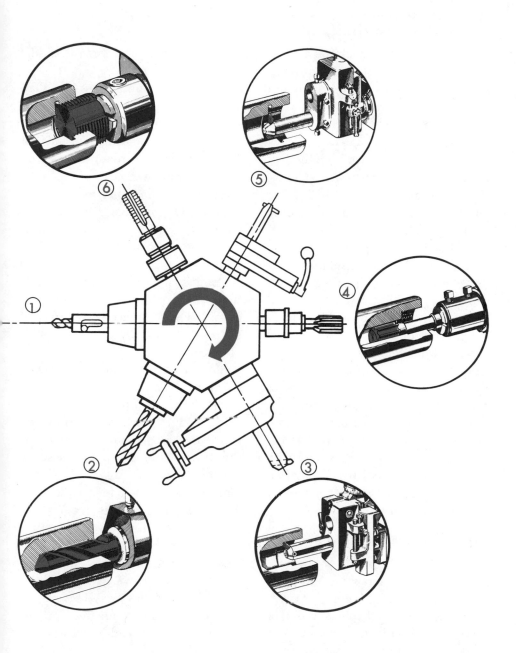

Automatic Screw Machines

Automatic screw machines are automatic turret lathes capable of mass producing precision turned and formed parts of endless variety. Reproductive accuracy is excellent, tolerances can be close, and the low per-unit cost makes this method of manufacture ideal for producing machined parts in large quantities. The automatic screw machine is the only standard automated machine designed to produce special parts.

Automatic screw machines may have one, four, five, six or eight work spindles. They can handle material as small as $\frac{1}{64}''$ dia or as large as $2''$ dia, but all of these machines have the following common features: (1) The workpiece material is in rod, bar or tube form; (2) production of consecutive pieces is fully automatic; (3) operations are similar to those performed on lathes; (4) automation of the machine is achieved by the use of cams; and (5) production rates are greater than for manually operated machines.

Types of Automatic Screw Machines

All automatic screw machines fall into one of three general classes: single-spindle, multiple-spindle or Swiss-type machines. Generally, but subject to many exceptions, parts $\frac{1}{8}''$ dia and smaller with complex contours are the province of Swiss-type machines. The broad middle range of parts from $\frac{1}{8}''$ to $1''$ dia, particularly for quantities less than 5000 pieces, are usually considered single-spindle work. Parts larger than $1''$ dia, in quantities of 5000 or more, are most often produced on multiple-spindle machines.

These distinctions are general guides only. Some parts can be made most efficiently on single-spindle machines regardless of quantity; others are most efficiently produced on multiple-spindle machines. For those parts which can be produced on either machine, the quantity of the run usually dictates the choice of machine.

Single-Spindle Automatic Screw Machine. This is a fully automated turret lathe, equipped with a bar-stock feeding mechanism. An example of the single-spindle machine is the Brown & Sharpe automatic screw machine shown in Fig. 33. The work material is gripped in a collet and rotated while the tools, mounted in a turret which revolves in a vertical plane, advance in sequence to machine the workpiece.

Cross slides move at right angles to the axis of the work material to machine outside diameters of the bar, while the turret tools perform end work (interior machining). One cross slide is always used to hold a cutoff tool, which separates the

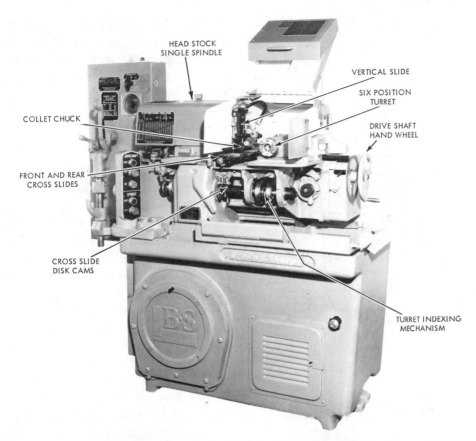

Fig. 33. Principal elements of the single-spindle automatic turret lathe. (Brown & Sharpe Mfg. Co.)

finished workpiece from the bar. Following cutoff, the work bar is automatically advanced by cam-actuated fingers and the machining cycle repeats.

Generally, parts made on single-spindle automatic screw machines are of the type shown in Fig. 34. The major diameter is smaller than 1″, the part requires profiling operations on its OD, and internal work (holes, counterbores, etc.) is such that the largest ID is outboard, i.e., opposite the machine's collet.

The part shown in Fig. 34 is also tapped and its OD is turned and knurled. Still other machining operations are possible. For example, a slot or a straddle milled area could be produced, or a recess could be cut in the hole at any depth. In addition, the burr left at the rear of the hole by the cutoff tool could be removed by unloading the part

$\frac{7}{32}$ DRILL THRU

$\frac{11}{32}$ C'DRILL $-\frac{3}{4}$ DEEP

$\frac{3}{8}$ -32 TAP $-\frac{11}{16}$ DEEP

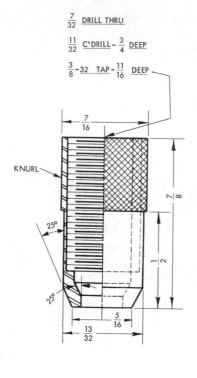

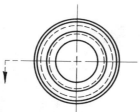

Fig. 34. Typical part produced on the single-spindle automatic.

with a transfer arm and presenting it to a rotating deburring tool while the next piece is being made on the machine.

A part like the one shown in Fig. 34 qualifies as single-spindle work because no more than 10 tools are needed for its machining. While the same part could be produced on

multiple-spindle equipment, these machines are usually used for parts which require 12 to 20 tools. However, if relatively simple parts are to be made in lots of 20,000 or more, the shorter cycle time of a multiple-spindle machine would offset the cost of its longer setup time and slightly higher tool charges.

Operating Features of the Single-Spindle Machine. A bar of stock or a workpiece is held firmly in the spindle by means of a spring collet or other suitable chucking device. Cutting operations are performed on the workpiece by cutting tools mounted in the vertical turret, on the cross slides and vertical slide and, on machines so equipped, on upper front and rear slides.

The various tool slides are moved to and from the work by cams and springs. The cross slide disc cams, shown in Fig. 33, which move the tool slides, are mounted on cam shafts driven from the constant speed driveshaft through change gears which provide different rates of production. These cams, custom made for each workpiece, control the sequence of operations and, in combination with the feed change gears selected, govern the feed rates for the cutting tools. The cams are of simple design, inexpensive and easy to make, yet offer a practical method for rapid setups, efficient production and accurate duplication of work. Return springs exert sufficient force to return the slides to the next starting positions and hold the lever rolls against the cams during the cutting operations.

The spring collets are closed by action of the chuck fork sleeve and chuck levers at the left end of the spindle. The bar of stock is advanced through the spindle and collet by means of a feeding finger affixed to the feeding tube. The feeding tube is moved forward, or backward, by the action of the feed slide.

Cross Slides. The two cross slides move horizontally at right angles to the spindle. Each of the slides operates independently. Cams advance

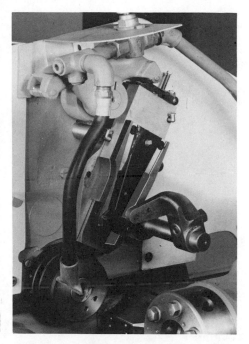

Fig. 35. In addition to cross slides, single-spindle automatics also have a vertical tool slide. (Brown & Sharpe Mfg. Co.)

the slides toward the work by means of levers, and springs return the slides when the cutting operations are finished. Each slide is adjustable along a rack. The cutting tools are held in tool holders mounted on top of the slides.

Vertical Slides. In addition to the two cross slides used to approach the work from the sides, some single spindle machines are equipped with a vertical slide with a downward motion for cutting, Fig. 35. The vertical slide, generally used for cutting-off, has a cutoff blade mounted directly in the tool holder which is an integral part of the unit. Cams for this slide are the same as those for operating the cross slides.

Multiple-Spindle Automatic Screw Machine. The multiple-spindle machine, Fig. 36, uses a transfer system to present a carousel of work bars to tooling mounted on slides which have only forward and reverse motion. Whether the multiple-spindle machine has four, five, six or eight bar stock spindles, the method of operation is the same. Bars are gripped in collets and fed forward as on the single-spindle machine.

Among the advantages of this type of screw machine is the overlap of operations, i.e., all tools are at work at the same time, and one finished part is cut-off for each indexing or transfer of a spindle to its next station.

Fig. 36. The multiple-spindle automatic turret lathe. (New Britain Machine Div., Litton Industries)

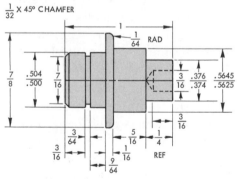

Fig. 37. Typical part produced on a multiple-spindle machine.

Even though its dimensions are within single-spindle capacity, the stud shown in Fig. 37 would generally be made on a multiple-spindle machine. The amount of work to be done on the OD and the close tolerances required necessitate both rough and finish turning. The tooling diagram for the stud, Fig. 38, suggests the use of five cross slides; the single-spindle machine, however, has only four at most. Another advantage of the multiple-spindle machine, therefore, is that operations can be broken down into a greater number of steps than possible on a single-spindle machine, reducing overall cycle time.

On parts like the stud shown in Fig. 37, considerably more internal

394

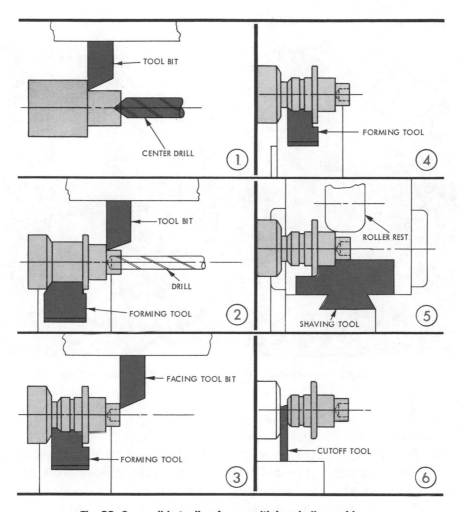

Fig. 38. Cross slide tooling for a multiple-spindle machine.

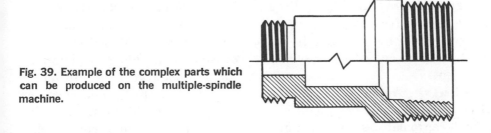

Fig. 39. Example of the complex parts which can be produced on the multiple-spindle machine.

Fig. 40. The Swiss-type automatic screw machine. (Bechler)

work can be done on a multiple spindle machine. Multi-diameter holes could be drilled, reamed, tapped and recessed, and the OD could be threaded on one or more diameters, including those behind a shoulder. Fig. 39 illustrates a fitting that requires some of these operations. The multiple-spindle machine has a standard provision for rolling threads. And if the rear of the part requires minor machining, this can be done by means of a pick-off attachment.

Swiss-Type Automatic Screw Machine. While single- and multiple-spindle automatic screw machines are of U.S. origin, the Swiss-type automatic screw machine, Fig. 40, was developed by the watch industry of Europe, to provide rapid production of miniature pinions, screws and studs. Widely used both in the U.S. and abroad, the Swiss-type has no exact counterpart in a domestic machine.

All Swiss-type machines have sliding-headstock mechanisms. The work material is gripped in a collet contained in a sliding spindle which can be actuated longitudinally. Nearly all of the tooling is of the cross-slide type. The Swiss machine uses four or five cross tools depend-

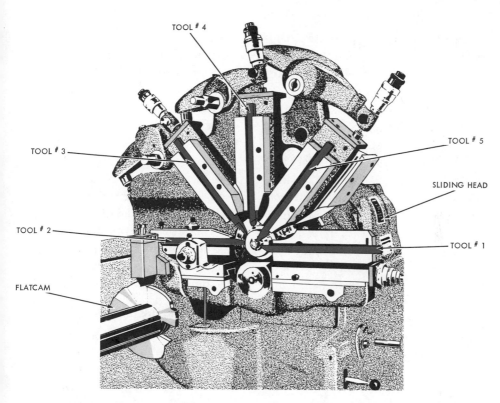

TOOL # 4

TOOL # 3

TOOL # 5

SLIDING HEAD

TOOL # 2

TOOL # 1

FLATCAM

Fig. 41. Sliding head mechanism and tooling setup for the Swiss automatic (Bechler)

ing on the type of part to be machined. The tools are fixed firmly in tool holders which are arranged radially around the workpiece, Fig. 41.

Each tool holder advances or retracts to bring the cutting tool closer to, or farther from, the center of the bar. Tool holder movements are controlled by flat cams, which are easily made from cast iron or steel.

Each of the tool holders can be adjusted individually in all directions by means of precision microm-

eter screws. These screws have a large graduated dial which can be set at zero or any desired division. No calculations are required for adjustment of the various tools for diameter or length of steps, or centering height relative to the axis of the barstock. The micrometer dials can be obtained with graduations in hundredths of a mm (metric system) or in thousandths of an inch.

Another advantage of the Swiss automatic, various devices such as cross-drilling, back-drilling and slotting attachments can be fixed to

397

a carefully finished surface, thus permitting certain additional operations to be performed, eliminating secondary operations.

Swiss machines make close-limit parts with diameters as small as 0.005″. Holes as small as 0.002″ dia can be drilled and bored. Machines of this type use a support bushing which follows the work material, supporting it within a few thousandths of the cutting point, thus making long, slim work practical. Although used for general work as well, Swiss machines excel on small

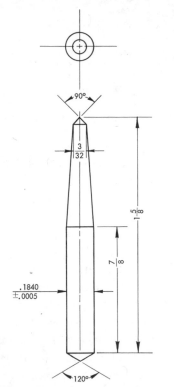

Fig. 42. Typical part produced on a Swiss automatic.

parts of complex shape, pinion-type work, and parts requiring turned diameters behind shoulders. Steps, shoulders, complex radii and configurations difficult to produce by other means are relatively simple on the Swiss.

A typical part suited for Swiss-type screw machines is shown in Fig. 42. The long, slim taper, the high length-to-diameter ratio, and the complex interrelationship of dimensions take this part out of the range of both the single- and multiple-spindle machines.

Production Considerations

Quantity. Production times on automatic screw machines range from less than one second to several minutes per part. Net production varies from 50-90% of theoretical production. Factors that contribute to this wide variation include the type of workpiece material, length of the part, number of close-tolerance dimensions to be produced and the type of machine used.

Because of the high production rates of these machines, and because the setup and teardown time must necessarily be charged against each job, small-lot jobs carry a proportionately higher downtime burden per piece than high-volume jobs. Similarly, special tools for each job are charged against it. (Standard drills, reamers, taps, knurls, threading rolls, die chasers, and other tools carried in stock are not charged to a specific job.) Where several lots of similar parts can be run end-to-end, costs usually can be reduced.

Number of pieces required has little bearing on the cost of this method of manufacture; rather, total machine hours needed is a better criterion. For example, 500 pieces, each requiring 5 min to produce, would use 2500 min of machine time. A 5-sec part would require a quantity of 30,000 to use the same number of machine hours.

The ability of screw machine tooling to be reset easily is an important advantage. If minor design changes are needed—even after the run has been started—they can be made more easily and at less cost than with many other methods of manufacture.

The standard rule of thumb in screw machine work is that the production run should be at least as long as the setup time. A quantity as low as 50 parts can be economically practical for certain parts but most orders are, of course, considerably larger.

Accuracy. Subject to limitations imposed by various types of equipment, the nature of the material and the part configuration, tolerances of $\pm 0.005''$ on overall length and $\pm 0.002''$ on diameters are typical and can be even closer when required. Screw machines are capable of continuously producing identical tolerances, subject, of course, to tool-wear deviations.

Under normal conditions, concentricity of a hole with a turned OD is usually within 0.002'' TIR (Total Indicator Reading), and concentricity of a hole with an unturned OD will be within 0.004'' TIR. Lengths between shoulders or the distance from a shoulder to an inside dimension can be within $\pm 0.003''$ without undue cost. In most cases much closer tolerances are possible but, they usually require added tooling attention. During design of the product, savings in manufacturing costs can be gained if parts are considered in the light of normal screw machine accuracies.

Shape. The work material may be of almost any shape: round, square, hexagonal, triangular, or any rolled or drawn form. Most shops have standard collets in stock for all but special shapes and even these can be had on short delivery.

Through the use of forming tools, which are accurately-ground reverse profiles of the desired part contour, diameters can be formed (turned) at low cost. Multiple steps, shoulders, chamfers, angles, radii, and tapers are not only simple to produce, but commonplace. For inside diameters, the general rule is that the outboard hole should be the largest. Multi-diameter holes can be drilled, reamed, tapped, bored, taper-reamed, profiled, or chamfered.

Turret Lathe Production Project

The following pages outline the tooling, setups and procedures required to produce the various parts of the center punch shown in Fig. 43. *Blueprints* for the center punch body, head, cap and point indicate the dimensions for the various parts. *Operation sheets* for each of the parts specify the type of turret lathe to be used, the tooling required, and the sequence of machining operations for each part.

The tooling is identified for each movable element of the machine by referring to the numbered layout on each operations sheet. By following this layout and the tool list sequence, the correct sequence for machining the part is established.

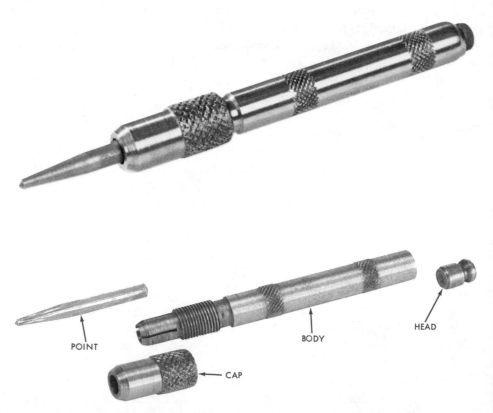

Fig. 43. The center punch and the components from which it is assembled.

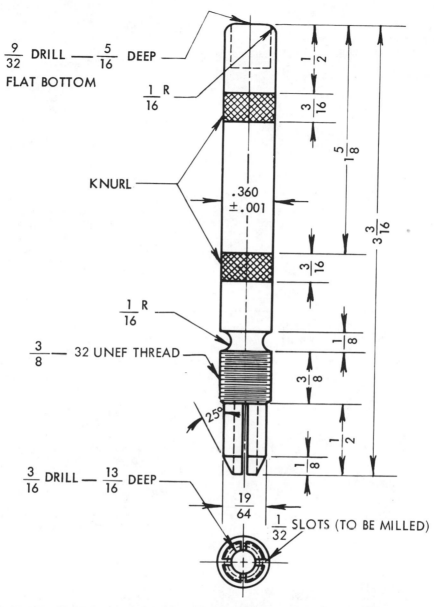

$\frac{9}{32}$ DRILL — $\frac{5}{16}$ DEEP

FLAT BOTTOM

$\frac{1}{16}$ R

KNURL

.360 ±.001

$\frac{1}{16}$ R

$\frac{3}{8}$ — 32 UNEF THREAD

25°

$\frac{3}{16}$ DRILL — $\frac{13}{16}$ DEEP

$\frac{19}{64}$

$\frac{1}{32}$ SLOTS (TO BE MILLED)

$\frac{1}{2}$

$\frac{3}{16}$

$5\frac{1}{8}$

$\frac{3}{16}$

$3\frac{3}{16}$

$\frac{1}{8}$

$3\frac{1}{8}$

$\frac{1}{2}$

$\frac{1}{8}$

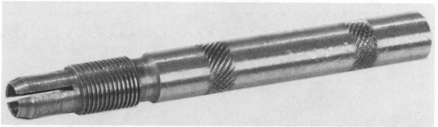

OPERATIONS SHEET

TOOLING SETUP: <u>CENTER PUNCH BODY</u>
(1st operation: Machining collet end of punch)

TOOLING
Hexagon Turret
   1   Stock Stop
   2   Center Drill
   3   3/16" Drill
   5   Stock Stop
   6   Automatic Die Head
Square Turret
   4   Circular Form Tool
Rear Tool Post
   7   Cutoff Tool

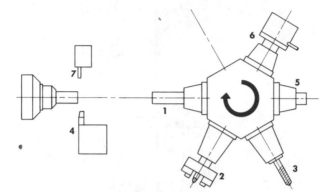

MACHINE TOOL: <u>RAM TYPE TURRET LATHE</u>　　MATERIAL: <u>AISI 1020 CRS Bar Stock</u>

RELATED INFORMATION: Collet chucks used for small diameter bar stock up to 2 1/2" are of the spring type. The bar can be fed through without stopping the spindle.
   Spring type collets are of three types: (a) pushout, (b) drawback and (c) stationary. A stationary type collet is recommended for this job to prevent the stock from being withdrawn when the collet is closed.

PROCEDURE	KEY POINTS
1 Feed stock to stop	1 Motor must be running in order for stock feed to work. Be sure collet is reclosed before taking cut.
2 Center drill	2 Spindle speed: High. Feed Center Drill manually to stop.
3 Drill 3/16" hole x 13/16" deep	3 Spindle speed: High. Withdraw Drill to clear chips several times during this cut.
4 Form thread relief, 19/64"dia. & 25° bevel	4 Spindle speed: High. Feed Circular Forming Tool to stop, withdrawing occasionally to clear chips and cool tool.
5 Feed stock	5 Use short Stock Stop and feed Stock two strokes. Make certain collet is reclosed.
6 Thread 3/8--32 UNEF Caution: Be sure there is ample clearance between cross slide tools before threading.	6 Spindle speed: Low. Close Automatic Die Head. Advance ram slide and follow through to stop. Die Head will release automatically.
7 Cutoff	7 Spindle speed: High. Feed Cutoff Tool very slowly as part is cut free in order to eliminate burr on finished part.
8 Check finished part	8 Inspection gages should be provided for checking depth of collet opening, overall length, and thread. Use a micrometer for checking the formed diameter.

OPERATIONS SHEET

TOOLING SETUP: <u>CENTER PUNCH BODY</u>
(2nd operation: Machining opposite end of
punch body)

TOOLING:
  Hexagon Turret
    1 Center Drill
    2 9/32" Drill
    3 9/32" Bottom Drill
    4 Facing & Chamfer Tool
    5 Working Support

  Rear Tool Post
    6 Knurling Tool

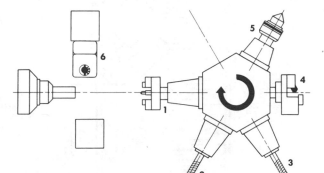

MACHINE TOOL: <u>RAM TYPE TURRET LATHE</u>
             OR
      <u>HAND SCREW MACHINE</u>

MATERIAL: <u>AISI 1020 CRS BAR STOCK</u>

RELATED INFORMATION: This operation illustrates one of the most important uses of the hand screw machine, that of secondary operations. Small-diameter stock previously cut to length can be mounted in a hand screw machine especially when the piece part requires internal machining operations.

PROCEDURE	KEY POINTS
Insert center punch body in collet. Be sure the 3/16" hole is free of burrs and chips so that the part is accurately located from the bottom of the hole.	
1 Center drill.	1 Spindle speed: high.
2 Drill 9/32" hole x 5/32" deep.	2 Spindle speed: high. Withdraw the drill periodically to clear chips.
3 Bottom drill the 9/32" hole.	3 Press bottom drill firmly against stop.
4 Face and chamfer.	4 Inspection gages should be provided. Check at least every fifth part.
5 Support work.	5 Too much overhang on workpiece requires support.
6 Knurl.	6 Lubricate lightly and use light feed.

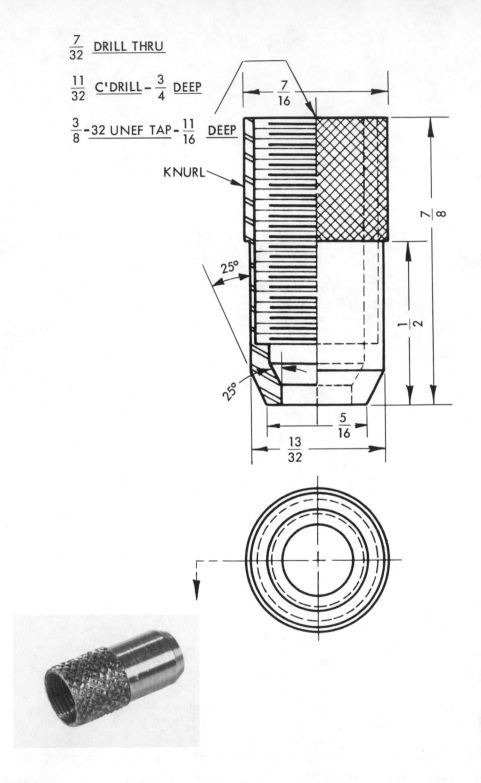

$\frac{7}{32}$ DRILL THRU

$\frac{11}{32}$ C'DRILL – $\frac{3}{4}$ DEEP

$\frac{3}{8}$ –32 UNEF TAP – $\frac{11}{16}$ DEEP

KNURL

$\frac{7}{16}$

$\frac{7}{8}$

$\frac{1}{2}$

25°

25°

$\frac{5}{16}$

$\frac{13}{32}$

OPERATIONS SHEET

TOOLING SETUP: <u>CENTER PUNCH CAP</u>

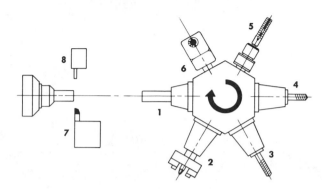

TOOLING
  Hexagon Turret
  1   Stock Stop
  2   Center Drill
  3   7/32" Drill
  4   11/32 Tap Drill
  5   3/8-32UNEF Tap(#2Tap)
  6   Knurl
  Square Turret
  7   Circular Form Tool
  Rear Tool Post
  8   Cutoff Tool

MACHINE TOOL: <u>RAM TYPE TURRET LATHE</u>   MATERIAL: <u>7/16" dia. AISI 1020 CSR Bar Stock</u>

RELATED INFORMATION: In calculating horsepower at the tool tip simplicity dictates that only the force required to push the tool ahead through the cut be used. This force is responsible for practically all work done and is sufficiently accurate in most cases. The approximate horsepower can be calculated using the formula:

$$HP = \frac{12 A_o V}{K}$$

where V = cutting speed in fpm, A_o = area of the cross section in sq. in., and K = cubic in. of metal removed per min. per horsepower.

PROCEDURE	KEY POINTS
1 Feed stock	1 Hold Stock Stop firmly in position while feeding stock.
2 Center drill	2 Feed manually to stop.
3 Drill 7/32" through hole	3 Use .004 feed. After feed stops, feed tool manually to dead stop.
4 Drill 11/32" x 3/4" deep	4 Use .004 feed.
5 Tap 3/8--32UNEF x 11/16" deep	5 Feed tap manually to stop. When tap holder revolves, reverse the spindle and back the tap out. After tap is out of hole, reset spindle to Forward.
6 Knurl	6 Feed manually to stop. Light Knurl pressure to prevent crushing of thin wall.
7 Form 13/32" stepped diameter	7 With carriage locked at carriage stop #1, feed manually to dial clip #1.
8 Cuttoff and chamfer	8 Feed slowly so that there will be no burrs on the finished part.

CENTER PUNCH HEAD

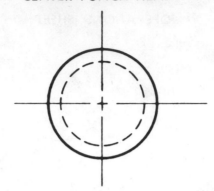

$\frac{1}{16}$ R

$\frac{1}{32}$ R

$\frac{13}{32}$

$\frac{1}{4}$

$\frac{9}{32}$

OPERATIONS SHEET

TOOLING SETUP: <u>CENTER PUNCH HEAD</u>

TOOLING:
<u>Hexagon Turret</u>
   1 Stock Stop
<u>Front Tool Post</u>
   2 Circular Form Tool
<u>Rear Tool Post</u>
   3 Cutoff Tool

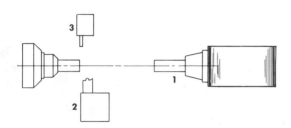

MACHINE TOOL: <u>SINGLE SPINDLE AUTOMATIC</u>   **MATERIAL:** <u>9/32 AISI 1113 Steel</u>

RELATED INFORMATION: The success of the automatic screw machine in producing work with a high degree of accuracy, good finish, and at a rapid rate, depends to a great extent on the selection and setting of the tools.

Circular forming tools on the cross slide are used as a rule for all form turning and for straight line turning behind shoulders, especially where the length of cut is not too great in proportion to the diameter.

Time per part can be calculated from the formula:

$$t = R/N$$

where t = time in min., N = spindle speed in rpm, and R = total revolutions to make one part.

Revolutions per work piece can be calculated from the formula:

$$r = T/f$$

where T = distance the tool moves in cutting plus approach distance, and f = feed in in. per revolution.

PROCEDURE	KEY POINTS
1 Feed stock to stop	
2 Form part contour	2 Start motor. Engage clutch to start camshaft.
3 Cutoff	3 Stop machine at end of cycle (when piece part is severed from bar stock). In an emergency, stop the machine immediately.

CENTER PUNCH POINT

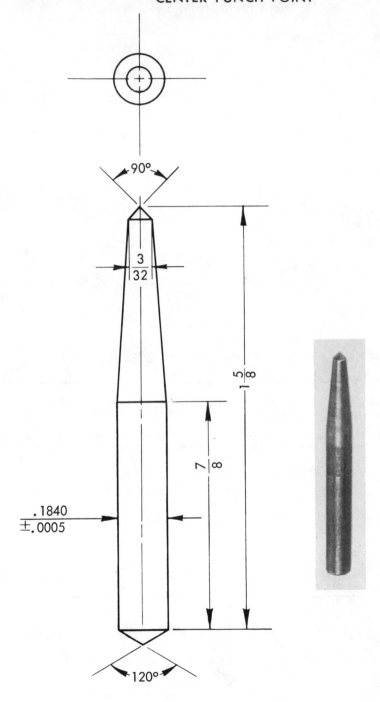

90°

$\frac{3}{32}$

$1\frac{5}{8}$

$\frac{7}{8}$

.1840
±.0005

120°

OPERATIONS SHEET

TOOLING SETUP: <u>CENTER PUNCH POINT</u>

TOOLING:
1 set of cams
1 collet
1 bushing
3 straight shank tool bits ground for the
following operations:

1 Cut off and form point on next
part. (Position $\underline{V}$)
2 Turn 3/4" taper .093" to .1840".
(Position $\underline{IV}$)
3 Chamfer the cut off end. (Position $\underline{III}$)

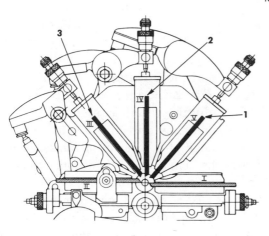

MACHINE TOOL: <u>SWISS TYPE AUTOMATIC</u>   MATERIAL: 0.184" dia. Drill Rod

RELATED INFORMATION: On the Swiss type automatic screw machine a taper may be generated by the
relative movement as directed by two cams, one for axial movement and one for radial movement.

PROCEDURE	KEY POINTS
1 Start spindle.	1 Spindle should be allowed to run for a few minutes before starting cycle.
2 Engage clutch to start cam shaft.	2 Watch tooling closely through first cycle to be sure cams and tooling are properly adjusted.

3   Automatic operations:
    A.   Form point
    B.   Generate taper
    C.   Feed through
    D.   Cut off and round end
	3 The cutoff tool serves as a stock stop and must remain across the stock until head stock is withdrawn and collet relocked.

4 Stop Machine.	4 The clutch should be disengaged when the head stock is moving toward rear of machine.

Safety Precautions—Turret Lathes, Automatic Screw Machines

When operating the turret lathe, automatic screw machine — or any other machine tool — the greatest hazard to the machine operator is his own carelessness. The operator should be alert to potential hazards at all times. The best assurance of safety — for the operator and those working near him—lies in developing responsible work habits and a cautious respect for the machine itself, each of which attitudes is embodied in the following safety suggestions:

1. Injuries frequently occur when the operator fails to move the turret back as far as possible when changing or gauging work. Be careful not to strike your hand, elbow, etc., on cutting tools when adjusting or setting up the job.

2. Never use the power of the machine to start a face plate or chuck onto the spindle.

3. Keep hands clear of the turret slides while the machine is running.

4. Exercise caution when turning odd-shaped or oversized workpieces.

5. Enclosures or guards over the chuck, which confine chips, oil splashes and also act as exhaust hoods for removal of fumes, should be kept in good condition, especially on automatic machines.

6. A chip produced in a continuous spiral is a frequent cause of hand and arm injuries. Wherever possible, chip breakers should be used.

7. Hazards encountered in turret lathes and screw machines are similar to those encountered in engine lathe work, and most of the same precautions apply. The student is urged to review the section on lathe safety precautions at the end of Chapter Six.

NOTE: Please see review questions at end of book.

Grinding Machines:

Surface, Cylindrical, and Internal Grinding

The grinding wheel and the grinding machine have become such important factors today that without them our mass production would be impossible. Only the grinder can produce the surface finishes and the close tolerances required on many jobs.

Because of the construction and accuracy of grinding machines today, workpieces can be held to extremely close tolerances in dimension and finish. With the aid of grinding machines, large-quantity production of parts of the same size, shape, and finish is possible. Pieces so produced in quantity can be used interchangeably in an assembled product.

The grinding operation depends upon the abrasive or cutting qualities of emery, corundum, carborundum, or other materials bonded and formed into a wheel. Each abrasive grain in the wheel can be considered a very minute sharp tool. As the wheel revolves, each grain cuts a small chip from the workpiece.

Grinding machines vary in design and construction depending on the kind of grinding work the machine is intended to handle. Some machines are designed to handle more than one type of work.

Surface Grinding

Surface grinding is done to produce flat surfaces on workpieces held in contact with a grinding wheel. Four different types of surface grinders are shown in Fig. 1.

A study of Fig. 1 will show two classes of machines, namely: those with horizontal spindles (see *Types I* and *II*) and those with vertical spindles (*Types III* and *IV*).

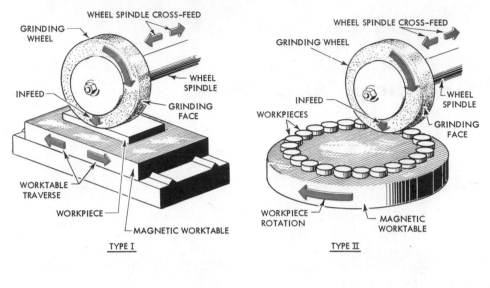

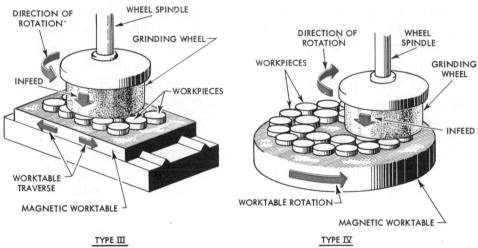

Fig. 1. There are two classes of surface grinders: those with horizontal spindles (types I and II) and those with vertical spindles (types III and IV).

Horizontal-spindle surface grinding machines are in turn divided into two classes. The first is the reciprocating grinder—one having a rectangular table that traverses under the revolving wheel (see *Type I* of Fig. 1). Another kind has a circular table that rotates under the wheel (see *Type II*).

Another type of horizontal-spindle surface grinder has a spindle which can be adjusted from a horizontal to an angular position to grind inside corners and dovetails.

Horizontal Reciprocating-Table Surface Grinder, Type I

The major parts of the horizontal-spindle surface grinder are discussed below, beginning with the base and proceeding in an order of assembly.

Base. The base of the horizontal surface grinder, Fig. 2, has a column at the back for supporting the wheelhead. The saddle which supports the table is mounted on carefully machined ways at the front of the machine.

Table. The table on most grinders is similar to the table on the milling machine. It is fitted to the saddle by carefully machined ways. It can be moved right or left (traverse travel) under the grinding wheel. This kind of table is called *reciprocating*. The table carries adjustable reversing dogs which may be set to limit table travel. Hand table feeds can be used as well as automatic power feeds.

On some machines, the table can be moved in or out from the vertical column which supports the wheelhead. This movement is known as cross-feed. On other machines, the wheel and driving motor (wheelhead) move on cross-slide ways.

Wheelhead. The wheelhead is mounted on the column. It moves on carefully machined ways by which it can be raised or lowered. A handwheel which turns a lead screw is used to move it up and down on the column.

A spindle for revolving the grinding wheel is mounted in the head in a horizontal position. Hence the machine is called *horizontal*. The heart of any precision grinding machine is the grinding wheel spindle and its bearings. Any wear in the bearings that permits the spindle to vibrate will make the machine useless for precision work.

A disk wheel is mounted on the horizontal spindle. The wheel grinds the workpiece as it feeds and reciprocates under the wheel.

An electric motor is built into the machine to furnish power to turn the grinding wheel. The wheel is usually driven by pulleys and belts or direct spindle drive.

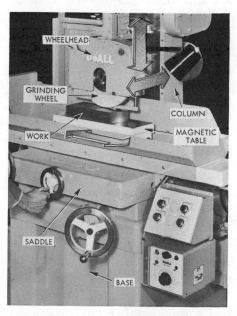

Fig. 2. The parts of a horizontal grinder. (DoAll Co.)

Coolant System. A coolant must be used to regulate the temperature of the work. The coolant is directed to the work by pipes from a pump and a tank.

Capacity. Grinders are described in terms of work capacity. This is usually given as the maximum working space provided by the workholding device, which usually is the magnetic chuck. A machine size of 6 inches by 18 inches designates the dimensions of the working area of the chuck or table.

Wheels. The wheels used in the horizontal machine are usually plain wheels which cut on the periphery. In grinding shoulders, a recessed wheel may be used. In grinding V's and other shapes, wheels with formed edges are used. For further information, see the section on wheels and wheel markings at the end of this chapter.

Setups and Operations— Horizontal Surface Grinding

To achieve good results, the operator must know how to set up a job, using the most suitable workholding device. The size and shape of the workpiece naturally determine the kind of holding device to use and the setup required.

The table of the horizontal surface grinder is usually provided with T-slots. The work can be bolted directly to the table top, or can be mounted in a vise or in V-blocks.

Where possible, the work is held on a magnetic chuck, Fig. 3, which is built into or attached to the machine table. Magnetic chucks are now in common use as a means of holding workpieces. Magnetic chucks are of two types: permanent and electric.

Only magnetic materials, such as iron or steel, will actually hold on the chuck. The magnetic chuck holds the work by exerting a magnetic force on it. The magnetic poles of the chuck are placed close together so that it is possible to hold very small pieces of flat work.

In clamping work to the table directly, the piece must be firmly held in place. Great care should be taken to avoid springing, warping, or other distortion of the workpiece. Ground work is expected to be true and accurate.

The operations performed on a horizontal surface grinder can be classified as *plain surface grinding*, *angular surface grinding*, and *form grinding*, using a form wheel.

Fig. 1 shows different examples of grinding a flat surface. A flat surface also can be ground on a cylindrical shaft. Fig. 3 shows still another example of plain grinding.

Work Speed. In surface grinding, the work speed is determined by the rate of movement of the machine table which supports the work and moves it back and forth past the grinding face of the wheel. Work

Fig. 3. The workpiece here is held on a permanent magnetic chuck. (Browne & Sharpe Mfg. Co.)

speed or table speed is also called the *rate of table traverse.* It is expressed in surface feet per minute (s.f.p.m.).

Cross-Feed Speed

The cross-feed speed is the *in* or *out* movement of the wheel on the work—a movement at right angles to the table as it reciprocates. The cross-feed should be in proportion to the width of the wheel and the finish desired. Cross-feed should not exceed, even in rough grinding, one-half the width of the wheel's face per table stroke. The lower the cross-feed for any given width of wheel face, the greater the depth of cut that can be taken.

Wheel Speed. For actual wheel speeds for safe operation, the recommendation of the wheel manufacturer should be followed at all times.

Infeed. The infeed, or depth of cut, is governed by the amount of metal to be removed and the finish desired. Other controlling factors are the desired accuracy of work and coolant used.

The handwheel at the top of the grinder, for use in setting the infeed, has a dial graduated in tenths of thousandths, thus making it possible to grind flat surfaces to an accuracy of .0001 inch.

Depth of cut determines to a great extent the pressure built up between the wheel and work during the grinding operation. High working pressure may create heat which will distort or burn the work. Thin

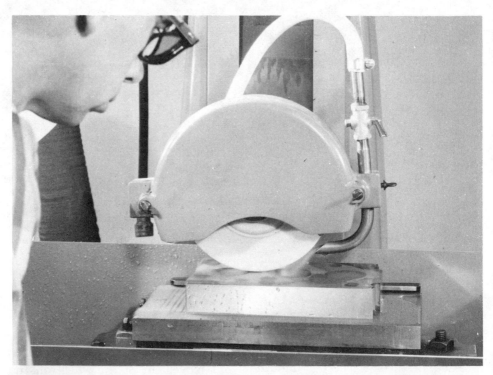

Fig. 4. Coolant should be directed to the contact area between the wheel and the work. (Harig Products Co.)

stock requires reduced infeeds and light cuts.

Dry grinding operations require lighter cuts than wet operations.

Coolant. If a coolant is used, the operator must regulate the direction and flow of it. Volume of coolant is more important than pressure. The nozzle of the coolant pipe should direct the flow of coolant to the contact area between the wheel and the work, see Fig. 4.

Dressing the Wheel. The removal of metal by grinding causes the grinding wheel's contact surface (face, in the case of the horizontal surface grinder) to wear and become irregular in shape. The metal particles "load" or glaze the wheel's contact surface. These metal particles must be removed and new abrasive grains exposed. This is usually done by using a hand wheel-dressing tool; it must be done by an experienced person. Some modern machines are equipped with an automatic wheel-truing device for dressing the grinding wheel. This arrangement makes it possible to true the wheel without disturbing the setup.

Cylindrical Center-Type Grinding

Grinding the outside diameter of a cylindrical piece, while it is revolving on its axis, is called *cylindrical grinding*. This operation reduces the workpiece to size and produces a fine finish. Typical cylindrical pieces ground on their outside diameters are: ordinary round metal parts, Fig. 5, *left;* tapered parts, Fig. 5, *center;* and pieces having two or more diameters, Fig. 5, *right*. Highly complicated shapes, such as crankshafts, can also be ground to precision accuracy.

For work that is not to be hardened, about .006 to .010 inch of

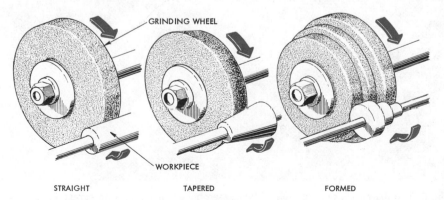

GRINDING WHEEL

WORKPIECE

STRAIGHT   TAPERED   FORMED

Fig. 5. Straight, tapered and formed parts can be cylindrically ground.

stock is left for grinding. A long piece of work that is apt to spring will need from .020 to .030 inch of grind stock allowed.

Two different types of machines are in use today for doing cylindrical grinding. The difference between these two types lies in the manner in which the work is held in the machine. In the center type, Fig. 6, the work is held on pointed centers. In the centerless type, the work is supported by means other than centers. In this part of the chapter, we will concern ourselves with center-type cylindrical grinding and the machines used to do this kind of work. Centerless grinding will be taken up later.

Center-Type Cylindrical Grinder

The center-type cylindrical grinder, Fig. 6, is designed to give three movements which are very important. These three movements, Fig. 7, are: (1) rotating the workpiece on its axis, (2) moving the work horizontally back and forth in front of the grinding wheel (*traverse* movement), and (3) moving the wheel into the workpiece and away (*in-feed*).

With the center-type grinder, the three movements can be accurately combined and controlled. The operator, however, must understand the machine and run it skillfully.

The three movements listed above must be performed without play, bind, vibration, or unevenness to insure good finish and accurate sizing. Any difficulty in cylindrical grinding, aside from wheel content or makeup, can be traced to one of the

Fig. 6. Two diameters of a differential box assembly are being ground simultaneously on this center-type cylindrical grinder. (Landis Tool Co.)

three movements mentioned above. It is important, therefore, that the machine be in good condition and the work be properly set up.

Base. The base gives rigidity and stability to the total machine and supports the parts mounted on it. On the top of the base are carefully machined ways for the table to slide on. At the rear of the machine is a column on which the grinding wheel, wheel spindle, and wheel power unit are mounted.

Table. The table has a machined surface on which the head and footstock that hold the workpiece are mounted. The table traverses back and forth on accurately machined ways to give movement of the workpiece past the grinding wheel, ex-

cept in the case of "plunge-cut" grinding. The length of table traverse is controlled by trip dogs. These can be positioned in different places along the table to reverse the direction of table movement when the grinding wheel reaches each end of the workpiece.

Headstock. The headstock, Fig. 6, is mounted on the left end of the table. The work drive motor is incorporated in the headstock. Some jobs must be held on a faceplate; others must be secured in a chuck or in a special fixture mounted on the spindle of the headstock. Most work to be ground cylindrically is held between centers. A drive plate is screwed on the spindle nose. A dog is fastened to the headstock

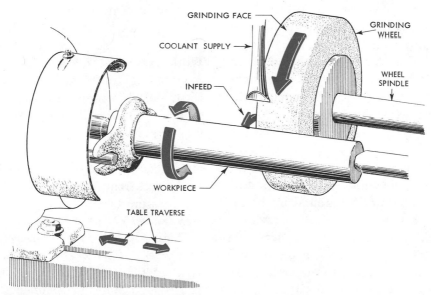

GRINDING FACE

COOLANT SUPPLY →

GRINDING WHEEL

INFEED

WHEEL SPINDLE

WORKPIECE

TABLE TRAVERSE

Fig. 7. The center-type cylindrical grinder rotates the workpiece, moves the workpiece back and forth horizontally, and moves the grinding wheel toward and away from the workpiece.

419

end of the workpiece, which is mounted between centers, and engages a pin in the drive plate. The drive plate revolves about the headstock dead center and gives the workpiece its rotary motion, see Fig. 7.

Footstock. The footstock is a unit similar to the one used on the lathe and can be adjusted and clamped in various positions along the table to take care of different lengths of workpieces. The dead center in the spindle supports the work.

Wheelhead. The grinding wheel is mounted on the end of a spindle which revolves in a carefully constructed head. It is driven by an electric motor. In the plain machine, the grinding-wheel unit (or wheelhead as it is usually called) is mounted on a cross-slide at right angles to the ways of the table. At each reversal of the table and the work, the wheel is fed toward the work either by hand or automatically. Sometimes it can be fed across the workpiece.

Coolant System. To control the sparks and abrasive dust of grinding so it does not fly around in the air of the room, and also to provide a coolant for the grinding action, a coolant storage tank, a pump, and piping are built into the machine to supply coolant to the wheel. An adjustable spout is supported near the face of the wheel above the point of grinding contact in order to deliver the coolant at the proper place.

Wheels. The wheels used on the cylindrical grinder are usually plain wheels which cut on the periphery. Occasionally, wheels are used with formed edges or faces for operations where the wheel is fed against the work and the work does not traverse.

Dressing the Wheel. On all cylindrical grinders, provision is made for dressing the wheel when it is necessary. Wheel dressing, or truing, is an important and delicate operation; its purpose is to restore the worn and glazed grinding wheel to its original shape and sharpness, see Fig. 8.

Work Rest. Centers alone are often insufficient for supporting long pieces of work against the heavy pressure of the grinding wheel. Work rests or back rests, Fig. 9, are mounted at suitable intervals along the table. The shoes of the work rest oppose the outward and downward pressure of the wheel. Work rests may be spaced six to ten work diameters apart to prevent long, slender workpieces from springing away from the pressure of the grinding wheel. Heavier work requires less support.

Plain and Universal Center-Type Machines. There are two variations of the center-type cylindrical grinding machine: (1) plain and (2) universal. They differ in construction and in kinds of work handled.

Fig. 8. A table-type diamond-wheel dresser can be used to true a grinding wheel. (Norton Co.)

Fig. 9. Work rests are used to support this eccentric shaft during grinding. (Landis Tool Co.)

Plain center-type cylindrical grinders are primarily production machines. They are designed for plain cylindrical work and have few attachments. The grinding wheel is mounted at right angles to the ways of the table, with the axis of the grinding wheel parallel to the axis of the work mounted between the centers. The plain machine has the wheelhead set permanently at right angles to the table travel and the headstock cannot be swiveled.

The principal difference in the case of the universal center-type grinder is that both the wheelhead and the headstock can be swiveled at an angle to the ways of the table. Thus, the universal can perform a wider variety of jobs, including internal grinding. Universal grinders are adaptable to many kinds of grinding jobs. They are usually equipped with special attachments and equipment.

Setups for Center-Type Cylindrical Grinding

When mounting the work in the center-type cylindrical grinder, the machine centers must be properly seated in the center holes of the work. The cylindrical grinding of any work held between centers cannot be more accurate than the center holes. Nor will it be more nearly round than the center points on which the work revolves.

Both the work centers and the center holes should have exactly 60-degree angles. Center holes should always be made with a slight relief at the bottom, just as in lathe work. Without this relief, the center hole will revolve on the rounded nose of the work centers. The hole is likely to wobble about on the work center. Care should be taken to see that no dirt particles get between the machine centers and the center holes in the ends of the workpiece.

Work Speeds. The general range of work speeds for the grinding of plain cylindrical work is from 60 to 100 s.f.p.m. (surface feet per minute).

In grinding cranks or other work that is out of balance, lower speeds are necessary. The rough grinding of automotive cams is usually done at 15 to 30 s.f.p.m. The finish grinding is done at half this speed.

Certain nonferrous alloys and soft metals are ground at higher work speeds — up to 200 s.f.p.m.

Lower speeds are necessary whenever the form of the wheel face is important. Form grinding by direct infeed, called *plunge grinding*, requires very low speeds. Thread grinding represents the extreme of these conditions. Here, work speeds as low as 2 to 6 s.f.p.m. are used.

Traverse Speeds. Traverse speeds should be in proportion to the width of the wheel face and the finish desired. Traverse should not exceed, even for rough grinding, three-quarters of the wheel face per work revolution.

The length of the table traverse should be set so as to enable the wheel to overrun the end of the work about one-fourth to one-third the width of the wheel face. This must be done to permit the wheel to finish the cut. If the wheel is not permitted to partly overrun the end of the work, the piece will be oversize at the end. Traverse should never be as far as to allow the wheel to fully extend beyond the work.

At the end of each traverse, the table stops momentarily. This permits the wheel to grind the work to size, to clear itself for the new cut, and to avoid the jarring motion which would result if the table came to a complete stop and reversed immediately.

Wheel Speeds. Most machines are designed to give a variable speed to the wheel. The wheel speed is measured in surface feet per minute (s.f.p.m.), never in revolutions per minute. Therefore, the larger the wheel, the greater the surface speed for the same r.p.m. Likewise, the smaller the diameter of the wheel, the less the surface speed for the same r.p.m. Table XVII in the Appendix provides wheel speeds in r.p.m. for desired peripheral speeds (s.f.p.m.) for various diameters.

The speed at which a grinding wheel revolves is important. If the speed is too slow, the abrasive is wasted without much work being done. If the speed is excessive, the grinding action will be hard and the danger of breakage is increased. Usually, it is best to operate the wheel at the speed recommended by the manufacturer of the grinder.

Infeed. Infeed, or depth of cut, is controlled by the amount of metal to be removed, the finish desired, and the power and rigidity of the grinding machine. Other controlling factors are the desired accuracy of work, coolants, and the work support provided.

The grinding wheel can be fed to the work by automatic feed or hand feed. It is not advisable to use the hand feed except to bring the wheel up to the work or to move it away or when taking very fine cuts.

The automatic feed takes from .00025 to .004 inch of each traverse of the table. It saves time and wear on the machine by taking more uniform cuts and gives longer life to the grinding wheel.

Using the Work Rest. When using the work rest, the jaws (shoes) must be kept properly adjusted to the height and diameter of the work. Otherwise the work might get caught between the lower jaw and the wheel. The workpiece may be thrown from the machine or the wheel may break. As the operation progresses, the work rest shoes must be adjusted to take up the reduction in diameter of the work.

Operations—Cylindrical Center-Type Grinding

On the cylindrical center-type grinder, numerous operations can be performed. Because both the wheel-head and the headstock of the universal machine can be swiveled at an angle to the table travel, a wide variety of operations can be performed on the workpiece. A few of the typical operations are described below.

Plain Cylindrical Grinding. A good illustration of plain cylindrical grinding is shown in Fig. 10. We see the workpiece mounted on a man-

Fig. 10. Work is often mounted between centers for plain cylindrical grinding. (Landis Tool Co.)

drel which, in turn, is mounted between the two dead centers. The dog mounted on the end of the mandrel engages with the headstock drive pin to revolve the work.

Form Grinding. Fig. 11 shows a close-up view of a form-grinding op-

eration. A formed wheel is grinding two diameters at once. Form grinding requires that the wheel be trued to the exact form desired.

Grinding Taper Work. Tapers can be ground on the cylindrical grinder. They vary from a slight ta-

Fig. 11. Formed grinding wheel is grinding two diameters, an adjacent face and adjoining radius of a front wheel spindle. (Landis Tool Co.)

per, as shown in Fig. 5, to a steep taper, as shown in Fig. 12. Either slight or steep tapers may be ground on a universal grinder. On the plain cylindrical grinder, taper grinding is limited.

For grinding slight external tapers on the universal grinder (up to 8 degrees), the usual procedure is to swivel the work table. On most machines, the table top can be swiveled on its reciprocating base from 0 degrees to a maximum of 8 degrees on either side of zero. This method is suitable for slight tapers only.

Steep tapers or angles can be ground by either of two methods: (1) by swiveling the headstock on the table, or (2) by swiveling the wheelhead on its base. The first method, swiveling the headstock, is used, for example, in grinding the 60-degree point on a machine center, Fig. 12. In this setup, the center has been placed in the live spindle of the headstock with the headstock

Fig. 12. A 60° conical point must be ground on a *universal* cylindrical grinder. So steep a taper could not be ground on a plain cylindrical grinder. (Macklin Co.)

swiveled through a 30-degree angle. In grinding machine centers, it is important that the center point have an included angle of exactly 60 degrees.

Where the headstock is swiveled, the workpiece is supported by the headstock only, see Fig. 12. It does not permit the use of the dead center in the footstock as a support for the end of the workpiece.

To grind a steep taper on a workpiece held between centers, the wheelhead is swiveled the desired an-gular amount. Headstock and footstock remain in perfect alignment.

Regardless of the method used, the grinding of an exterior taper should be done, if possible, with the wheel pressure toward the headstock.

Plunge Grinding. Sometimes the length of the workpiece to be ground is less than the width of the wheel face. In these cases, it is not necessary to move the work across the wheel. The grinding wheel can be fed straight into the work until

Fig. 13. Plunge grinding of a driven gear. (Landis Tool Co.)

the desired diameter is secured. This is called "plunge-cut" grinding. Plunge wheels may be as large as 9 inches across the face.

For grinding a shoulder or stepped cylinder, plunge-cut grinding is convenient. It maintains a square corner between the cylinder and the adjoining shoulder on the workpiece.

With suitable forming attach-ments, the face of the grinding wheel can be shaped to many contours. In this way, work with an irregular contour can be plunge-cut accurately to shape.

Fig. 13 illustrates the plunge grinding of an end diameter, adjacent face and radius of a driven gear. The gear is then turned end for end to grind the opposite end diameter, adjacent face and radius.

Cylindrical Centerless Grinding

The centerless grinder, Fig. 14, as its name implies, does not make use of center points. In this type of grinding, there is no need for locating and drilling center holes in the workpiece.

The machine elements of a centerless grinding machine are the grinding wheel, regulating wheel, and work-rest blade. All three of these are necessary to support the workpiece in the grinding position, see Fig. 15.

The work passes between the wheels, one a grinding wheel and the other a regulating wheel. The grinding wheel rotates at high speed. Its purpose is to grind material off the surface of the cylindrically shaped workpiece. The regulating wheel is opposite the grinding wheel and it rotates at a low speed. The regulating wheel has three functions: (1) to rotate the work so the grinding wheel can grind over the entire surface, (2) to support the work against the horizontal thrust of the grinding wheel, and (3) to feed the work between the grinding and regulating wheels as it rests on the work-rest blade.

All centerless machines employ so-called *negative work speed.* In other words, the angular rotation of the work and that of the grinding wheel are in different directions. The grinding wheel revolves at standard grinding speeds, approximately 6,000 feet per minute. The operating speeds of the regulating wheel can be varied at will, within a range of from 50 to 200 feet per minute peripheral speeds.

For best straightening-out effect of long slender-diameter work, the center of the piece should be placed below the center line of the wheels, and the rate of traverse should be high. Grinding in this position is primarily for straightening the work.

A. Grinding wheel	6. Grinding wheel truing rate adjustment
B. Work rest	7. Cutting fluid valve
C. Regulating wheel	8. Regulating wheel truing device
1. Micrometer adjustment	9. Micrometer adjustment
2. Loadmeter	10. Booster lever
3. Grinding wheel truing device	11. Regulating wheel speed range lever
4. Booster lever	12. Infeed lever
5. Grinding wheel truing engaging lever	

13. Regulating wheel truing traverse control	
14. Upper slide clamp	
15. Regulating wheel speed change handwheel	
16. Tachometer dial for regulating wheel speeds	
17. Infeed handwheel	
18. One shot lubrication plunger	
19. Master start-stop buttons	

Fig. 14. Centerless grinder for cylindrical work. (Cincinnati Milling Machine Co.)

Subsequent passes can be made with normal setup for corrective rounding action.

Classes of Centerless Grinding

These centerless grinding principles lend themselves to almost unlimited applications through the use of machine setups involving special relationships between the grinding wheel, the regulating wheel, and the work-rest blade, combined with various types of work guides and feeding mechanisms.

Through-feed centerless grinding is accomplished by passing the workpiece between the grinding wheel and the regulating wheel. The longitudinal or actual movement of the work past the face of the grinding wheel (a movement corresponding to traverse in center-type grinding) is imparted by the regulating wheel.

The infeed method is usually employed when grinding work which has a shoulder, head, or some portion larger than the ground diame-

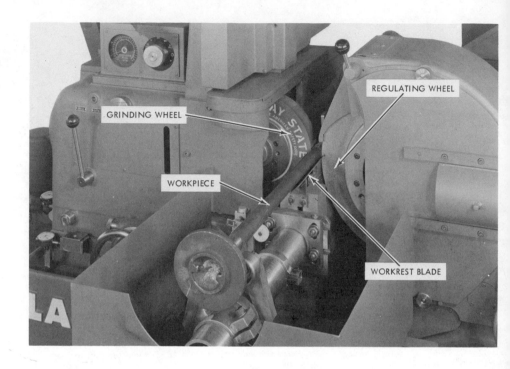

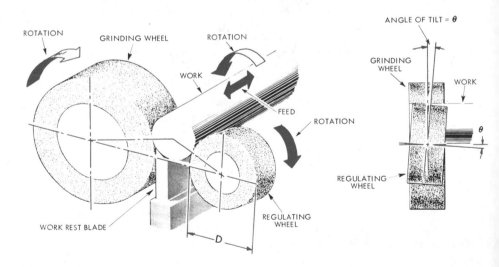

Fig. 15. *Top:* The grinding wheel, regulating wheel and workrest blade support the workpiece in the grinding position. (Landis Tool Co.) *Bottom:* The operating principle of the centerless grinder.

ter. The same method is used for the simultaneous grinding of several diameters of the work, as well as for finishing pieces with taper, spherical, or any other irregular profile. In general, this method corresponds to plunge-cut or form grinding on the center-type grinder.

End-feed grinding is used only on taper work. The grinding wheel, the regulating wheel, and the blade are set in a fixed relation to each other. The work is fed in from the front, manually or mechanically, to a fixed end stop. Either the grinding wheel or the regulating wheel, or both, are dressed to the proper taper for the work being ground.

Internal Grinding

The grinding of internal surfaces, or holes (inside diameters), is called *internal grinding*, Fig. 16. The application of internal grinding is extensive. The range of hole sizes and types of pieces that can be worked is limited only by the capacity of the grinding machine. Both general-purpose and special-purpose grinders have been developed so that internal grinding can be performed rapidly and economically on a wide range of hole sizes. The holes may be straight, tapered, or formed. Through holes, blind holes, or holes with more than one diameter may all be ground by these machines.

Internal grinding is the most widely used method of finishing holes. It is accurate and economical, and produces a very satisfactory surface. Reaming, in many instances, has given way to internal grinding. Very often in the heat-treatment process, there is a certain amount of distortion of the workpiece. The holes in workpieces can be finished by internal grinding to secure accurate diameters and true surfaces.

Types of Internal Grinding Machines

There are many types of internal grinding machines, each designed to meet certain specifications. Nevertheless, internal grinding machines may be classified into three main groups according to the way the work is held:

1. The work-rotating type, which uses a three- or four-jaw chuck, a faceplate, or fixtures to hold and turn the work, see Fig. 16 *A*.

2. The planetary grinder type, in which the workpiece is reciprocated to obtain traverse, but is not rotated. The eccentric travel of the wheel spindle generates the correct size hole, see Fig. 16*B*.

3. The centerless internal-grinder type, which uses a set of rollers to hold and turn the work.

The work-rotating type of grinder is the kind most commonly found in

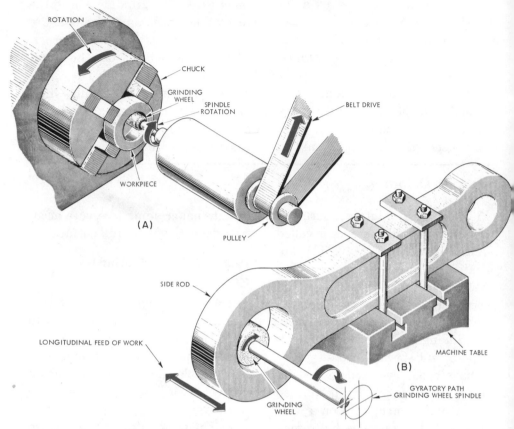

Fig. 16. At A—the work-rotating internal grinder. At B—the planetary grinder.

toolrooms. The work is held by a chuck, faceplate, or fixture in the headstock. The wheel and work center lines are in the same horizontal plane, see Fig. 17. They are not in the same vertical plane, however, because the grinding wheel is smaller than the hole. As the work turns slowly, the rapidly rotating grinding wheel is reciprocated along one side of the bore.

Wheelhead Arrangement. The wheelhead is mounted differently on

different machines for internal grinding. One type has the wheelhead mounted on a cross-slide which can be fed horizontally across the ways. Another has the wheelhead mounted on a swinging arm suspended from an overhead crossbar. In both types, the wheelhead can be traversed parallel to the ways to finish the full length of the surface.

Grinding Wheels. The wheels used in internal grinding are generally much smaller than those used in external grinding. They range from the smallest mounted points to wheels of 10 to 12 inches in diameter. The wheel is, of course, smaller than the hole to be ground, see Fig. 17.

Because of the small wheel size, the spindle of an internal machine must run at a very high rate to give the wheel its proper surface speed of 4,000 to 6,000 surface feet per minute. New operators sometimes find these high speeds a little frightening. There is no need for fear, however, as a properly operated machine is thoroughly safe.

Work speeds on internal grinders are low because of the large area of contact which the wheel makes with the surface being ground. Work speeds range from 50 to 100 surface feet per minute.

Setups and Operations— Internal Grinding

Work is frequently held in a chuck or fixture mounted on the headstock spindle, see Fig. 18. In high production, special chuck jaws are often prepared for holding each job. For irregular-shape pieces, a special fixture is made in the toolroom. Sometimes, work can be held best on a magnetic chuck mounted on the headstock spindle.

Most modern internal grinders are highly automatic in operation. After the job is set up, the operator has only to start and stop the ma-

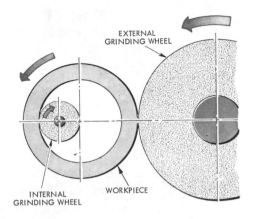

Fig. 17. Grinding wheels used for internal grinders are much smaller than those used for external grinding.

chine, and load and unload the chuck or fixture holding the workpiece. A typical cycle might run as follows:

1. The work is mounted in the machine and rotated.

2. The wheel is traversed rapidly up to the work.

3. The traverse is slowed to grinding speed.

Fig. 18. Work may be mounted in a chuck for internal grinding operations.

4. The wheel rough grinds the hole.

5. The wheel is withdrawn to the automatic dressing tool and resurfaced for the finish cut.

6. The wheel re-enters and finish grinds the hole.

7. The wheel is withdrawn to starting position and stopped.

Sizing or Measuring the Work. Sizing may be done by hand measurement with a hand plug gage. Taper may be checked with a taper plug gage. Various gages are made for this purpose. The size of the run and the production requirements generally control the method used. Telescoping gages are also used for taking exact inside sizes of holes. The size is then transferred to a caliper or micrometer caliper.

Size of Wheels. The relation between the sizes of the hole and the wheel is not of great significance except as it affects the strength and rigidity of the grinding-wheel spindle extension, the allowable wheel wear, and power economy. If the diameters of the wheel and hole are too close, the area of contact may become so great that difficulties would be encountered, especially in large holes. The difference between the diameter of the wheel and the hole varies according to the size of the hole. The difference is small when the hole is small, but is greater when the hole is large. For instance, in grinding a ½-inch hole, a ⁷⁄₁₆-inch wheel would be in order. A 2-inch hole would normally be ground with a 1¾-inch wheel.

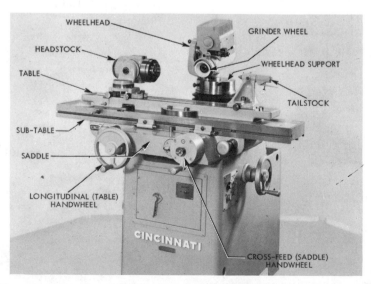

Fig. 19. The universal cutter and tool grinder has a wheelhead that can be swiveled and positioned for a variety of setups.

Universal Cutter and Tool Grinder

The universal cutter and tool grinder, Fig. 19, can be set up for a wide variety of grinding operations. With various attachments, it can do internal, gear cutter, cylindrical, and surface grinding. It is used to sharpen milling-machine cutters. Many attachments and specially formed grinding wheels are available.

Because of the wide variety of work performed, no attempt will be made to cover completely all the possible setups.

Construction of Universal Cutter and Tool Grinder

The machine is constructed with a heavy, rugged, box-type bed or base, see Fig. 19. A saddle is mounted directly on top of the bed. It moves on antifriction ball bearings on hardened ways. Mounted on the saddle is the column supporting the wheelhead. The column is movable up and down and swivels to either side. The wheel angles required for various kinds of work are thus made possible. The saddle also provides the means for moving the wheelhead forward and back. The grinding wheel can then be brought in contact with the work.

Over this saddle is mounted the top base containing the gears and mechanism which control the table movement. The top base supports the ways for the bearings on which the table rests and moves.

The sub-table moves longitudinally on these ways and provides the traverse of the work before the grinding wheel.

Table. The worktable is mounted on the sub-table. It has T-slots for mounting the work and the attachments used on the machines.

The worktable is designed to swivel. With the table slightly turned, the operator can grind tapers.

Traverse of the table is controlled by a handwheel. Handwheels also control the movement of the column and saddle.

The table carries the headstock and tailstock, which, in turn, hold the workpiece.

Headstock and Tailstock. A headstock and a tailstock similar to those on an ordinary lathe are mounted on the table, see Fig. 19. For cylindrical grinding, the workpiece is positioned between the centers and driven exactly as in the ordinary lathe. The work rotates toward the operator, and the grinding wheel also turns toward the operator. At the point of contact, the wheel and the work are moving in opposite directions.

The headstock rotates a chuck for internal grinding operations.

Wheelhead. Back of the headstock and tailstock is the wheelhead. The abrasive wheel or wheels are driven by a separate motor in the wheelhead. The wheelhead can be swiveled and positioned on the base for varied setups.

Cutter Grinder Wheels. Fig. 20 shows the common wheels used on cutter grinders. At *A* is the plain wheel; at *B* is the dish or saucer; and at *C* is the cup. Wheel shape depends on the cutter to be sharpened.

Accessories for Universal Cutter and Tool Grinder

The range of operations that can be performed with the universal cutter and tool grinder is broadened by means of suitable accessories. Among such accessories are the center gage and two types of tooth rests, which will be discussed here.

Center Gage. The center gage is useful in making setups for cutter and tool grinding. It is used for setting the center of the head, the foot-

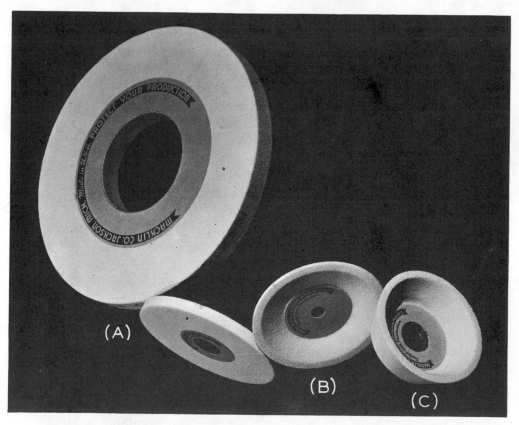

Fig. 20. Wheels used on cutter grinders; At *A*—a plain wheel, at *B*—a dish or saucer wheel, at *C*—a cup grinder. (Macklin Co.)

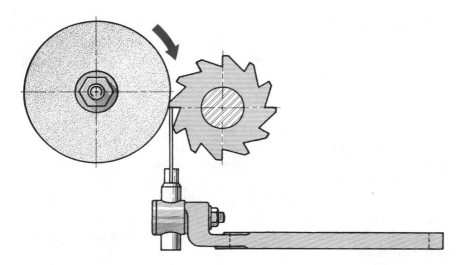

Fig. 21. A tooth rest is used to support a cutter tooth.

stock, and the tooth rest in line with the center of the wheel spindle. Alignment is necessary in grinding cylindrical work and in setting milling cutters and reamers for proper tooth clearance.

A slot is milled in the wheelhead in line with the center of the spindle. The center gage, Fig. 28, can be set at the same height as the center of the spindle of the head and footstock. The knee of the machine is adjusted vertically until the center gage touches the bottom of the slot. The two centers can then be brought into perfect alignment with the center of the wheel spindle.

Tooth Rests. A tooth rest is an important tool-holding device used to support the tooth while the cutter is in the grinding machine. It can be mounted either on the machine table or on the wheelhead,

depending on the cutter to be sharpened. The distance between the end of the rest and the base should be short to support the tooth rigidly. Fig. 21 shows the tooth rest set to support the tooth from the bottom. In Fig. 22, the tooth rest is holding

Fig. 22. A tooth rest can be used to support a helical mill when it is ground with a cup wheel. (Cincinnati Milling Machine Co.)

437

the tooth of a helical mill being ground.

Usually, there are two types of tooth rests furnished with the machine: the plain and micrometer. The micrometer tooth rest has a micrometer adjustment, graduated in thousandths of an inch for close adjustment.

The tooth rest can be mounted on the top or bottom of the wheelhead, or on the table. The extension bar and universal joint make the various positions possible.

Setups and Operations

The working efficiency of a cutter is determined by the keenness of its cutting edge. A keen edge lasts much longer if the cutter is ground with the proper clearance and rake.

An improperly ground edge is soon dulled and leaves a poorly finished surface on machine parts. The

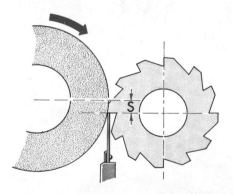

Fig. 23. The wheel can be rotated away from the cutting edge when a plain wheel is used to grind mill cutters. Burrs must be removed with a stone after grinding.

edge of a milling cutter breaks down in time. Then it is necessary to grind away a large portion of each tooth to restore the cutting edges. Very little metal is ground away if cutters are sharpened at the first sign of dullness.

In general, cutters can be divided into two groups according to design or method of sharpening:

Group 1. Cutters in this group are sharpened by grinding the clearance angle behind the cutting edge of each tooth. This grinding is done on the periphery of the cutter or on the sides. Plain milling cutters, cut-off saws, reamers, and helical mills are ground on the periphery. See Fig. 22. Face milling cutters, shell mills, and end mills are ground on the end or side.

Group 2. Cutters in this group are sharpened by grinding the cutting faces of the teeth. The profile of the teeth is not altered, as these cutter teeth have a definite profile for producing a given contour or form in the workpiece. Cutters of this type include those used for machining taps, form tools, involute gear tooth cutters, etc.

There are two methods of grinding cutters and reamers. They are based on the direction of rotation of the grinding wheel in relation to the cutting edges of the teeth. Fig. 23 illustrates a setup for sharpening a plain milling cutter. Here a plain wheel is used. The direction of rotation is away from the cutting edge. The rotating wheel tends to hold the front face of the tooth being ground in contact with the tooth rest as the cutter is fed past the wheel. There is no tendency for the tooth to dig into the wheel. This method, however, leaves a slight burr on the cutting edge, which must be removed by stoning. It also has a tendency to draw the temper of the steel.

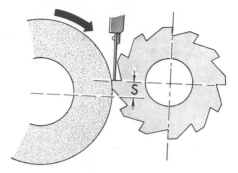

Fig. 24. A plain wheel can be rotated against the cutter.

With the cutter position reversed, Fig. 24, the direction of wheel rotation is toward the cutting edge. The centers of the wheel and cutter are offset so as to provide the desired relief as shown in Figs. 23 thru 26. There is a tendency for the cutter to dig into the wheel. Great care must be used to hold the cutter on the tooth rest, since the rotation of the wheel is toward the cutter edge, it tends to turn the cutter away from the rest. If the cutter turns, the tooth is ruined. However, when this method is used by a careful, skilled workman, it produces a keener cutting edge without burrs.

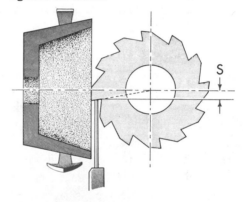

Fig. 25. Cutters can be ground with a cup wheel. The "far" side of the cup wheel contacts the cutter and grinds away from the cutting edge.

Cup wheels are also used for grinding cutters and reamers, Figs. 25 and 26. The two methods of using cup wheels are similar to those used with plain wheels. The comments on plain grinding wheels apply to cup wheels. More care, however, should be taken in using cup wheels because of the greater area of contact. The cuts should be light.

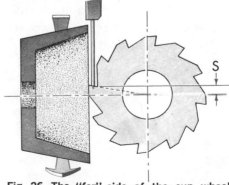

Fig. 26. The "far" side of the cup wheel can be rotated against the cutting edge.

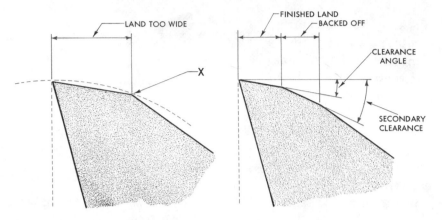

Fig. 27. To control the width of a land a secondary clearance must be provided on milling cutter teeth.

Clearance Grinding. Correct clearance back of the cutting edge is important. Insufficient clearance will make the teeth drag over the work and will result in friction and slow cutting. Too much clearance will cause the teeth to wear rapidly and produce chatter. The clearance or relief in a cutting tool is that area removed from the teeth behind the cutting edges, see Fig. 27. Each edge must be sharp and the clearance angle correct.

The type and diameter of the cutter and the metal to be machined determine the proper clearance an-

gle. Small cutters require greater clearance than large cutters. Cutters used on relatively soft metals— bronze and aluminum — require greater clearance than cutters used on hard metals, such as steel and iron. Table I shows the angles for cutters of average diameter.

Suggestions on Grinding Proper Clearance Angle. The clearance angle is produced on the cutter by properly locating the wheel, the cutter, and the tooth rest. The kind of wheel used, the shape of the cutter, and the location of the rest help determine the method to use. The

TABLE I. CLEARANCE ANGLES FOR TEETH OF AVERAGE-DIAMETER CUTTERS

WORKPIECE METAL	ANGLE DEGREES
ALUMINUM	10 To 12
BRONZE, CAST	10 To 15
BRASS AND SOFT BRONZE	10 To 12
CAST IRON, FAST FEEDS	3 To 7
COPPER	12 To 15
HIGH-CARBON AND ALLOY STEELS	3 To 5
ORDINARY LOW-CARBON STEEL	0 To 7
STEEL CASTINGS	6 To 7
TOBIN BRONZE, VERY TOUGH	4 To 7

440

wheel may be either a plain wheel or a cup wheel. The cutter may be plain (straight) or tapered, with straight or spiral teeth. The tooth rest may be located on the wheel-head or on the table.

Plain Wheel Used in Clearance Grinding. When using a plain wheel, the clearance angle depends upon the diameter of the wheel. With a cup wheel, the diameter of the cutter is the determining factor. In general, when grinding a cutter with the teeth facing downward, the center of the work is below the plane of the axis of the wheel; and when the teeth face upward, the center of the work should be above the center of the wheel.

Cup Wheel Used in Clearance Grinding. In using a cup wheel, the cutter center, the wheel center, and the tooth rest when mounted on the wheelhead are brought into the same plane. The proper height setting of the tooth rest will provide the required clearance.

Width of Land on Milling Cutters. The land on each tooth of a milling cutter is the narrow surface immediately behind the cutting edge, see Fig. 27. The land is ground to the clearance angle. It varies in width from about $\frac{1}{64}$ inch to $\frac{1}{16}$ inch, depending upon the type and size of the cutter. After repeated grinding, the land may become so wide as to cause the heel of the tooth to drag on the work, Fig. 27.

Fig. 28. A center gage should be used to position a shell end mill for grinding. (Cincinnati Milling Machine Co.)

Fig. 29. Grinding a radius on a shell end mill. (Cincinnati Milling Machine Co.)

Machine Shop Operations and Setups

To control the width of the land, a secondary clearance, usually from 20 to 30 degrees, is ground, Fig. 27.

The clearance of the end teeth of end mills should be about 3 to 5 degrees. On formed cutters or involute gear cutters, clearance need not be considered in resharpening. The teeth are so formed that when ground radially on the face, the clearance remains the same.

Fig. 22 shows a cup wheel being used to sharpen a helical mill.

Fig. 28 shows a shell end mill being positioned at proper height in relation to a grinding wheel by the use of a center gage. Fig. 29 shows the operation of grinding a radius on a shell end mill. Fig. 30 shows a close-up view of a cup wheel being used to grind a large-diameter saw on a cutter and tool grinder.

Fig. 30. A cup wheel can be used to sharpen a circular saw. (Cincinnati Milling Machine Co.)

Grinding Wheels

The removal of stock when accomplished by grinding is done with a grinding wheel. Its action upon a metal surface may be compared to that of a milling cutter, in that each abrasive particle becomes a cutting tool or edge.

Grinding wheels are made of small, sharp crystals of very hard material held together by strong, porous bonds. As each grain whirls past the work, it cuts a small chip. The accumulated effect of this action by many grains gives the work its smooth, straight surface. As each grain becomes dull, it is torn from the bond and fresh crystals are presented to the work.

Grinding Wheel Construction

Five important factors determine the nature of a particular wheel: the kind of abrasive, grain size, grade, bond, and structure.

Kinds of Abrasive. Modern grinding wheels are formed of two kinds of material: silicon carbide and aluminum oxide. Silicon carbide has harder and sharper crystals, but aluminum oxide is very hard and tough. Aluminum oxide does not break down so readily when used on high-strength materials. Silicon-carbide abrasive wheels are preferred for grinding materials low in tensile strength, such as cast iron, brass, aluminum, hard rubber, and building stone. Aluminum-oxide abrasive wheels are better used on most steels, malleable iron, and tough bronzes. Most manufacturers make a few variants of each type of abrasive to meet special job conditions.

Grain Size. The size of the abrasive grains in modern wheels varies widely. Grains are carefully screen-graded. They vary from grains that will pass through a screen having 8 meshes to a linear inch (but not through one having ten meshes) to grains that will pass through a 240-mesh screen. Even finer "flour" sizes are made. In general, coarse grains are used for rapid stock removal; fine grains are used for fine finish; coarse wheels for soft ductile materials; fine wheels for hard brittle work. Many jobs, however, can be both roughed and finished with a medium-grain wheel.

Bonding Material. The grains are held in the wheel by a bonding material. There are six chief types: vitrified, silicate, resinoid, rubber, magnesite, and shellac. About 75 per cent of all grinding wheels have a vitrified bond. Vitrified-bond wheels are strong and porous and are not affected by water, acid, or oils at ordinary temperatures. Silicate-bond wheels release grains more readily, and so produce a better cutting action, as new sharp grains are constantly being exposed. Usually, silicate is the bond used in making large wheels. Shellac wheels are light-duty, high-finish wheels frequently used in cutlery work. Resinoid and rubber bonds are used for high-speed wheels, such as cutoff wheels.

Grade. The strength with which the grains are held by the bond is called the grade. Grade might also be described as the hardness of the wheel. It depends, usually, on the size of the bond posts between abrasive grains. The ideal grade for a job would be one in which the cutting grains break free at the exact moment when they become so dull that they no longer produce the desired surface in the desired time. There are other factors than the manufacturer's grade which influence this condition however.

Structure. Abrasive grains are not packed tightly in the wheel, but are distributed through the bond. The relative spacing is referred to as the structure. This is an important factor in the cutting action of the wheel. The finest finishes are produced with a closely spaced structure. However, close-grained wheels allow very little clearance for chips and therefore tend to clog easily. Moreover, close-grained wheels do not release dulled grains as readily as wheels with more open structure. Fine structures are best for brittle materials and finishing cuts; coarse structures are best for heavy cuts and ductile materials.

Wheel Markings

At one time, each manufacturer used his own system of indicating the nature of a grinding wheel. Now, however, a standard system has been set up. This has been accepted by the Grinding Wheel Manufacturers Association. The marking consists of six parts: (1) abrasive, (2) grain size, (3) grade, (4) structure, (5) bond and (6) manufacturer's record. A typical marking might be:

$$B\ 46\ M2\ 5\ V\ E$$

This marking would designate a vitrified bonded wheel (V) of refined aluminum oxide (B) of medium grain (46), grade (M), and structure (2). The letter E is the manufacturer's record.

The abrasives naturally fall into two groups: the aluminum-oxide group (A) and the silicon-carbide group (C). The abrasive is represented by a letter as follows:

(A) Aluminum oxide, regular
(B) Aluminum oxide, refined
(AB) Mixture of regular and refined aluminum oxide
(C) Silicon carbide, regular
(CD) Silicon carbide, refined
(D) Corundum
(E) Emery
(F) Garnet

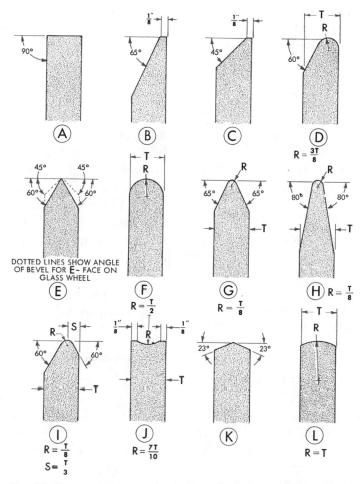

Fig. 31. Standard shapes of grinding wheel faces. (Norton Co.)

The grain size is given by the number of meshes per linear inch of the grading screen, through which the grains are sifted as they are sorted for particle size. The screen sizes are standard throughout the industry and provide a standardization of grinding wheel grain size.

The numbers run as follows: 10, 12, 14, 16, 20, 24, 30, 36, 46, 54, 60, 70, 80, 90, 100, 120, 150, and 220. Additional sizes occasionally used are 240, 280, 320, 400, 500, 600.

The grade is given by a letter of the alphabet, A to Z, soft to hard in all bonds or processes.

The structure of the wheel is indicated by a number of from 1 to 15, where 1 is the most dense and 15 is the most open.

445

The bond is given as follows:
V—Vitrified
S—Silicate
E—Shellac or elastic
R—Rubber
B—Resinoid (synthetic resin)
O—Oxychloride

The sixth position is reserved for any private factory records of the individual manufacturer.

Wheel Shapes and Faces

Manufacturers of grinding wheels have standardized on nine wheel shapes and twelve faces. Fig. 20 shows the wheel shapes. You will see these different wheels in use in the illustrations of setups and operations for grinding. In purchasing different shaped wheels, the manufacturer's catalogue should be studied carefully. Often, it will be more advantageous to use one type of grinding wheel rather than another. Proper selection, of course, comes with experience.

The standard shapes of grinding wheel faces are shown in Fig. 31. It is necessary that these faces be retained during the lifetime of the wheel in order to form the proper shape or contour for given grinding operations. This is accomplished by truing the wheel with the proper wheel dresser.

Storage of Wheels

Great care should be used in the storage of wheels. Suitable racks or bins should be provided to accommodate the various types of wheels carried in stock.

Most straight and tapered wheels can be best supported on edge in racks. Thin rubber, shellac, and other organic bonded wheels should be laid flat on a plane surface to prevent warpage. Cylindrical wheels and large cup wheels should be stocked on the flat sides with corrugated paper or other cushioning material between them. Small cup and other shape wheels and small internal grinding wheels may be stored in boxes, bins, or drawers. Very large wheels can be stored in their original containers.

Inspection of Wheels

As soon as wheels are received, they should be closely inspected to make sure they have not been injured in handling or shipping. Cracks can be detected by tapping the wheel gently with the handle of a screwdriver while the wheel is suspended. If the wheel sounds cracked, it should not be used.

All wheels do not produce the same tone when rung; nor does a low tone signify a cracked wheel. Vitrified and silicate wheels emit a clear metallic ring; organic bonded wheels give a less clear sound, but the sound of a cracked wheel will be perfectly apparent. Oil- or water-soaked wheels do not ring clearly.

Operating and Maintenance Tips

Before Starting the Machine

Check the Wheels. Use the widest possible wheel for most economical operation. When straight infeed grinding, the wheel face should be ⅛ to ¼ in. wider than the surface to be ground.

Examine the face of the wheel: a loaded wheel will mar the finish of the work; a glazed wheel will rub, burn and ruin work.

Check the Machine. Check the stops, feed trips, and levers to make sure the wheel will not run into the machine or the work. For traverse grinding, set the table dogs so that no more than one-third of the width of the wheel runs off the work.

Check the Workpiece and the Workholding Devices. The angle of the work center holes must match the angle of the work center points. To grind work round, true work center points and round center holes in the work are essential. Taper of centers should fit spindle tapers. With hardened work, it is desirable to lap the center holes to be sure they are accurate and clean.

Clean and oil the work centers and clean the work center holes before putting the work on the machine. Lubricate the work center holes with "Center Lube" or red or white lead mixed in oil.

Always clamp the work drive dog securely to the work. Clamp it on the largest diameter. The driving pin must not fit the driving dog too tightly. A small amount of play is desirable. To avoid distorting slender stock during the grinding operation, adjust the footstock center to hold the work firmly but without excessive pressure.

During the Grinding Operation

Run new wheels at full operating speeds for at least one minute before applying to work. Step safely aside after pressing the start button. Flush the wheel heavily with coolant after starting.

When engaging the wheel to the work, apply a light pressure. Undue pressure is harmful to the wheel, work supports and spindle bearings. Don't feed the wheel in too fast. This will cause uneven wear, inaccurate work and poor finish. After the first cut has been taken, measure the work for straightness.

Use the automatic wheelfeed mechanism wherever possible. It speeds production and ensures greater uniformity of grinding action.

Always use plenty of coolant. An

insufficient amount will cause the wheel to cut out of round; will permit the workpiece to heat up; will result in a rubbing rather than a grinding action.

Keep ample tension on the belt drive at all times. Check the wheel spindle and bearings often for proper clearance and lubrication. However, do not tamper with the bearings unless absolutely necessary. If the spindle runs hot, check the level of the oil reservoir. Use only the oils recommended by the manufacturer for the spindle bearings. A cool-running spindle may be an indication of too much clearance between the spindle and the bearing.

When grinding hollow or tubular work, feed the wheel slowly to prevent distortion from heat. Dress the wheel often, especially when internal grinding.

Tips for Fine Finishes. Use a fine grain wheel of resinoid, rubber or shellac bond. Reduce wheel speed, work speed and infeed. Use plenty of coolant. The coolant should be filtered and cleaned (changed) frequently. True the wheel by moving the diamond slowly across the face.

Tips for Fast Stock Removal. Use a coarse grain wheel. The speed of the work carriage should be a bit less than the width of the wheel for each work revolution. Reduce work speed and traverse speed and increase wheel speed. True the wheel by moving the diamond rapidly across the face.

Grinding Machine Maintenance

Make certain that the machine is level and solidly positioned. There should be no machine vibration. Check all V-belt drives to see that sheaves are in line and that belts have ample tension and are free of oil and moisture.

Always clean the top of the swivel table before changing the position of the headstock or footstock. Keep dirt from oil holes and bearings. Clean the inside of the wheel guard periodically. This will help in producing a finer finish.

Work centers should be ground periodically on the center grinding attachment. Use a center gage against a strong light to check the accuracy of the angle. The point must be concentric with the taper.

Be sure that plenty of lubricating oil is delivered to the table and wheel slide ways at all times. Follow the grinder manufacturer's instructions regarding lubricating and hydraulic oils.

Drain oil frequently if the machine is used regularly: every six months is practical. Use good, clean oil for flushing. Examine the oil to check whether coolant has seeped into the system. Keep oil at the level recommended by the manufacturer and clean oil filters regularly.

Safety Precautions—Grinding Machines

1. Wear safety glasses at all times when performing any grinding operation, even though the wheel has a glass shield.

2. Always test grinding wheels for possible cracks. Tap the wheel with a wood object such as a screw driver handle. A solid wheel will sound a clear ring; a damaged wheel will have a dull sound.

3. Guard against excessive wheel speed. Check the Table of Grinding Wheel Speeds for the correct speed. This speed should not be exceeded with new wheels. As wheels wear to smaller diameter, the rpm may be safely increased.

4. Do not operate the grinder unless all guards are in place and the wheel is secure on the spindle.

5. Before starting the machine, make sure the workpiece is properly mounted, that the footstock is locked in position and that the drive plate and dog are correctly adjusted.

6. Use the proper wheel specified for the job.

7. After starting the machine, always step back from the rotating wheel.

8. Don't feed the wheel too fast; otherwise, the workpiece may be forced from between centers.

9. When shoulder grinding, do not permit the side of the wheel to press too hard against the work. The work may be forced from between centers or the wheel may break from excessive side pressure.

10. Stop all motors before making adjustments to the machine.

11. When dressing the wheel, make sure the diamond is securely locked in position.

12. Never hand gage the work while it is turning.

13. When mounting or removing work, back the wheel far enough away to provide adequate clearance for your hands.

14. Never wear long sleeved shirts, neckties, jewelry, or anything else which might get caught in the machine when operating a grinder.

15. When leaving the grinder for an indefinite period, stop the machine.

NOTE: Please see review questions at end of book.

Steel and Its Alloys:

Types and Characteristics

A great many people in industry today are involved in designing and making tools. Tools like those shown in Fig. 1 are made by toolmakers and are used daily in all phases of manufacturing. Our high standard of living today is partly a result of mass production methods, which in turn are made possible by tools of superior design and construction.

If a tool is to operate efficiently and give maximum production, the following steps must be taken:

1. The tool designer must study the part to be machined and design a tool which will best do the job.
2. He must specify a steel that will meet the requirements of strength, hardness, etc., of the machining operation.
3. The toolmaker must make the tool accurately in accordance with the designer's plans.
4. The tool must be properly heat treated.

Metals for Tools

To be a toolmaker or good student of machine shop practices, it isn't necessary to have a thorough knowledge of all the technicalities of steel manufacture. On the other hand, ignorance of common terms which are likely to be encountered —both in reading and in conversation — will prove embarrassing if your job is selecting, purchasing, or working tool steel. In this chapter and those following we will learn some of these terms.

Specifically, in this chapter we will

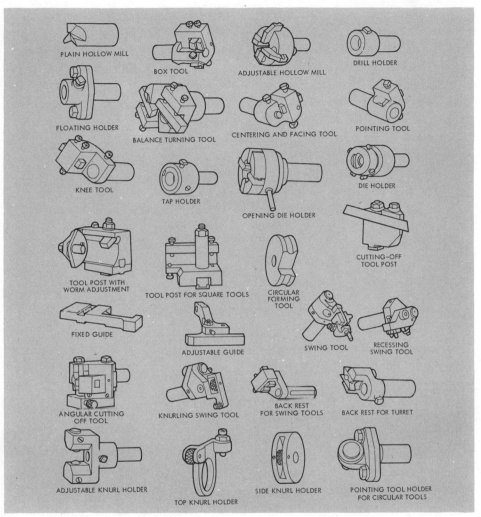

PLAIN HOLLOW MILL

BOX TOOL

ADJUSTABLE HOLLOW MILL

DRILL HOLDER

FLOATING HOLDER

BALANCE TURNING TOOL

CENTERING AND FACING TOOL

POINTING TOOL

KNEE TOOL

TAP HOLDER

OPENING DIE HOLDER

DIE HOLDER

TOOL POST WITH WORM ADJUSTMENT

TOOL POST FOR SQUARE TOOLS

CIRCULAR FORMING TOOL

CUTTING-OFF TOOL POST

FIXED GUIDE

ADJUSTABLE GUIDE

SWING TOOL

RECESSING SWING TOOL

ANGULAR CUTTING OFF TOOL

KNURLING SWING TOOL

BACK REST FOR SWING TOOLS

BACK REST FOR TURRET

ADJUSTABLE KNURL HOLDER

TOP KNURL HOLDER

SIDE KNURL HOLDER

POINTING TOOL HOLDER FOR CIRCULAR TOOLS

Fig. 1. These tools and attachments for standard use on automatic screw machines were made by toolmakers. The steel used must meet the requirements of strength, hardness and endurance of the machining operation it was designed to perform. (Browne & Sharpe Mfg. Co.)

study the characteristics of different kinds of metals, especially those related to tools. In the next chapter we will study the various operations carried out in order to change the characteristics of a metal by the application of heat, called *heat treating.*

Iron

About one-twentieth of the earth's crust is iron; only oxygen, silicon,

451

and aluminum are present in greater quantities. Iron ore is found on every continent and in nearly every country. In the United States it has been found in every state and is obtained commercially in 28 of them. In this country, however, most ore is obtained from those sections in which the ore yields at least 40% iron.

Pig Iron. The type of iron ore most abundant in the United States is a heavy reddish, stone-like material called *hematite*. Before this ore can be used commercially it must be processed in a blast furnace charged with iron ore, coke, and limestone in the proper proportions. The mixture which results from the smelting operation is called *pig iron*. The pigs contain about 92% iron and 3.5% to 4% carbon, with small quantities of silicon, manganese, and phosphorus.

Cast Iron. When pig iron is remelted in a cupola with scrap iron, flux, and other materials, and cast into molds, the end product is *cast iron*. Fundamentally cast iron is an alloy whose chief elements are iron, silicon, and carbon. The general classification of cast iron includes gray cast iron, white cast iron, and mottled, chilled, malleable, and alloy cast irons.

Malleable Iron. Malleable iron is made from white iron castings, which are produced by melting a charge of ordinary pig iron and scrap

under controlled conditions to give, on solidification, a hard, brittle white iron. To produce malleable iron it is necessary to anneal these castings approximately 48 to 60 hours in a furnace at a temperature of about 1600°F. Malleable iron can be machined with great ease and hammered into different shapes without cracking. The metal is used for parts which are subject to continual shock and vibration, such as railroad and automobile equipment, and farm tools.

Carbon Steels

Steel is the basic metal of our industrial economy. An automobile built in 1910 used 11 kinds of metal; today's car uses over 160 kinds of steel. Steel is made from cast iron by removing excess carbon and some of the other impurities. Molten cast iron placed in a converter or open-hearth furnace has heated air blown through it to burn out undesired impurities.

The percentage of carbon is the most important factor relative to the physical properties of steel, although other materials, such as silicon and manganese, are also present. Steel exists from about 0.08% carbon to about 2.00% carbon. Although all steel contains carbon, the terms *carbon steel* and *plain carbon steel* are used to distinguish a steel to which no special alloying element, such as nickel, tungsten, or chromium has been added in appreciable amounts.

Steel is considered one of our strongest materials. Steel needles can be made hard enough to scratch glass; also, the metal can be made soft enough so that a common needle will scratch it. Steel can be given a hard surface to resist wear and a soft tough core to resist breaking. The metal has magnetic properties that makes it valuable for use in electrical equipment.

Low Carbon Steel. As its name implies, low carbon steel contains from 0.05% to 0.30% carbon. It is made in the Bessemer converter, open-hearth furnace, or crucible furnace. If this steel is rolled, while cold, between highly finished rollers under great pressure, it comes out with a very smooth finish and exact dimensions and is referred to as *cold-rolled steel*. Low carbon steel is also known as machine steel, machinery steel, and mild steel. It does not contain enough carbon to cause the steel to harden to any great extent when heated to the critical point and quenched in oil, water or brine. It may be heat treated to increase its resistance to wear.

Low carbon steel has many uses, such as in nails, screws, wire fencing, bolts, nuts, washers and similar items where the surfaces do not receive much wear. It is also used in forge equipment, rivets, chains, and machine parts which do not require great strength. Low carbon steel is used for almost every purpose for which wrought iron is used and has gradually replaced wrought iron.

Medium Carbon Steel. Medium carbon steel, SAE 1030 through 1050 (0.30% to 0.50% carbon), contains enough carbon to cause the steel to become partially hard upon proper heat treating. These steels can be used whenever requirements of higher strength and wear resistance than can be obtained in the low carbon groups are desired. Uses include gears, shafting, hand tools such as pliers and screw drivers, etc., and a few edge tools such as tin snips and bush knives.

High Carbon Steel. Steels from SAE 1055 through 1095 (0.55% to 1.70% carbon) are classified roughly as high carbon steel. These steels are used chiefly for heavy machinery parts, agricultural cutting tools, springs, dies and punches, and tools of various kinds. High carbon steels possess greater strength than low or medium carbon steels and can be given a thorough and uniform hardness to increase wear resistance and edge cutting service.

Hot-Rolled Steel. Hot-rolled steel is usually a low carbon steel that is rolled to shape while hot. After pig iron has been made into steel in a furnace, the molten metal is poured into large molds; the cooled shapes are called *ingots*. These ingots are rolled while hot

453

between heavy, powerful steel rollers which are moved closer together as the hot steel travels back and forth between them, producing what is known as *hot-rolled steel*. The surface produced is comparatively rough and must be machined before the steel is usable in various types of tools.

Cold-Rolled Steel. Cold-rolled steel is made by taking hot-rolled steel after it is cold and rolling it between highly finished rollers under great pressure. First it is necessary to remove the black skin or scale from the hot-rolled steel by a process known as pickling with sulphuric acid. The rolling gives the steel a smooth bright finish and also permits exact sizes to be reached. Because of this, the steel can be used for many purposes without additional finishing or machining.

Cold-Drawn Steel. Cold-drawn steel is bar steel that is drawn through a smooth hole in a block of hardened steel called a die. After hot-rolled steel cools and is pickled, it is drawn through the die, which reduces the bar to the desired size. After the drawing process the steel is passed through straightening rolls. This same process is used in making wire.

Drill Rod

Drill rod is a carbon tool steel which is described under alloy steels, and may be purchased in round or square lengths. It is cold drawn and in some instances is ground or polished to very accurate limits. Drill rod has a great many uses. The tool designer specifies drill rod for various types of small cutters, pins of all types, and shafts where it is necessary for them to be hardened and ground. Typical sizes available for drill rod are $\frac{1}{8}$ to 1″. Larger diameters up to 2″ can also be supplied on special request. It is usually supplied in lengths of 3 feet.

Cast Steel

To secure the many irregular shapes needed in manufacturing today, molten steel is poured into molds in the same way that cast iron is. The resulting metal is called *cast steel*, and the molded shape is called a steel casting. At times, casting methods are the only feasible way to produce certain large, irregularly-shaped products.

The composition of cast steels is similar to that of wrought steels; however, the silicon and manganese content differs. These elements are usually present in greater amounts in cast steels to insure thorough deoxidation and to prevent gassing of the metal in the mold. Cast steels, like wrought steels, may contain different alloying elements such as nickel, chromium, vanadium, and copper to give desirable combinations of hardness, tensile strength, and toughness not readily secured in plain carbon steels. The carbon content of steel castings varies from 0.25% to 0.65% carbon.

Steel castings may be heat treated for purposes of improving the as-cast structure and the mechanical properties. The customary heat-treating procedures are annealing, normalizing, annealing and normalizing, annealing and tempering, and spheroidizing. It is not uncommon, however, to quench and temper steel castings where the size, shape, and composition are not subject to serious distortion and cracking due to quenching.

Alloy Steels

Alloy steels have some element or elements other than carbon which have been added to change the characteristics and properties of the basic steel. Some of the more common alloying elements are: chromium, manganese, vanadium, silicon, nickel, molybdenum, and tungsten. The effect of each of these elements will be explained when the steel alloy containing the elements is discussed.

Although alloy steels are relatively more expensive than plain carbon steel of the same nominal carbon content, their added cost may be justified in many instances by the advantages gained using the alloyed product. As a rule, the additional cost of the alloyed steel is roughly proportional to the kind and the amount of alloying elements used. On the other hand, it would be unwise to select, on the basis of hardenability and other characteristics, a steel of relatively high alloy content when a less expensive, low alloy steel would adequately meet the hardenability or other requirements.

Reasons for Adding Alloying Elements

Alloying elements are added to ordinary steel for the purpose of changing the behavior of the alloy during heat treatment and altering the mechanical and sometimes the physical properties. The addition of these elements are made for one or a combination of the following reasons:

1. To improve the tensile strength without lowering the ductility appreciably.
2. To improve the hardenability.
3. To improve the toughness, particularly at subnormal temperatures.
4. To improve the wear resistance.
5. To enable successful quenching with less danger of distortion and cracks from cooling.
6. To retard the softening rate of

fully hardened steel during tempering or during exposure at elevated temperatures.

7. To improve the corrosion-resisting properties.
8. To bring about fine grain size in the steel.

Alloying Elements

Chromium. When alloyed with steel, chromium causes the hardness to penetrate deeper and when present in sufficient quantity will give oil-hardening properties. Chromium contributes wear resistance and toughness characteristics to the alloy and raises the temperature necessary for hardening.

Manganese. Manganese is a constituent of all steels and where it does not exceed 0.80% is used as a purifier to help offset the effects of impurities in the metal. When added to steel in the amounts from 1% to 15% an alloy with hardness and resistance-to-wear properties is produced.

Vanadium. Vanadium is alloyed with steel to act as a deoxidizer and cleanser and produces a fine-grained steel. When used with tool steel it resists shock and jarring better than straight carbon steel.

Silicon. Silicon has about the same effect on steel as manganese, in that it acts as a purifier and cleanser. When silicon is added to steel in large quantities it imparts definite magnetic properties. In tool steel silicon imparts hardening and toughening qualities.

Nickel. Nickel is used as an alloying element in steel to increase its strength and toughness and to resist certain heat-treatment strains. For general use the amount of nickel varies from 1% to 4%. When added to steel in quantities of from 24% to 36% it causes the steel to become practically non-magnetic and reduces the coefficient of expansion.

Molybdenum. Molybdenum increases the endurance limit and yield point of steel and is added in quantities of from 0.10% to 0.40%.

Tungsten. Tungsten is added to straight carbon steel to produce a fine and dense grain structure, and when used in small quantities helps cutting tools to retain a sharp cutting edge. When tungsten is added in amounts from 17% to 20%, with certain other elements, it produces a steel that is capable of retaining its hardness at speeds far in excess of those for which tool steel cutters can be used.

Tool Steel

The name *tool steel* is rather generally applied to any steel used to cut, form, or otherwise change the shape of materials. Originally, tool steels were mainly composed of iron and carbon with small amounts of manganese, silicon and other impurities.

The development of alloy tool

steels, especially the high speed types, make it possible for the tool-maker to meet a great variety of specific production problems, and undoubtedly has been responsible to a large measure for the rapid growth of industry since the turn of the century.

Tool steel in America is made by a small group of specialty steel mills, many of which have experience dating back into the last century. All these companies have brand names and each manufacturer jealously guards the quality of the product that carries his brand and backs it up with his guarantee.

Practically all carbon tool steel is made either by the crucible or electric furnace process. Great care is exercised to keep the impurities, such as phosphorus and sulphur, and the non-metallic inclusions, including dirt, at a minimum. These steels may contain 0.65% to 1.40% carbon, depending upon the use to which they are put.

Carbon and Low Alloy Tool Steels. Carbon tool steels were used exclusively for all tools and dies before the development of high speed and alloy tool steels. Even with high speed and alloy steels, carbon tool steels are still widely used in industry today. Carbon tool steels are manufactured in a number of carbon ranges or tempers and in different grades. The low carbon range (0.70% to 0.90%) is most

suitable for tools subject to shock, and the higher carbons (1.10% to 1.30%) are most suitable for cutting tools where a keen edge is required. By adding a small amount of chromium, a low alloy steel is developed for use in making tools where a higher degree of hardness and greater depth of hardness penetration are desired. This kind of steel is necessary for drawing dies and rolls in cold work operations, for drills, reamers, and cutlery dies. Fig. 2 shows a penetration-fracture classification chart for carbon tool steels.

Cold Work Steels. In industry today many of the manufacturing processes include blanking and cold-forming operations. Alloy die steels have been found to be better than carbon tool steels for tools and dies where both longer life and better nondeforming qualities are required. Dies such as blanking, forming, thread rolling, coining and trimming dies, and bushings, gages, taps, and broaches are good examples of tools made from cold work steels. High carbon, high chrome steels, manganese oil-hardening steels, and chrome-tungsten steels are included in this group.

Shock-Resisting Steels. This group of steels was developed to replace carbon steels in those applications where resistance to shock is most important, and where hardness and wear resistance is secondary.

457

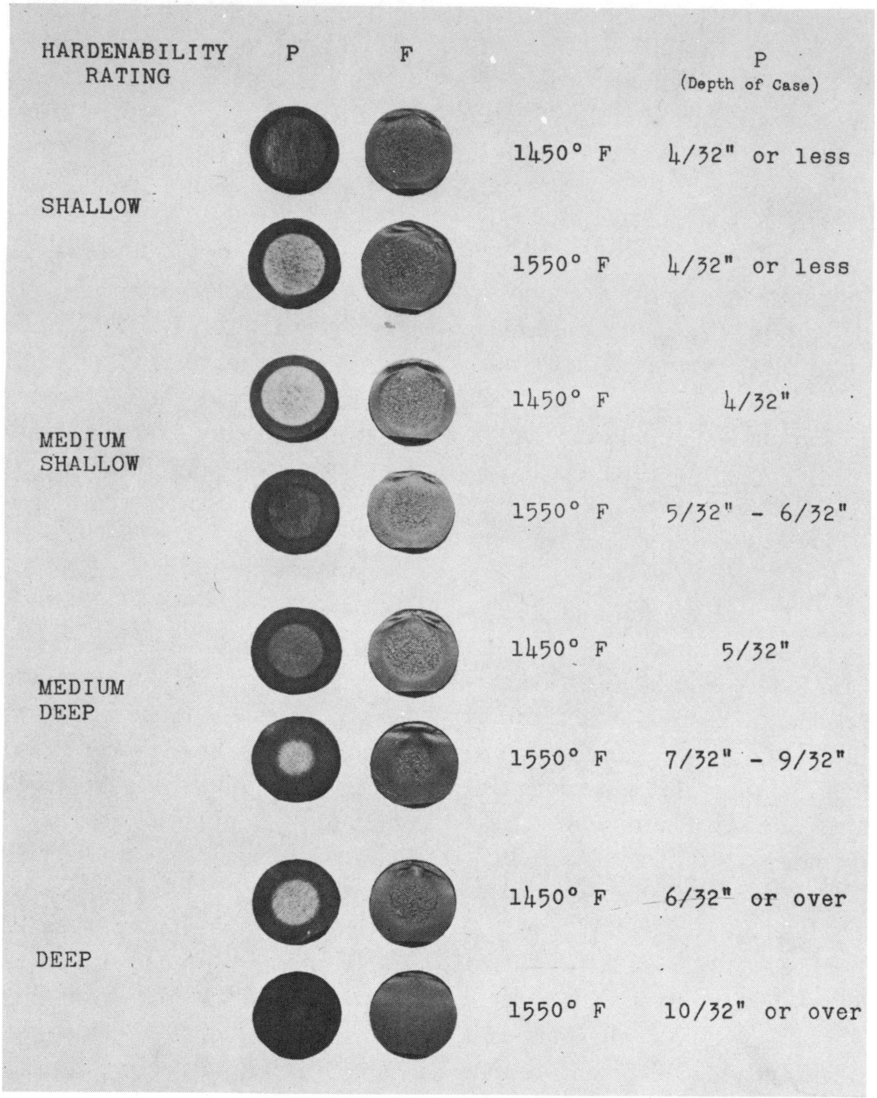

HARDENABILITY RATING	P	F		P (Depth of Case)
SHALLOW			1450° F	4/32" or less
			1550° F	4/32" or less
MEDIUM SHALLOW			1450° F	4/32"
			1550° F	5/32" – 6/32"
MEDIUM DEEP			1450° F	5/32"
			1550° F	7/32" – 9/32"
DEEP			1450° F	6/32" or over
			1550° F	10/32" or over

Fig. 2. Penetration-fracture classification chart for carbon tool steels. (Allegheny Ludlum Steel Corp.)

There are four primary groups: a carbon-vanadium type, a chrome-molybdenum type, a tungsten-chrome type, and a silicon-manganese type, all of which have excellent resistance to shock. These steels are particularly suitable for pneumatic tools such as chisels, rivet sets, rivet busters, backing out punches, and boilermakers' tools in

458

general. Heavy duty shear blades and punches are also made from shock-resisting steel.

Hot Work Steels. Many manufacturing operations involve the punching, shearing or forming of metals at comparatively high temperatures. As the name implies, hot work steels are used in operations where the material being worked is usually in a temperature range of 1100° F to 2000° F with a tool temperature range from 600° F to 1200° F. Tools used for this purpose must meet a great variety of physical requirements. They must have the quality of red hardness in order to retain their hardness at high temperatures. Wear resistance is necessary to overcome abrasion and washing action, and shock resistance to withstand the impact load. Hot work steels are usually alloys of tungsten, chromium, vanadium, and/or molybdenum in varying amounts, with relatively low carbon content. These steels are used for tools such as header dies, gripper dies, hot punches, shear blades, trimmer dies, extrusion dies, permanent molding dies and dies for die-casting.

High Speed Steels. High speed steel is a form of alloy steel. It is used to make tool bits, milling machine cutters, reamers, broaches, forming tools, etc. The steel contains a high percentage of tungsten (17.50% to 19.00%), and chromium (0.35% to 0.45%), vanadium (0.90% to 1.20%), and carbon (0.65% to 0.75%). The chromium imparts hardness to the alloy and the vanadium acts as a purifier and fatigue resister; therefore, high speed steel is often used in cutting tools. High speed steel with this composition still remains a major tool material despite the development of carbides, Stellites, and ceramics.

High speed steels possess the characteristic known as *red hardness*, meaning the ability to retain hardness at high temperatures. These steels derive their characteristics through the addition of alloying elements such as tungsten, molybdenum, chromium, and vanadium, along with carbon. Sometimes cobalt is also added. Steels alloyed with these elements have high abrasion resistance coupled with a comparable degree of shock resistance.

Although they have numerous other applications, high speed steels are used mostly for cutting tools, such as milling cutters, hobs, twist drills, reamers, shapers, broaches; lathe, planer and boring tools; woodworking tools; and punches and dies of a special nature such as lamination dies.

While high-speed steel is more costly than plain carbon steels, its high cost is repaid in the performance of tools made of this steel, which offer twice the normal tool life even at very high temperatures.

459

How Steels Are Classified

Two systems for coding steels are in use today. The Society of Automotive Engineers sponsor a system which has been widely used and is known as the SAE method of classification. A second system, the AISI method of classification is sponsored by the American Iron and Steel Institute, which is somewhat more comprehensive and yet not in conflict with the SAE system. Table I shows the SAE system.

TABLE I. CLASSIFICATION OF CARBON ALLOY STEELS

Type of Steel	Grade of Steel within Type	Digits
Carbon		1XXX
	Basic open hearth and acid Bessemer (non-sulphurized)	10XX
	Basic open hearth and acid Bessemer (sulphurized)	11XX
	Basic open hearth carbon steels (phosphorized)	12XX
Manganese 1.60% to 1.90% Man.)		13XX
Nickel		2XXX
Nickel-Chromium		3XXX
Molybdenum		40XX
	Chromium-Molybdenum	41XX
	Nickel-Chromium-Molybdenum	43XX
	Nickel-Molybdenum	46XX
Chromium		50XX
	Low Chromium	51XX
	Medium Chromium	52XX
Chromium-Vanadium		61XX
Tungsten		7XXX
Silicon-Manganese		92XX

TABLE II. CLASSIFICATION OF CARBON AND ALLOY STEELS

Type of Steel	AISI No. (1942)	SAE No. (1942)	Special Characteristics	Common Uses
Carbon	C 1010	1010	Low tensile strength Machines rough	Welding steel, sheet iron, tacks, nails, etc.
	C 1020	1020	Very tough Machines fair	Fan blades, sheet steel, pipe, structural steel.
	C 1030	1030	Heat treats well Stronger than 1020	Seamless tubing, shafting, gears.
	C 1040	1040	Heat treats average Machines fair	Auto axles, bolts, crankshafts, connecting rods.
	C 1045	1045	Thin sections must be carefully quenched	Some coil springs, auger bits, screw drivers, forge.
	C 1055	1055	May be oil tempered	Miscellaneous coil springs.
	C 1060	1060	Soft tool steel does not hold edge	Valve springs, lock washers, cushion springs, non-edged tools.
	C 1070	1070	Stands severe shocks Very tough and hard	Wrenches, anvils, dies, cold-rolled forms, knives.
	C 1080	1080	Holds edge well Medium hard	Shovels, hammers, chisels, shear blades, vice jaws.
	C 1085	1085	Hard tool steel	Music wire, knife blades, auto bumpers, taps, saws.
	C 1090	1090	Very hard tool steel Thin edges will be brittle	Coil and leaf auto springs, taps, hacksaw blades, milling cutters.
Free Cutting	B 1112	1112	Excellent machining characteristics	Screw machine stock, studs, screws, bolts.
	C 1115	1115	Stronger and tougher but machines slowly	Any of the above items where more strength is desired.
	C 1117	X1314	Machines well, case hardens very well	Used where surface hardness and strength is desired.
	C 1132	X1330	Machines very well	Used where better quality hardness is desired.

TABLE II (*continued*)

Type of Steel	AISI No. (1942)	SAE No. (1942)	Special Characteristics	Common Uses
Mang.	A 1330	**1330**	Withstands hard wear, hammering and shock	Burglar proof safes, railroad rails (curved).
Nickel	A 2317		Withstands vibration, shocks, jolts, wear	Flexible tapes, armor plate, wire cables, steel rails.
Nickel-Chromium	A 3115 A 3130 A 3140 A 3150	3115 3130 3140 3150	Very hard and strong " " " " " " " " " " " "	Armor plate Gears Springs, axles Shafts
Molybdenum	A 4130 A 4140 A 4150 A 4320 A 4615 A 4815	4130 4140 4150 4320 4615 4815	Withstands high heat and hard blows Same as above " " " " " " " " "	Very fine wire, ball and roller bearings, high grade automobile and machinery parts. Same as above. " " "
Chromium	A 5120 A 5140 E 52100	5120 5140 52100	Hard and tough Resists rust, stains and scratches Hard and very tough	Burglar proof safes, springs, cutting tools, bolts and rollers for bearings. Same as above.
Chromium-Vanadium	A 6120 E 6150		Hard and very strong Resists corrosion	Auto parts, steering, gears Springs, frames and axles, unbreakable-tools-chisels.
Stainless-Chromium	410 414 420 430	51210 51310 51335 51710	Not heat treatable Can be heat treated Can be heat treated Does not corrode Same as above	Stainless cooking utensils, ornaments, sinks. Same as above. Stainless steel cutlery, tableware, ball bearings, aircraft valves—used for form and pressed parts. Same as above.

NOTE: Tungsten and high speed steel are considered "special steels" and therefore, are not included on this list.

TABLE III. COLOR CODE FOR MARKING STEEL BARS °

Type of Steel	SAE No.	Color
Carbon	1010, 1015, X1015 1020, X1020 1025, X1025 1030, 1035 1040, X1040 1045, X1045 1050 1095	White Brown Red Blue Green Orange Bronze Aluminum
Free Cutting	1112, X1112 1120 X1314 X1315 X1335, X1340	Yellow Yellow and brown Yellow and blue Yellow and red Yellow and black
Manganese	T1330, T1335, T1340 T1345, T1350	Orange and green Orange and red
Nickel	2015 2115 2315, 2320 2330, 2335 2340, 2345 2350 2515	Red and brown Red and bronze Red and blue Red and white Red and green Red and aluminum Red and black

Each system makes use of a series of digits. Each series has associated with it a range of percentage values for the elements which are definitely to be expected in the steel designated by the number. Therefore, when steel is ordered today the purchase order specifies the chemical analysis of the steel. By consulting SAE or AISI tables in a handbook, we can determine what the steel is composed of. Table II shows both systems of classification.

The SAE System. The first digit of the four or five digits in the num-ber designates the general type of steel. The second digit generally indicates the relative amount of the principal alloying element. The last two digits (in some cases the last three digits) indicate the average carbon content in hundredths of 1%.

Thus, an SAE 1050 steel has the following analysis:

The first digit (1) indicates a plain carbon steel.

The second digit (0) indicates no other alloying agent is present.

The last two digits (50) indicate

463

TABLE III (*continued*)

Type of Steel	SAE No.	Color
Molybdenum	4130	Green and white
	X4130	Green and bronze
	4135	Green and yellow
	4140, 4150	Green and brown
	4340, 4345	Green and aluminum
	4615, 4620	Green and black
	4640	Green and pink
	4815, 4820	Green and purple
Chromium	5120	Black
	5140, 5150	Black and white
	52100	Black and brown
Chromium-Vanadium	6115, 6120	White and brown
	6125	White and aluminum
	6130, 6135	White and yellow
	6140	White and bronze
	6145, 6150	White and orange
	6195	White and purple
Tungsten	71360	Bronze and orange
	71660	Brown and bronze
	7260	Brown and aluminum
Silicon-Manganese	9255, 9260	Bronze and aluminum

* Simplified Practice Recommendation, R 166–37, April 1, 1937, U.S. Dept. of Commerce.

a carbon content of 0.50%. This content may range from about 0.46% to 0.54%; the figure of 0.50% is an *average* amount.

The AISI System. The AISI method of classification of steel makes use of the same designations of the elements as that described for the SAE system excepting letter prefixes are added to the designations to indicate the manufacturing process employed in producing the steel. The following prefixes are explained:

The prefix *C* denotes a basic open-hearth carbon steel. In the *open-hearth* process for making steel an intense flame is applied to the surface of a large trough of iron, gradually burning out the excess carbon which results in a high quality steel. This process, while slower than the Bessemer process, can be controlled more closely. The steel is used for bolts, bridge members, shafting, rails, etc.

The prefix *B* denotes an acid Bessemer carbon steel. The *acid Besse-*

TABLE IV. SPARK CHARACTERISTICS OF METALS

Metal	Color of Spark	Description of Spark
Wrought Iron	Straw yellow	Long shafts ending in forks and arrowlike appendages.
Low Carbon Steel (.15% to .25% Carbon)	White	Shafts shorter than wrought iron and end in forks and appendages. Forks become more numerous and sprigs appear as carbon content increases.
Medium Carbon Steel (.40% to .50% Carbon)	White	Explosions are more numerous and more brilliant than low carbon steel.
High Carbon Steel (.80% to .90% Carbon)	White	Large volume of short, bushy clusters of sparks beginning close to the wheel.
High Speed Steel	Dull Red	Streaks with explosions.
Manganese Steel	White	Numerous explosions that branch off like bushes.

mer process entails the use of a Bessemer converter, a large, pear-shaped, open top container, which is filled with molten iron. This iron is changed to steel very quickly by forcing cold air through the iron from the bottom. Combustion, resulting from the uniting of oxygen from the air with carbon in the iron, removes most of the carbon and produces a cheap low grade steel. It is used for building steel structures, nails, screws, etc.

The prefix *A* denotes an open-hearth alloy steel.

The prefix *CB* denotes that either the Bessemer or open hearth process may be used at option of the mill.

The prefix *E* denotes an electric furnace alloy steel. The *electric furnace* process, by making use of electrical heat, which is the purest form of heat, is able to produce a very high grade of steel for tools, dies, etc.

In Table II, some of the more prominent characteristics of each steel are indicated in the fourth column. Listed in the fifth column are the more common uses of these steels. As you can see from this table the selection of steel for a particular job is not arbitrarily made by the engineer or designer. There

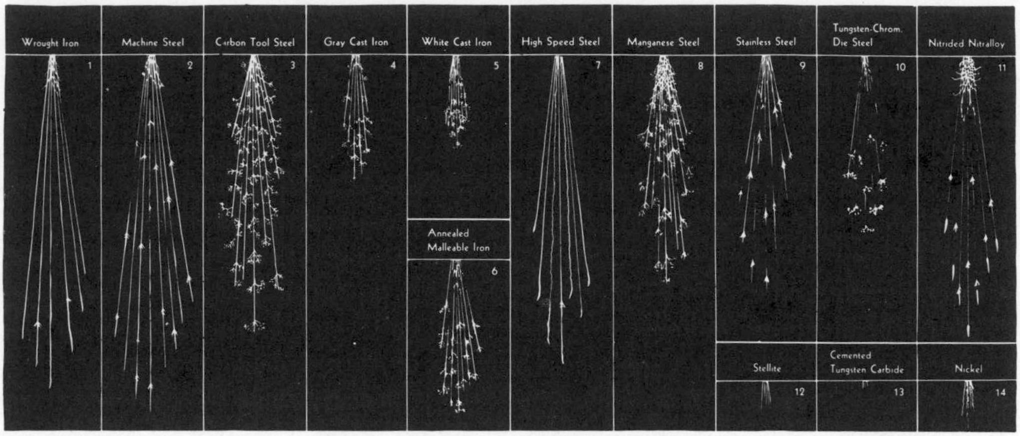

Metal	*Volume of Stream*	*Relative Length of Stream, Inches†*	*Color of Stream Close to Wheel*	*Color of Streaks Near End of Stream*	*Quantity of Spurts*	*Nature of Spurts*
1. Wrought iron	Large	65	Straw	White	Very few	Forked
2. Machine steel	Large	70	White	White	Few	Forked
3. Carbon tool steel	Moderately large	55	White	White	Very many	Fine, repeating
4. Gray cast iron	Small	25	Red	Straw	Many	Fine, repeating
5. White cast iron	Very small	20	Red	Straw	Few	Fine, repeating
6. Annealed mall. iron	Moderate	30	Red	Straw	Many	Fine, repeating
7. High speed steel	Small	60	Red	Straw	Extremely few	Forked
8. Manganese steel	Moderately large	45	White	White	Many	Fine, repeating
9. Stainless steel	Moderate	50	Straw	White	Moderate	Forked
10. Tungsten-chromium die steel	Small	35	Red	Straw*	Many	Fine, repeating*
11. Nitrided Nitralloy	Large (curved)	55	White	White	Moderate	Forked
12. Stellite	Very small	10	Orange	Orange	None	
13. Cemented tungsten carbide	Extremely small	2	Light Orange	Light Orange	None	
14. Nickel	Very small**	10	Orange	Orange	None	
15. Copper, brass, aluminum	None				None	

†Figures obtained with 12" wheel on ben stand and are relative only. Actual length in each instance will vary with grinding wheel, pressure, etc. *Blue-white spurts. **Some wavy streaks.

Fig. 3. Metals can be identified by the spark pattern which is generated by grinding.

is a best type for each of the many products which we make of steel.

Standard Color Code for Steel

A quick and simple means of identification of steel is afforded by means of a color code. The color marking is applied in the form of paint to the ends of bars one inch in diameter or larger. Smaller bars are usually painted across the ends of the bundles. It is always good practice to cut off pieces of steel from the bar of steel from the same end so the paint on the other end remains to the very last. Table III shows the color code for various kinds of steel.

Spark Test

When the kind of steel is unknown it is often possible to determine its identity by means of a spark test. When any kind of iron or steel is held against a grinding wheel, small particles, heated to red or yellow heat, are released from the metal and hurled into the air. Upon contact with oxygen in the air they oxidize or burn. If an element such as carbon is present rapid burning occurs, resulting in a bursting of the particles. This requires very careful observation and some practice before one can see the characteristics of the spark when the metal is brought in contact with a rapidly revolving grinding wheel. The "spark picture" of each type of steel should be studied carefully. The best way to master this method of identification is to obtain a series of sample pieces of steel of known identity and study the sparks from these. By comparing the sparks emitted from a piece of unknown steel a fairly accurate identification can be made. The data in Table IV gives these characteristics. Fig. 3 shows typical sparks generated by the grinding of metals.

NOTE: Please see review questions at end of book.

<table>
<tr><td>

Chapter

14

</td></tr>
</table>

Heat Treating:

*Methods, Equipment,
and Hardness Testing*

In preceding chapters we saw that a wide variety of metals are machined in lathes, milling machines, shapers, and other machines. These metals are carefully chosen because of physical properties, chemical composition, the design of the workpiece and the function it is intended to fulfill. Some metals require additional improvements in their microstructure in order to meet physical specifications. They usually have to go through another process, that of heat treatment, in which the metal is given greater toughness, hardness, and wear resistance.

In this chapter, the various phases of heat treatment are fully explained, with a recommended procedure given for each phase. The heat treatment of tools will be stressed.

Heat Treatment Defined

Heat treatment can be defined as "a combination of heating and cooling operations timed and applied to a metal or alloy in the solid state in a way that will produce desired properties. Heating for the sole purpose of hot working is excluded from the meaning of this definition."[1]

Grain Structure of Steel

Like stone and wood, steel has a grain structure. With the naked eye we can see the grain in a polished piece of wood; with steel, however, the grain structure must be magnified many times to be visible. In order to see and study grains in steel, we prepare a sample by giving its surface a very high polish. After the sample is polished, it is etched

[1]ASM Metals Handbook.

in acid. The action of the acid makes the grain visible under a powerful microscope.

The grain structure of steel is composed of iron and carbon. Some of the iron and carbon in steel forms a chemical compound. The iron that does not go into a compound with carbon also has its part in grain formation. The grains of steel we find then are composed of alternate layers of almost pure iron and the iron and carbon that has joined chemically to form a compound.

When steel is in a molten state, there is no grain structure. As the steel cools and begins to harden, grains start to grow around the edges of the mold. As the metal continues to cool, grains continue to form until the metal all becomes solid. The grain structure of a piece of steel may be very coarse or very fine. Good quality tools have a fine grain structure.

Controlling Grain Size. The grain size in steel can be altered and controlled by several methods. Hot forging, or hammering while hot, will refine the grain structure, as will the addition of certain elements. Heat treating operations will also give the grain structure desired.

When a piece of steel is being heated, the grain structure starts to change and continues to change until the steel melts. When the steel being heated reaches the critical temperature, the carbon in the iron carbide compound dissolves and goes into solution within the steel. This happens much in the same way that a drop of die would mix in a glass of water.

When steel at a temperature slightly above the critical temperature is quenched, the steel becomes very hard and has a hard needle-like grain structure. The carbon that was in solution is mixed throughout the steel in the form of tiny globules. In this hardened state, no grain structure pattern is evident.

During the beginning of the drawing or tempering operation, as the metal starts to heat, the grain structure begins to return to its original pattern, but is allowed to progress only far enough to give the desired hardness and toughness to the piece being tempered. By controlling the growth and size of the structure, through controlled heating and cooling processes, we can obtain a good temper in any given piece of steel.

In machine shop operations, before one can become proficient as a machinist, he must possess some knowledge of heat treating, especially in regard to the various types of steels used in tools. The toolmaker is frequently called upon to recommend and select proper material for the tools he must make. Tool steels are the materials most frequently used in tool making and machine shop work.

Heat-Treating Methods

Tools and other machined parts are hardened to increase their strength and wear resistance. The properties of all steels may be changed very markedly by heating and cooling under definite, controlled conditions. The object of heat treatment is to make the steel better suited, structurally and physically, for some specific application. The most important heat treatment is that of hardening. In heat treating steel, the temperature to which the steel is heated is very important since the metal must be heated above a certain temperature before it will become hardened.

Critical Temperature

The *critical temperature* (or critical temperature range) is the temperature above which a steel must be heated in order that it will harden when quenched. Steels have temperatures at which some definite change takes place in their physical properties. These temperatures vary with different kinds of steel. The more carbon steel contains, the lower will be its critical temperature and the less it should be heated for hardening.

By knowing the critical temperature of the steel being treated, the operator can accordingly regulate the heat supplied to the furnace through use of an instrument known as a *pyrometer*.

Hardening

The purpose of hardening tools is to develop the maximum physical properties, such as tensile or compressive strength and wear resistance, of the tools' materials. The process consists of heating these materials, usually steel, to a temperature well over the critical range and cooling rapidly in water, oil or air, whichever is most suitable for the grade being hardened.

The rate of heating for alloy steels, as the alloy content increases, should be proportionately slower than for straight carbon. Slow, uniform heating helps to avoid warpage. It is good practice to preheat slowly in one furnace and then transfer to another furnace held at the hardening temperature. With the highly alloyed types, such as high speed steel, this is the usual procedure except for small sizes. Of the various methods of heating, the rate of heat transfer is most rapid in molten baths, somewhat slower in open or semi-muffle, fuel-fired furnaces, and the slowest in electric or complete muffle-type furnaces. Rate of heat transfer is also affected by different furnace atmospheres.

Hardening Carbon and Low Alloy Types

Cold tools should be charged at comparatively low furnace temperatures, preferably below 1000° F, except for small sizes. Small sections may be placed directly in a furnace or liquid bath maintained at the hardening temperature. When tools are charged in the furnace and equalized at a temperature of about 1000° F, they can be brought up with the furnace heat to the hardening temperature. Slow heating reduces the danger of warpage.

Preheating in a separate furnace at about 1000-1200° F, then transferring to another furnace held at the hardening temperature, is good practice. The steel should be held at the hardening temperature until uniformly heated through, allowing sufficient time to insure proper solution of the carbides. *Unduly long soaking should be avoided* as it may cause decarburization, scaling, and grain coarsening. Furnace atmospheres should be neutral to slightly oxidizing.

When no protective atmosphere or heating bath is available and it is necessary to preserve the surface, *pack hardening* may be resorted to. The process consists of packing the tools in a container with cast iron chips, spent carburizing compound or other carbonaceous material, the combination dry and free from dirt and scale. The entire charge should be brought up to the hardening temperature and, when heated through, the tools should be removed quickly and quenched in the proper medium.

Hardening High Alloy Types

Due to the rather high alloy content of these steels, slow heating is important in hardening to insure proper solution of the carbides. Slow preheating at about 1200° F in controlled atmosphere furnaces, followed by heating in hardening furnaces maintained at the proper temperature is a desirable heat-treating operation. Pack hardening of these steels, as described previously, is common practice in ordinary furnaces when the surface must be preserved. In this case, preheating to the lower part of the hardening range is recommended. Overheating should be avoided, otherwise low hardness and shrinkage will result.

Flame Hardening

Flame hardening consists of heating a selected area rapidly, usually with a high-temperature gas burner or oxyacetylene torch, to or above the hardening range of the steel. Flush quenching in water applies to water-hardening tool steels; there is some risk of cracking if applied to oil-hardening tool steels. Air-hardening steels should be preheated and cooled in air. The process is not used very often with tool steels owing to a lack of close control over temperature.

Induction Hardening

Like flame hardening, induction hardening is used where selective or localized hardening of the parts is required. In general this method of hardening can be used to advantage with carbon or low alloy tool steels when a number of similar tools or parts are to be heat treated. To the present time it has not been found suitable for high speed and other highly-alloyed tool steels.,

Tempering should follow flame or induction-hardening operations, following the same practice as when furnace hardened.

Case Hardening

Many parts made in the machine shop need only their surfaces hardened. It is possible to give a wearing surface, like gears or sprockets, a hard surface and still maintain a tough interior. The operation by which this is achieved is known as *case hardening*. Sealing the machined part in a box containing material rich in carbon and heating it to a high temperature for many hours will effectively harden the surface. The carbon is absorbed into the surface layer $\frac{1}{16}$ to $\frac{1}{8}$ of an inch, the depth depending upon the length of time of *carburizing* (see following paragraph). The part is then quenched and tempered. Certain

parts which are specified to be finish ground should be case hardened first. There are several processes used in case hardening, notably carburizing and cyaniding.

Carburizing. Carburizing is a process whereby carbon is absorbed into the surface of steel alloys. The steel is heated in contact with a carbonaceous material to a temperature below the melting point of steel. Alloy steels containing not more than 0.20% carbon (low carbon steel) are used. These low carbon steels cannot be hardened by the usual method of heating and quenching but are suitable for case hardening.

Carburizing Material. The carburizing material may be any one of a number of charcoal base, carbonaceous materials now on the market. It may be coke, bone black, etc., and can be used over and over again. The material should be run through a sieve before using to eliminate the fine particles. Suggested procedure in carburizing workpieces:

1. Pack workpieces in a steel box, entirely surrounding them with the carburizing material. The pieces should be so placed that they will not touch each other.
2. Place cover on box and seal with fire clay.
3. Place box in furnace; set temperature at 1650° F.
4. Heat for about six hours. This will give a case of about $\frac{1}{16}$ of

an inch in depth.

5. When the time limit is reached, shut off the gas or electricity and let box remain in furnace until cold.
6. Remove pieces from box, reheat in furnace to 1650° F, and quench in water.
7. Reheat to 1450° F and quench in water or oil.
8. Temper by heating to 250-325° F and quench.

Cyaniding. Another case-hardening compound on the market is a material known as cyanide of potassium. It is from this material that the process of *cyaniding* gets its name.

Cyaniding is achieved by immersing the piece in a molten bath of potassium cyanide from 5 to 30 minutes, depending on the size of the piece and the depth of penetration desired. This is a common way to case harden steel where the thickness of the case seldom exceeds .008". Low carbon steels can be satisfactorily case hardened by this method but it must be kept in mind that with low carbon steel the cyanide will penetrate to a depth of not more than .015". Therefore, parts made from low carbon steel cannot be ground below this depth or there will be no case-hardened material left. Suggested procedure in cyanide hardening:

1. Fill the pot with cyanide of potassium, then light the furnace and regulate the flame. When the temperature is up to between 1500° and 1600° F, the cyanide will be in a molten state.
2. Fasten a soft wire to each of the workpieces, leaving approximately 6 or 8 inches to form a hook.
3. Preheat the pieces to be case hardened. This is done to remove all moisture. On some furnaces the pieces can be preheated by laying them around the top of the furnace near the pot.
4. The pieces are now ready to be suspended in the molten cyanide. Leave them in the cyanide until the desired depth of case is produced. Cyanide penetrates from .001" to .0015" each minute of soaking.
5. Remove pieces and quench in either water or oil to produce hardness.

CAUTION: Cyanide is a deadly, gaseous poison and care must be exercised in its use. Wearing safety glasses or a face shield would be advisable; an exhaust fan is a necessity. When water comes in contact with hot cyanide it causes a violent "spatter." If using tongs or hooks to remove work from pot, make sure that all moisture is removed from the work before bringing it in contact with the molten bath.

Tempering

After quenching and before tempering, all tools are in a more or less highly strained condition, and there is often danger of cracking if hardening strains are not promptly relieved. It is good practice not to allow tools to become cold before tempering.

Tempering (often called drawing) consists of reheating tools, which have previously been hardened, to a comparatively low temperature in order to relieve hardening strains and increase toughness. The temperature varies with the type of steel and the nature of the tool.

Tempering Steel According to Color. Where modern equipment is not available in tempering tools,

the old-fashioned method of observing color is still used. The piece to be tempered is hardened in the usual manner and then a piece of emery cloth is used to polish the surface of the steel so that the color variations that occur during the process can be observed readily. The colors progress as follows: faint straw, dark straw, bronze, purple, dark blue, light blue, steel gray. Table I shows the color that is used for tempering certain tools and other items.

As an experiment, suppose we polish a sample of SAE 1090 steel which has been hardened to 65Rc (Rockwell) and heat it to 400° F. The oxide coating which forms due to the reaction of oxygen in the air on the iron, will be a very light yellow. At

TABLE I. TEMPERING OR DRAWING CHART

°F.	Oxide Color	Suggested Use for Carbon Steels
430	Faint Straw	Tools for metal cutting that must be of maximum hardness, drills, taps, paper knives, lathe tools, etc.
460	Dark Straw	Tools that need both hardness and toughness, rolled-thread dies, punches and dies.
500	Bronze	Rock drills, hammer faces, shear blades, and tools where toughness is required.
540	Purple	Axes, wood-carving tools and tools that may be sharpened and shaped by use of a file.
570	Dark Blue	Knives, iron and steel chisels.
610	Light Blue	Springs, screw drivers, saws for wood.

this temperature, the hardness will be nearly the same as when fully hardened but the internal stresses in the steel will be relieved.

If we now heat the steel to 450° F, the steel will have an oxide coating that is straw in color. If cooled and then checked for hardness, it will be found that the steel is about 62Rc. It is still very hard but the stresses are further relieved and the toughness of the steel has been increased.

When tempered at 500° F the steel will have an oxide coating that is bronze in color and the hardness now is about 60Rc. The part is tougher and has less stresses. At 540° F, the oxide is purple and the hardness is about 55Rc. A temperature of 570° F will produce a dark blue oxide coating and the hardness will read about 50Rc.

In using the temper color method for tempering, care should be taken not to heat beyond the temperature desired. The hardened and polished steel is slowly heated over a fire or any hot medium until the color corresponding to the desired temperature is seen on the polished surface. Cooling may be done in oil or air.

Annealing

Tool steel is annealed to soften it for machining, forming, and grain refinement. As processed by the mill, the metal is fully annealed, free from internal stress and of proper hardness for ready machining.

The annealing operation consists of heating the steel to slightly above its critical range and cooling very slowly. In order to avoid excessive scale and surface decarburization, the steel should be packed in a tight container, surrounded by cast iron chips, lime, mica, or any other neutral material. The charge should not be in contact with the container, and the presence of scale should be avoided.

The steel should be heated slowly and uniformly to the annealing temperature range and held within that range long enough for complete penetration of the heat and readjustment of the grain. The length of time will depend on the size of the charge and the alloy content of the steel; 1 to 4 hours is generally sufficient. The steel is cooled slowly, preferably in a furnace. A maximum cooling rate of 50° F per hour down to 1000° F is suggested.

High speed steels are usually annealed at 1600-1650° F, the charge held in that range until uniformly heated through, followed by a slow cool in the furnace for maximum softness. In the packing operation, an excess of carbonaceous packing material must be avoided, as carburizing may occur.

Normalizing

Normalizing consists of heating steel well above its critical range and cooling in air. The purpose of normalizing is to refine the grain and relieve forging strains in those steels that are known to have poor grain structures. In the building of tools, jigs, and fixtures which are of welded construction requiring that close tolerances be maintained, normalizing is necessary before finish machining.

Heat Treating Small Tools

How to Harden and Temper a Cold Chisel. First heat the end of the chisel to dark red, two or three inches back from the cutting edge. Then cool about half of this heated part by dipping it in clean water, being sure to move it about in the water until the end is cooled down enough to hold it in the hands. Quickly polish one side of the cutting end by rubbing it with emery cloth or filing lightly. Carefully watch the colors pass toward the cutting end as the chisel continues to cool. The first color to pass down will be yellow, followed by straw, brown, purple, dark blue, and light blue. The color desired for the cold chisel is dark blue.

When the dark blue reaches the cutting edge, immediately dip the end into water and move it about rapidly. To prevent the shank of the cold chisel from cooling too rapidly and becoming too hard and brittle, only the cutting edge is placed in the water. The heat in the shank slowly dissipates into the air. When the tool is first dipped, it is important that it

be moved up and down in the water to prevent the formation of a sharp line between the hardened end and the shank. Such a line might cause the tool to break at that point when in use later.

After the heated cutting end is dipped in the water, the end becomes very hard. The heat left in the shank of the chisel gradually moves down to the cutting end and softens it. When the cutting end has been softened to the desired degree of hardness, in this case a dark blue color, the end is quickly quenched to prevent any further softening. The various colors indicate different temperatures. To see these colors it is necessary to clean the surface with an emery cloth or a file.

How to Temper Punches, Screw Drivers and Similar Tools. Punches, screw drivers, scratch awls and tools of this kind may be tempered in the same manner as a chisel except that the degree of softness or hardness will depend upon the use to be made of the tool.

A scratch awl should be made somewhat harder than a cold chisel, a center punch just a little harder, a screw driver somewhat softer. In case a tool proves to be too hard and the edge chips or crumbles as a chisel or screw driver could easily do under heavy usage, it should be retempered and the colors allowed to go out a little further. It is important to remember what color you allowed to appear at the tip before quickly putting it in water.

Quenching

Why Quenching is Necessary. For thousands of years man has known that to produce fully hardened steel rapid cooling was required. Different kinds of steel require different quenching mediums. Carbon steels, in general, require a relatively fast rate of cooling such as obtained by a water quench and are often classified as *water-hardening steels*. Many of the low alloy steels require a slower rate of cooling than obtained in a water quench and are often quenched in oil. Such steels are sometimes classified as *oil-hardening steels*. Some of the high alloy steels have a very slow rate of transformation and are therefore allowed to cool in still air. Such steels are often referred to as *air-hardening steels*.

Quenching baths may contain fresh water, salt brine, caustic soda solution, or various types of oils. Fresh water is not the most desirable medium for quenching tool steel because it dissolves large quantities of gas from the atmosphere. When a hot tool is immersed in a bath of this kind the boiling action throws the gas out of solution, causing it to settle in the form of a bubble on the surface of the work, particularly if the piece has pocket-forming holes or recesses in it. Also, water-hardening steels are most likely to crack when quenched in fresh water. For this reason, a salt brine is a better medium for quenching water-hardening tool steels. The action of the salt prevents the water from dissolving atmospheric gas.

Although considerable hardening is still done in fresh water, better results are usually obtained by adding salt. The more salt added the less dissolution occurs, but to add too much salt slows the action of quenching. Therefore, it is well to learn the proper amount of salt to add to the quenching bath. A 5% solution of caustic soda makes an efficient quenching bath. The main objection to this solution is that it has a harmful effect on the hands and damages clothing.

In winter the water to be used for quenching tools should be warmed to approximately 70° F before immersing the tools. If considerable quenching is to be done the temperature of the bath should not exceed 100° F; therefore, it is essential that the quenching tanks be large enough to dissipate the heat.

Oil Quenching. To quench high speed steel and other hardening tool steels, it is necessary to have quenching tanks filled with one or more oils. Although almost every kind of oil has been tried, mineral oils are now used almost exclusively in modern heat-treating methods. These prepared mineral oils are manufactured under strict technical control and the correct type of oil for the job can be secured from any reputable oil manufacturer. When considerable heat treating is done, the oil is sometimes kept cool by a series of coils circling the tanks. Cold water is piped through the coils to keep the temperature within the range desired.

Quenching High Speed Steels. High speed steels may be quenched in several ways, depending upon the type of tool. The medium is usually oil, air, or a molten bath. Small parts may be quenched in oil and brought down to about 150° F. Larger parts, especially those of molybdenum high speed steels, should not be allowed to overcool in the oil. They should be removed when a little below a red heat and cooled down in air. Cooling in a current of dry air is suitable in some cases where toughness is required and where scaling is not objectionable. Full hardness will not be developed in large sections. Quenching in a molten bath at 900° to 1200° F, equalizing, then cooling down in air to about 150° F before tempering is good practice. This is sometimes called *interrupted quenching*.

Grinding

Grinding is an important operation that follows hardening in making a great many tools. The effect of grinding is to produce intense local heat. This heat, of course, is rapidly absorbed by the surrounding metal and coolant. With the expansion and contraction caused by heating and rapid cooling, severe strains are developed momentarily in the surface of steel being ground. If a sufficiently heavy cut is taken when grinding a hardened tool, very small cracks, called *grinding cracks*, will form in the surface. Usually these cracks are accompanied by temper colors. As long as the heat is insufficient to produce the colors there should be no danger. Considerably heavier cuts can be taken when grinding with a coolant than with dry grinding. Some types of tool steel, such as high carbon-high chrome and high speed steel, are more difficult to grind than others — and more susceptible to grinding cracks.

Straight carbon and manganese oil-hardening grades of steel grind the most readily. Of the high speed group 18-4-1 is the most foolproof in this respect. With increasing vanadium content grinding becomes more difficult. The amount removed per pass on such steels should not be over half that removed on straight carbon or low alloy tool steels.

Excessive wheel speeds should be avoided on steels which are harder to grind. Usually wheel speeds run about 5000 to 6000 surface feet per minute. Sometimes by dropping the wheel speed as low as 3000 sfpm it is possible to use relatively harder wheels without visibly burning the work.

The Grinding Wheel. Frequent and proper sharpening of tools lengthens their life and enhances their capacity to produce accurate work. Excessively dull tools slow up production and sometimes cause failure by breakage; also, much more of the tool has to be ground away to renew the cutting edge. In selecting grinding wheels for toolroom work, the proper abrasive, grain, and grade should be chosen so the wheel will grind rapidly without unduly heating the work. To avoid drawing the temper from the tool's cutting edge the cut should be light and not forced. This type of cut may be done at a comparatively low temperature with carbon and low alloy tool steels. The wheel should never be too hard or too fine in grain size or it is apt to burn the tool. If the wheel is too soft, it will wear away rapidly and not leave a satisfactory edge on the work.

Brittle steel, caused by faulty heat treatment, is much more susceptible to developing grinding cracks. Often a combination of two conditions produces these cracks: overheating in hardening and improper grinding practices. Properly hardened tool steel will stand considerable abuse in grinding before cracks develop. So-called abusive grinding may cause excessive heat by (1) a feed of too heavy a pressure, or (2) the action of a glazed or loaded wheel. Glazing may result if a wheel that is too hard or wheel speeds that are too high are used. Surface discoloration of the work is an indication of poor grinding practice.

In grinding any kind of tool, heavy pressures sufficient to produce temper colors should be avoided. This applies to single point cutting tools which are hand ground and cooled in water. When grinding high speed steel tools still retaining some scale from heat treating, it is desirable to get down under the scale on the first grinding pass. Too light a pass may result in glazing over instead of cutting into the scale. After the scale is removed, lighter cuts should be taken.

When grinding tools of thin section, whether using a coolant or not, light cuts should be taken to avoid excessive heating which may draw the temper. Heavy cuts or the removal of unequal amounts of metal may result in warpage of tools which have thin sections.

In general, grinding sets up minor strains in a tool even though no grinding cracks are produced. Under some conditions, particularly when the design involves sharp corners and the tool is subject to a degree of shock, it is helpful to relieve any strains by retempering after the grinding operation.

All grinding wheels are aluminum oxide abrasive with vitrified bond, except wheels for soft steel dies, which are resinoid bonded.

Furnaces

There are three types of furnaces commonly used in heat treating—gas, oil, and electric furnaces. Since oil and gas furnaces are similar in design and operation, we will consider them together.

Fig. 1. The laboratory-type gas furnace shown has a controlled temperature range of 300° to 2400°F.

Gas and Oil Furnaces

Fig. 1 shows a laboratory-type gas furnace. It has a capacity of 300° to 2400° F, even heat distribution, and close temperature control, thus combining in one unit what would normally require two units of different temperature ranges. This furnace is used with temperatures as low as 400° F for drawing, 1475° to 1600° F for hardening, and at 1400° F for annealing brass.

The gas used in this furnace can be natural, artificial, mixed, butane, or propane, at 3 to 8 ounces psi pressure. Air is supplied at 12 ounces to 2 pounds pressure.

Electric Furnaces

The electric furnace, see Figs. 2 and 3, is very clean and simple to operate and adapts readily to automatic regulation if so equipped. Some furnaces are heated by resistor elements, which may take the form of coils located in the sides of the heating chamber or, where higher temperatures are required, of a bar situated in the heating space to provide heat by radiation and convection. Heating elements in electric furnaces are usually made from nichrome wire, ribbons, or bars.

Fig. 2. The electric furnace shown is clean and simple to operate. (Lindberg Hevi-Duty Div., Sola Basic Industries)

Controlling Temperature of Furnace

Just a few years ago it was quite common for the person who handled the heat treating to also watch the color of the workpiece in the furnace. As the color changed he was able, from experience, to determine the work's temperature. For example, cherry red was thought to indicate the proper temperature of tool steel. It was soon found that there was a large element of chance in this procedure and much depended upon the

Fig. 3. Some electric furnaces are heated by resistor elements which are often formed into coils. (Electric Hotpack Co., Inc.)

color vision and judgment of the operator. Science and research soon proved that this early method could be off as much as 200° F from the correct temperature. It was necessary, therefore, to develop temperature controls that would take as much as possible of this guesswork out of heat treating.

Furnace Atmosphere. Proper control of furnace atmosphere is important to avoid excessive scaling and maintain a good surface on the work. Atmospheres may be *oxidizing*, *neutral*, or *reducing*. How they react on the steel depends on the steel's composition, the temperature in the furnace, and the circulation of the atmosphere.

In a fuel-fired furnace the atmosphere is controlled by regulating the air-fuel ratio. In an electric or complete, muffle type, fuel-fired furnace a prepared atmosphere is introduced independently of the fuel gases. When more air is introduced than is necessary to burn the fuel, the hot burned gas contains an excess of oxygen, and the furnace atmosphere is oxidizing.

On the other hand, if there is insufficient air for complete combustion of the fuel, the excess will burn on contact with the oxygen of the outside air; the furnace atmosphere is *reducing*. A *neutral* atmosphere represents a balanced condition between oxidizing and reducing, where there is just the proper amount of air (oxygen) to consume the fuel.

Actual analysis of furnace atmospheres by means of gas analysis equipment is the best way to maintain uniformity. However, an experienced operator can control furnace atmosphere quite well by close observation of the flame from the furnace parts and how carbonaceous material such as charcoal or wood is consumed on the furnace hearth.

In general, tool steels that require low hardening temperature (under 1600° F), such as straight carbon and manganese oil-hardening steels, are best handled in a slightly oxidizing atmosphere which will not scale heavily. Those hardened at higher temperatures (over 1600° F) are less heavily scaled with a reducing atmosphere which also does not decarburize. Such steels would include the more highly alloyed group such as high carbon chrome, hot die, and high speed steels.

Measuring Temperatures. In metallurgical work, temperatures are usually measured with a thermoelectric pyrometer, Fig. 4. Such an instrument may consist of two dissimilar wires welded together at one end, called the thermocouple, with the opposite ends connected to a millivoltmeter. The welded ends of the thermocouple are placed in the furnace where temperature is to be measured. As the end becomes heated, a simplifying electromotive force (emf) or voltage is generated and a current flows through the thermocouple wires which in turn gives a reading on a graduated scale.

The use of the proper temperature-indicating and control equipment is an important factor in obtaining the required properties and longevity from heat-treated parts, tools, dies, etc. The pyrometer in Fig. 4 is known as a controlling and indicating pyrometer, which indicates temperature on a direct reading scale and regulates it automatically. No temperature recording is made with this type, although many modern indicating devices do also have attachments for recording temperatures versus times.

Two other types of surface temperature-indicating devices in common use are the optical and ratio pyrometers. Detection and controlling features of some of these pyrometers are fully automatic in operation, while others still require adjustments to be made manually on the basis of visual comparisons of the color brightness of a standard source (such as a special electric lamp filament) with the color brightness of the workpiece being heated. Important considerations in the use

Fig. 4. The pyrometer is used to measure, indicate and control temperatures in metallurgical work. (Minneapolis Honeywell Corp.)

of optical and ratio pyrometers as surface temperature measuring devices are: (1) proper calibration by means of suitable standards, (2) that the pyrometer has an unhindered 'line-of-sight' view of the workpiece, and (3) that measurements are made during uniform or 'steady-state' temperature conditions, so the surface of the heated workpiece can be considered to be at the same temperature as the interior.

Testing for Hardness

The hardness of a material can be measured only by comparing it with some other material. For example, you have probably tested a material for hardness by trying to scratch it with your fingernail. Some material we know can be scratched with a knife but not a fingernail. Still other material can be scratched only with a diamond.

What you think hardness means depends a great deal on who you are. The metallurgist thinks of hardness as the ability of a material to resist indentation or penetration. The machinist considers hardness as an index of machinability.

Testing Hardness with a File

Occasionally it is necessary to estimate the hardness of a metal when no hardness testing machine is available. The simplest procedure is the file test used by shop men to predict the machinability of a metal. Table II gives a comparison between the hardness found by using the Brinell hardness testing machine and the action received with a machinist's new hand file. Remember that this is only an approximate method but is helpful if a hardness machine is not available.

Hardness Testing Machines

Several machines are now on the market which measure the hardness of a material very accurately. Of these, the Brinell, Rockwell, and Shore Scleroscope testers are the most common.

Brinell Hardness Tester. With the Brinell Hardness Tester, Fig. 5, the hardness of the material under test is determined by measuring the resistance it offers to the penetration of a steel ball under pressure. The Brinell tester is useful in testing soft and medium-hard materials and large pieces. When testing hard steel the impression is so small that a low-power microscope must be used.

The Brinell hardness number is found by measuring the diameter of the impression caused by the steel

TABLE II. FILE HARDNESS TEST DATA

Brinell Hardness No.	File Action
100	File bites into surface very easily.
200	File removes metal with slightly more pressure.
300	File meets its first real resistance to the metal.
400	File removes metal with difficulty.
500	File just barely removes metal.
600	File slides over surface without removing metal. File teeth become dulled.

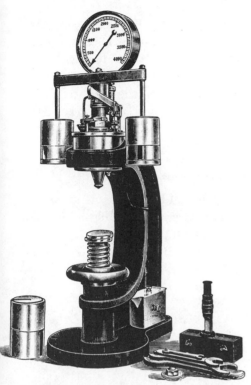

Fig. 5. The Brinell Hardness Tester is useful in testing soft and medium-hard materials.

ing is also based on a resistance-to-penetration measurement and is similar to the Brinell system; however, the hardness number is read directly from a dial on the tester, see Fig. 6, and the depth of impression is measured instead of the diameter. When testing hardened steel, the impression is made by forcing a diamond cone (usually with a 120° included angle) into the steel under pressure. The depth of penetration is measured and indicated in Rockwell units on the dial. SAE 1090 steel will show a hardness of about 63Rc on the *C* scale.

A typical example of the hardness information that would be given on a blueprint or drawing would be: Material SAE 3140 heat treated to 29-31 Rockwell *C*. In testing softer materials with the Rockwell instrument a $\frac{1}{16}''$ steel ball is used and the hardness is read on the *B* scale.

ball's penetration. The greater this diameter, the softer the metal, and the lower the Brinell number. After measuring the diameter with a microscope, the hardness number corresponding to this measurement is found by consulting a standard chart. Brinell numbers range from about 150 for annealed, high carbon steel to 750 for fully-hardened, high carbon steel.

Rockwell Hardness Tester. The Rockwell method for hardness test-

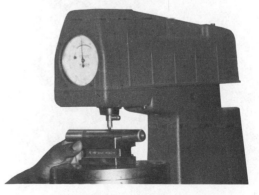

Fig. 6. The Rockwell Hardness Tester measures the depth of impression. (Chicago Public Schools, Manpower Training Div.)

The Shore Scleroscope. In the scleroscope test, Fig. 7, a diamond-pointed hammer is dropped through a guiding glass tube onto the test piece and the hammer's rebound checked on a scale. The harder the steel the higher the hammer will rebound, since the rebound is directly proportional to the resilience of the test piece. Scleroscope numbers range from approximately 20Scl for annealed tool steel to about 95Scl for fully hardened tool steel.

The chief advantages of the Scleroscope are: (1) it is easily portable, (2) it can be used for large sections, and (3) the diamond hammer leaves an invisible mark and does not harm the surface finish. Fig. 8 shows a direct-reading Scleroscope mounted on a swing arm.

Typical Heat Treatments

A typical heat treatment for tool steel is to heat it in a closed furnace to 1400-1450° F, quench in oil, and temper in oil at 350-375° F. A typical heat treatment for high speed

Fig. 7. The Scleroscope is a non-destructive testing instrument. (Shore Instrument & Mfg. Co., Inc.)

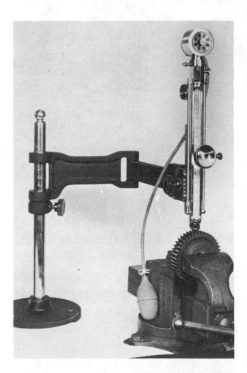

Fig. 8. The Scleroscope can be mounted on a swing arm for testing the hardness of wheels and gears. (Shore Instrument & Mfg. Co., Inc.)

steel is to heat it to 1400-1450° F, superheat to a minimum of 2350° F, quench in oil, and temper in a furnace at 1050-1100° F; the steel should test at 62-64 on the Rockwell C scale.

Hardening temperatures for carbon tool steels depend upon the type of tool and its intended service. Wherever possible the recommendation of the tool steel manufacturer should be followed. The temperature ranges which follow are usually recommended:

1450-1550° F for carbon tool steel of 0.65-0.80% carbon,

1410-1465° F for carbon tool steel of 0.80-0.95% carbon,

1390-1430° F for carbon tool steel of 0.95-1.10% carbon,

1380-1420° F for carbon tool steel of 1.10% and over.

A higher temperature tends to produce deeper hardness penetration and increased strength. A lower temperature results in shallow hardness penetration but increased resistance to splitting or cracking.

Safety Precautions—Heat Treating

1. Make sure furnaces and other equipment are safe and have all the necessary safety devices. Review each piece of equipment to make sure in the event of any utility flame, temperature, or instrument failure the safety device will operate.

2. When furnaces or heat-treating equipment are purchased, make sure that they are adequately protected with safety devices. A safety device should be considered as an integral and necessary part of the equipment, not as an extra.

3. Be sure you are following approved procedure. Make certain that furnaces and ovens are adequately purged before lighting and that proper methods are followed for introducing and removing furnace atmospheres.

4. Make sure all piping and valves are adequately identified. Piping should be identified by some standard color code. Valves should be readily accessible and their open and shut positions clearly visible from the floor.

NOTE: Please see review questions at end of book.

Machinability:

Variables and Ratings

Today's machine tools, because of their design, are capable of producing far more work than the average operator produces with them. Machine tools are capital investments, and the optimum performance must be obtained from them if the investment is to be redeemed. Because of improvements in machine tool designs and tooling, rates of speeds and feeds are attainable today which were thought impossible a few years ago. Therefore, the present-day operator must learn all of the facets of tooling in order to achieve the level of performance possible with each machine tool. Attention to the following components will help to boost machine performances and cut costs.

1. *Tooling*
   a. Use optimum speeds and feeds when machining.
   b. Select the best cutting tool for the material being machined.
   c. Grind proper tool geometry on cutting tools; relate grinding to material being machined.
2. *Dimensional Control*
   a. Machine to realistic tolerances.
   b. Select proper measuring tool; most suitable for gaging the tolerance specified.
   c. Keep machine adjustments in top shape.
   d. Take advantage of developments in free-machining materials and new cutting tool materials.
3. *Selection of work materials.*
   a. Use precision castings, forgings, seamless tubing, and other forms of stock that require a minimum of machining.

If the machine tool is in proper condition and has adequate horsepower, a number of practical observations can be quickly made by the alert operator. Many operators can correct and compensate for unpredictable variations by careful observation and by making the necessary adjustments. While observing the operation in progress, check for the following:

1. Do you have the proper setup? Is it rigid? Are tool overhang, work overhang, tool grade, geometry, etc., correct?

2. Chip colors—what do they tell? Fig. 1 suggests the proper color (speed) ranges for four basic tool materials. As cutting speed increases to about 300 sfpm, the temperature of the chip goes up—then levels off. The color of the chip produced at a given speed provides a rough indication of the temperature at the tool point, and is itself a reasonably accurate indication of the cutting efficiency of the tool at that speed.

For example, the preferred or most efficient workpiece speed when using a carbide cutting tool is that speed at which a blue or gray chip is produced.

3. Chip forms—what do they tell? In Fig. 2A, the Discontinuous Chip is shown. With this type of chip, proper speed and feed give more uniform tool pressure and a better work finish. Normal practice: Reduce feed and take a deeper cut. Speed here is critical; excessive speed will produce a splinter chip.

Fig. 2B shows heavy-duty chips. When roughing large workpieces on heavy-duty equipment, carbides produce good results at moderate speeds, with feed rates of 0.025 to 0.060 ipr and cuts up to 1″ deep. Higher cutting speeds result in greater removal rates, less feed pressure and greater accuracy.

The chatter chip is shown in Fig. 2C. Faults that produce these chips are: poor or obsolete equipment, excessive tool overhang, overly small tool shank, non-uniform feed rate, variations in the material, over-broken chips, too much feed for cut depth, poor length-to-diameter ratio (indicating the need for a steady

CUTTING
TOOL

	NOT COLORED	LIGHT STRAW	BROWN	PURPLE	BLUE	GRAY
H.S.STEEL	ACCEPTABLE	ACCEPTABLE	PREFERRED	PREFERRED		
CAST (NONFERROUS)			ACCEPTABLE	PREFERRED	PREFERRED	
CARBIDE				ACCEPTABLE	PREFERRED	PREFERRED
CERAMIC					ACCEPTABLE	PREFERRED

Fig. 1. The experienced operator can gage the best speed for the operation simply by observing the color of the chips produced. (General Electric Co., Machining Development Laboratory)

STEEL CHIPS PARTIALLY DISCONTINUOUS
BECAUSE OF NONUNIFORM SPEED

CAST IRON CHIPS DISCONTINUOUS
BECAUSE MATERIAL IS BRITTLE

A   DISCONTINUOUS CHIP

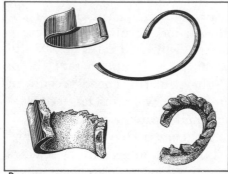

B   HEAVY DUTY CHIPS

	TOP	BOTTOM
TOOL MATERIAL	CARBIDE	HSS
SPEED (sfpm)	200	37
REMOVAL RATE (cu in/min)	105	65
FEED (In./Rev.)	0.060	0.149
CUT DEPTH (Inches)	0.750	1.000

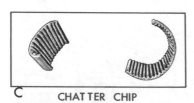

C   CHATTER CHIP

SPEED	250
FEED (In./Rev.)	0.018
CUT DEPTH (Inches)	0.250

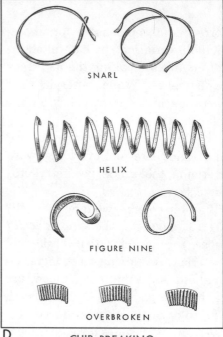

SNARL

HELIX

FIGURE NINE

OVERBROKEN

D   CHIP BREAKING

SPEED (sfpm)	.550
FEED (In./Rev.)	0.018
CUT DEPTH (In.)	0.250

TYPE OF CHIP	WIDTH BREAKER (Inches)	COMMENT
SNARL	0.125	BREAKER TOO WIDE
HELIX	(CRATER)	NORMALLY DUE TO WORN TOOL
FIGURE NINE	0.093	CORRECT CHIP CONTROL
OVERBROKEN	0.057	BREAKER TOO NARROW

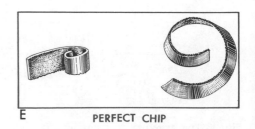

E   PERFECT CHIP

Fig. 2. **Chip forms and what they indicate about the quality of the machining operation. (Monarch Machine Tool Co.)**

490

Fig. 3. In a turning operation, be sure you have chosen a properly ground cutting tool of the best material and that you are using optimum speeds and feeds. Shown is a ceramic cutting tool removing metal at a high cutting speed. (Purdue University)

rest), non-rigid setup, excessive cutting speed, or dull tools.

Fig. 2D shows chip breaking. See the comments beneath the illustration.

Fig. 2E is an example of the perfect chip. When machining steel, a continuous *figure 9* chip is considered ideal. The ratio of feed to cut depth must be controlled to get this chip. Feed should be lowered and speed increased as depth of cut is reduced.

4. Did you plan tool life before beginning the operation? Fig. 3 shows the use of an extremely durable ceramic cutting tool.

What Machinability Means

The term *machinability* does not lend itself to an exact definition acceptable to all authorities. Generally speaking, however, machinability is a study of the relative ease (or difficulty) with which different materials can be machined and the many factors which contribute to ease of machining. Therefore, the following information is presented to acquaint the skilled operator with the machine and work material variables and some of the theory involved in metal cutting.

Ease of Metal Removal

The ease of metal removal is dependent upon two factors: (1) the power that is required for the removal of a given amount of metal, and (2) the rate at which the cutting tool is worn away by the machining operation. The second factor is important since excessive rates of wear result in higher cost because of the need to stop and sharpen or replace the tool. The lowering of machine output and the cost of tool maintenance have been a concern of industries for many years.

The ease with which a given material may be worked with a cutting tool changes with certain variables. Common variables affecting ease of cutting are[1]:

A. Machine variables.
 1. Cutting speed
 2. Dimension of cut (feed, depth, etc.)
 3. Tool geometry
 4. Tool material
 5. Cutting fluid
 6. Machine rigidity and set up
 7. Shape and dimensions of work

1. Adapted by permission from *Tool Engineers' Handbook,* edited by Frank W. Wilson; published by McGraw-Hill Book Company, Inc.

8. Nature of engagement of tool with work
B. Work material variables
   1. Hardness
   2. Tensile properties
   3. Chemical composition
   4. Microstructure
   5. Degree of cold work
   6. Strain hardenability

Ability to Develop an Adequate Finish

Machinability is further concerned with the ability to develop an adequate *finish* on the metal being machined. The element of cost also figures in here. While it is possible to produce a high type of finish on most metals, frequently the cost may be prohibitive.

A metal is said to have good machinability if portions of it can be removed easily by machining in such a way as to produce a good surface finish. The way to find out whether or not a certain metal has good machinability is actually to perform the particular operation on a sample piece. It must be remembered, however, that a metal showing good machinability for one operation may not show equally good machinability for a different operation.

Chip Formation

Another criterion by which to judge the machinability of a metal is the type of chip which is produced during machining. Chip formation is important as a means of determining the degree of finish the workpiece will have, besides revealing the efficiency of the machining operation.

The type of chips formed in the machining of metal fall into three classes. Chip types vary in their effect on surface finish and tool wear.

Discontinuous Chip Type. The discontinuous type of chip is shown in Fig. 4. When this chip is formed, the metal which is forced upward over the tool face is broken into short segments. This type of chip is caused by brittleness in the metal being machined and is present in the machining of cast iron and other brittle metals.

Fig. 4. Discontinuous Chip formed when machining brittle metals.

When machining brittle metal, a discontinuous chip usually means a fair surface finish, low power requirements and reasonable tool life. When

493

it occurs during machining of ductile materials, however, a discontinuous chip usually is accompanied by a poor surface finish and heavy tool wear.

Thick chips, low cutting speeds, and a small rake angle also cause discontinuous chips. A good surface finish is secured if the pitch angle of the chips is small. Tool failure occurs because of a gradual wearing away and rounding over of the cutting edge of the tool. However, this type of chip causes the least fouling of machine tools because, of the three chip types, it is the most easily removed from the cutting zone.

Continuous Chip Type. The continuous type of chip is shown in Fig. 5. This chip does not produce a built-up edge on the tool face. It is formed by continuous deformation of the metal being machined, the metal ahead of the tool deforming without fracture, producing a chip that travels smoothly up the tool face.

Thin chips, high cutting speeds, and a large rake angle are favorable to the formation of the continuous type of chip. The formation of the chip seems to be encouraged by circumstances which reduce friction between the chip-tool interface. The use of a tool material which does not weld to the work, highly polished steel faces, and suitable cutting fluids all tend to reduce this friction. With the continuous type of chip

Fig. 5. The continuous chip formed when machining ductile metals.

the best surface finish is produced and, for removal of a given amount of metal, the greatest efficiency in power consumption is yielded. The temperature at the cutting edge tends to rise less for this type of chip than for any other. As in the case of discontinuous chip formation, the face of the tool becomes worn by the abrasive action of the chip sliding over it so that the cutting edge is slowly rounded and worn away.

Continuous Chip Type with Built-up Edge. A third type of chip is the continuous type with a built-up edge, and is shown in Fig. 6. This type chip forms on metals which have good ductility. A compressed mass of metal adheres to the face of the cutting tool. Portions of the built-up edge break off from time to time and are carried away by the chip. Still other portions of the built-

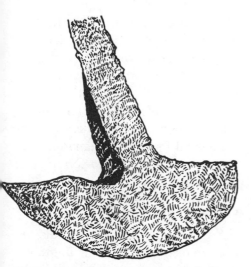

Fig. 6. The continuous chip with built-up edge, formed when machining ductile, wear-resistant metals.

up edge adhere to the workpiece and cause roughening of the machined surface. Although friction ap-

pears to be a partial cause of the built-up edge, there is also evidence that some welding may occur due to atomic bonding. Atomic forces become very great when two clean metal surfaces are squeezed together under high pressure. This type of chip will cause poor surface finish and severe wear on the tool. The flank of the tool is worn by the abrasive action of the portions of the built-up edge which adhere to the workpiece. The face is worn away by the action of the chip passing over it, producing a crater, which moves closer to the cutting edge as it grows. This has the effect of increasing the rake angle and finally results in breakage of the cutting edge by fracture or spalling.

Factors Affecting Machinability

Many factors have a pronounced influence on machinability. One is the *material characteristics* of the workpiece. Another factor of importance is the *type of cutting tool* used and the shape to which it has been ground.

A third factor is the *power required to produce the cut*. This is covered under the topic of feeds and speeds in each of the chapters. A fourth factor is *length of service of the cutting tool* before it becomes necessary to regrind it. A fifth factor, and certainly one that should not be overlooked, is the *kinds of coolants, lubricants, and cutting fluids used*. These factors will be explored in the following pages.

Workpiece Characteristics

Microstructure. The microstructure of a metal refers to its crystal or grain structure, visible through examination of etched and polished surfaces under a microscope. As a rule of thumb, metals whose microstructures are similar have similar machining properties. But there can be variations in the microstructure of the same workpiece that will make it difficult to machine.

Grain Size. Grain size and structure serve as a general indicator of a metal's machinability. A metal with small, undistorted grains tends to cut and finish easily. Such a metal is ductile, but it is also "gummy." While less easy to machine, metals with an intermediate grain size represent a compromise that permits cutting and finishing machinability.

Hardness Rating. Hardness is a measure of a metal's resistance to deformation. There are a number of testing devices for measuring hardness, most of which use indenters of precise shape and size to "dent" the metal being tested. The size of the indentation is then measured to determine the relative hardness of the metal. Common methods of making indentation hardness tests include the Brinell, Rockwell, Knoop, Vickers and Monotron tests.

Hardness alone is not an index of machinability. The common assumption that the harder the metal, the lower its machinability, does not hold true in the case of steel, for example. Steels of high hardness usually have lower machinability than steels of medium hardness, and very soft steels sometimes have lower machinability than steels of medium hardness.

Chemical Composition. Chemical composition of a metal is a major factor in machinability. The effects of composition are not always apparent, however, because the elements that make up a given alloy work both singly and in combination. Certain generalizations about the chemical composition of steels in relation to machinability can be made. But nonferrous alloys are too numerous and varied in composition to permit such generalizations.

Cutting Tool Materials

Six principal kinds of metals are used to make machine cutting tools: carbon, alloy and high speed steels, cast nonferrous alloys, cemented carbides, titanium carbides, ceramics, and diamonds. Which of these cutting tool materials to use for a specific operation is determined by the material to be machined, the type of finish required, the number of pieces to be produced and other job requirements.

Characteristics of cutting tool materials which can be compared when selecting the proper tool for a specific operation include: hardness, strength, high-temperature performance, rigidity, resistance to chemical action, abrasion resistance, thermal conductivity and coefficient of friction. The final factor, cost, represents the optimum selection of physical properties combined in a tool that will do the job at the lowest cost per piece.

Carbon and Alloy Steels. One of the oldest cutting tool materials, carbon tool steel is made of iron and carbon with small amounts of impurities such as manganese and silicon. Chromium and vanadium are sometimes added to increase hardness and to refine the grain.

By changing the carbon content, a range of properties can be obtained. Carbon tool steels in the low carbon range possess toughness and resistance to shock, while those in the higher range possess abrasion and wear resistance, hardness and the ability to hold a good cutting edge. Because they are often water quenched to attain the required hardness, warpage and dimensional changes incurred during hardening are problems with this grade.

High-Speed Steels. These steels are used principally for cutting tools and besides carbon may contain such alloying elements as tungsten, molybdenum, chromium, vanadium and cobalt. They usually contain enough carbon to harden uniformly throughout. Four general classifications of high-speed steel, named for their principal alloying materials, are (1) molybdenum, (2) molybdenum-cobalt, (3) tungsten, and (4) tungsten-cobalt. The addition of cobalt gives high-speed steels their ability to withstand high temperatures before cutting edges soften. They hold a high degree of hardness and strength at operating temperatures around 1100°F., a low-red heat.

High-speed steels find across-the-board use as cutting tools for lathes, milling machines, shapers and planers, reamers, drills, taps, etc. However, because of today's high machining speeds and the operating heats generated, the use of steel cutting tools is limited due to their inability to hold hardness at high-red heats.

Nonferrous Cast Alloys. This group of general-purpose cutting tool materials can be used for machining steel, iron, copper, brass, bronze, and aluminum and its alloys. These tools are used at surface speeds above those of high-speed steel and below those of carbides. There is always some overlapping, though, at the high and low ends of the speed ranges.

A characteristic of nonferrous cast alloys is their ability to retain hardness at high-red operating temperatures. They contain, usually, varying amounts of chromium, tungsten, boron, iron, vanadium, molybdenum, cobalt and carbon.

Cemented Carbides. Generally, "carbide" is a term applied to the chemical combination of a metal and carbon. Specifically, it is a term used to describe cemented carbide cutting tools made of various combinations of tungsten carbide, titanium carbide, and cobalt.

Cemented carbides are made in a variety of compositions to suit most machining requirements. They are extremely hard, high-speed cutting

497

Fig. 7. Edge wear along the friction faces of the tool results from friction or abrasion as well as from continuous tearing away of a built-up edge.

tools with high abrasion resistance, high transverse rupture strength, and a low coefficient of heat transmission so that there is no deformation of the tool at high operating temperatures. Their high wear resistance also permits holding to close tolerances. Cemented carbides have high compressive strength, excellent anti-weld properties, and retain their high hardness at elevated cutting temperatures.

The term "cemented carbides" implies that the carbides are held together by some cementing agent, such as cobalt. This is only partly true since the carbides themselves form a self-supporting skeleton. The cobalt facilitates the formation of the skeleton and reinforces it. The properties of cemented carbides depend on their composition, grain size

and processing techniques. A guide to the selection of carbide cutting tools for various workpiece materials and operations is presented in Table *XVIII* of the *Appendix*.

Tool Wear. In machining operations, two basic types of wear occur. These are commonly known as edge wear and crater. Edge wear, Fig. 7, occurs along the clearance faces of the tool and is the result of friction and abrasion. Edge wear can also occur when minute particles of carbide are removed from the cutting edge because of welding of the chip to the tool, and the gradual building up and tearing away of a built-up edge. Thermal cracking along the cutting edge also causes removal of small pieces of this edge.

Crater, Fig. 8, occurs on the rake face of the tool at the point of im-

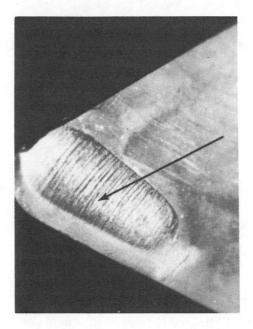

Fig. 8. High temperatures and pressures cause the chip to weld to carbide particles in the tool. Continuous tearing away of the chip eventually produces the cratering shown.

pingement of the chip with the tool, and is the result of high temperatures and pressures. These temperatures and pressures create a tendency for the chip to weld to microscopic particles of carbide present in the tool face. As they build up, pressures between these weld points and the chip increase until they are torn out and washed away. Experiments with workpieces containing radioactive tracers have shown this welding to be brought about by diffusion of the chip material into the carbide tool.

Edge wear is the prime concern when machining a material which has a brittle, flaky chip, or in any other operation where a continuous, tough chip is not generated. Non-

ferrous metals, non-metallics and most cast irons fall into this category. To cut these materials strong, wear-resistant grades are desirable. Such grades would be compounds of tungsten carbide and cobalt in varying percentages which fall within the C-1 to C-4 range of *Appendix Table XVIII*.

Force System

The force system acting on a cutting tool has three principal components, Fig. 9. The vertical component, which pushes down on the tool, is called the tangential force because it is exerted in a plane tangent to the workpiece. Another force is longitudinal, or parallel to the workpiece, and acts in the opposite direction to

499

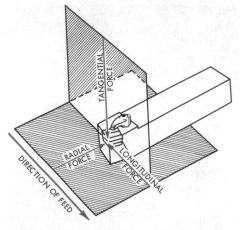

Fig. 9. Components of the force system acting on a single-point cutting tool during machining.

which the tool traverses the work. The third component is a radial force which also acts in a horizontal plane.

The longitudinal force and, in particular, the tangential force have a strong effect on the power required to make the tool cut. In general, most of the power is required to overcome the tangential force in a properly balanced operation. An exception to this is when the feed rate is in excess of good machining practice in which case the longitudinal force will exceed the tangential force.

Where cutting tool meets workpiece, the metal of the workpiece is placed in shear. This stress creates a shear plane which reaches maximum force when perpendicular to the face of the tool. Generally, the smaller the shear angle, the longer the path of shear and the thicker the chip. The power required to remove the metal in this case will be high. Conversely, the higher the angle of shear, the shorter the path of shear and the thinner the chip. The power required to remove the metal is less. The amount of compression and deformation that the metal undergoes during chip formation is measured in terms of shearing strain.

Chip formation is simplest when a

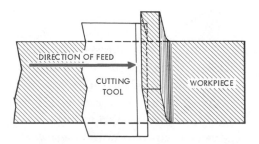

Fig. 10. In orthogonal cutting, the edge of the tool is perpendicular to the line of travel and only one cutting edge is working.

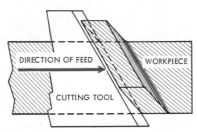

Fig. 11. In oblique cutting, the edge of the tool is inclined, imparting a helical curl to the chip.

continuous chip is formed in orthogonal cutting, Fig. 10. Here the cutting edge of the tool is perpendicular to the line of tool travel; tangential, longitudinal and radial forces are in the same plane, and only a single cutting edge is active. In oblique cutting, Fig. 11, the single, straight cutting edge is inclined in the direction of tool travel. This inclination causes changes in the direction of

chip flow up the face of the tool. When the cutting edge is inclined, the chip flows across the tool face with a sideways movement that produces a helical form of chip.

Milling Cutter Geometry

Because each tooth of a milling cutter is in and out of the work during every cutter revolution, each tooth takes an interrupted cut. For

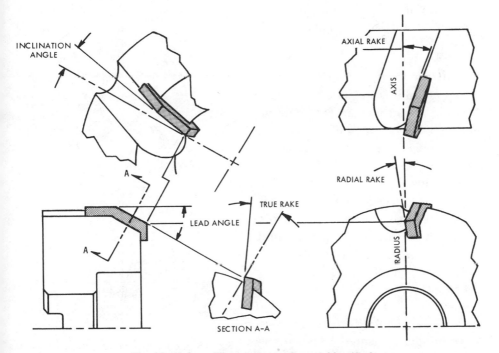

Fig. 12. Major milling cutter angles are identified.

501

this reason, the same basic principles which apply to single-point tools taking interrupted cuts apply equally to milling cutters. When measured relative to the workpiece, the cutting angles of the teeth of a milling cutter are comparable to those for a single-point tool. Thus the radial rake angle of a milling cutter compares to the side-rake angle of a single-point tool; the axial rake angle compares to the single-point tool's back rake angle; and the lead angle or bevel of the milling cutter tooth compares to the lead angle or side cutting edge angle of a single-point tool. But before continuing with milling cutter geometry, the student may find it helpful to review the section on single-point cutting tool angles in Chapter 7.

Milling Cutter Angles. The types of milling cutters covered in this discussion include face milling, end milling and side milling cutters. These tools have either indexible insert blades, or brazed-tip or solid carbide insertable blades. Indexible blades are discarded after their cutting edges are worn. Insertable blades, however, can be ground and re-used. The cutting edges of insertable blades, and their related angles, are illustrated in Fig. 12 and defined below.

Radial Rake Angle: Radial rake is the angle measured between the blade face and a radial line or reference plane drawn from the cutter axis to the cutting edge.

Axial Rake Angle: Axial rake is the angle measured between the blade face and an axial line or plane. The same reference plane is used for measuring radial rake and all other cutter rake angles.

Lead Angle: The cutting edge angle starting at the periphery of the cutter and converging toward the cutter axis.

True Rake Angle: The true rake angle is measured from, and perpendicular to, the lead angle and is the

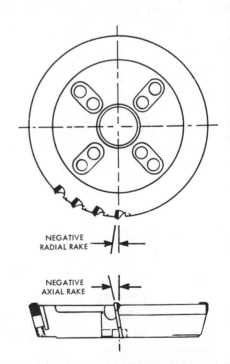

Fig. 13. A double-negative milling cutter has negative axial and radial rake angles.

angle between the blade face and the reference plane.

Inclination Angle: The inclination angle is similar to a helix angle or spiral. It is measured from the reference plane to the face of the blade on a line parallel to the lead angle cutting edge. The inclination angle determines the chip flow direction. A positive inclination pulls the chip up along the lead angle away from the cutter and work surface. A negative inclination angle pulls the chip down toward the work. A change in any of the cutting edge angles (radial rake, axial rake or lead angle) will change the inclination angle and the true rakes.

Basic Cutter Geometries. Milling cutters come from the tool-maker in three basic geometries: (1) double-positive (for positive axial and radial rake angles), (2) double-negative (for negative axial and radial rake angles), and (3) positive-negative (for positive axial and negative radial rake angles).

Double-negative geometry, Fig. 13, is more conventional and is used for milling hard materials at high speeds—350 sfpm or higher—where there is sufficient power and where part configuration permits. For milling free-machining metals at slower speeds, if the power and speed are not sufficient, double-positive cutters

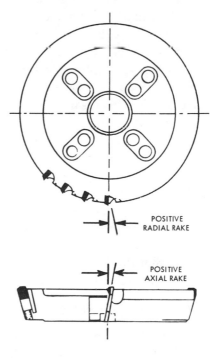

Fig. 14. A double-positive milling cutter has a positive axial and radial rake angle.

are used. A positive-negative cutter provides good control of chip flow and has a stronger cutting edge than the double-positive cutter, Fig. 14. The positive-negative cutter also generates a smoother surface finish.

Factors Affecting Cutter Selection. Each tooth in a milling cutter is designed to remove a certain amount of metal per revolution. Therefore, the greater the *number of blades* of the cutter, the more metal which can be removed per revolution. Other factors being equal, faster feeds and higher speeds can be used with fine-pitch cutters, but the horsepower requirements are accordingly greater. Fine-pitch cutters also produce a better finish. However, while higher speeds improve the surface finish, faster feeds generally cause a rougher finish.

Another factor that enters into the selection of milling cutters is *chip clearance*, or the space between the teeth. When necessary, chips can be removed by means of brushes attached to the cutters, or blown out by compressed air. When coolants are used, the fluid will wash away the chips.

When milling it is essential that the entire machine be as rigid as possible. *Rigidity of the arbor* and cutter have a direct bearing on tool life and surface finish.

Speed, Feed and Depth of Cut

Single-Point Tools. Each opera-

tion is an individual problem and such factors as rigidity, speed, tool geometry, feed and depth of cut must be determined for each specific operation. Experience shows that certain generalities can be made, however.

On long straight turning operations, speeds and feeds can be at the recommended maximum, consistent with the capability of the cutting tool and the capacity of the machine. Tables of recommended speeds and feeds for machining various materials using specific cutting tool materials are readily available. Table XIX of the *Appendix*, for example, indicates speed and feed ranges for turning with titanium carbide cutting tools.

Intricate shapes necessarily require slower feeds and speeds. Size of the workpiece also has a bearing on speed. For example, a large unbalanced casting may dictate turning at a relatively slow rpm; but because the work surface involved is large, the resultant surface speed may nevertheless be high.

Normally, a workpiece with small cross-sectional area will lack rigidity and be deflected easily. In such a case work supports are indicated, especially when the workpiece is of considerable length. In such cases a work rest or similar support should be used.

Chip Load. As a general practice, the maximum depth of cut and greatest feed, consistent with the capacity of the tool and the machine, are taken. The depth of cut is usually determined by the design of the workpiece and the time allowed for machining. Strength and size of the cutting tool are other factors.

Feeds are adjusted to provide the desired finish at the maximum metal removal rate. The slower the feed, the finer the finish.

The rate of speed (rpm) controls interface temperatures and determines tangential forces which will cause cutting edge failure. Speed also has an effect on finish: the higher the speed, the finer the finish. Speeds that are too slow will cause chatter. As the speed decreases, forces working against the tool increase and, unless the tool is sufficiently rigid, chatter and a herringbone finish will result.

Chip Control. The type and size of chip is controlled in several ways: by changes in depth of cut, by tool geometry, changes in speeds and feeds, and by the use of some type of chip breaker.

Milling Cutters. The relationship of feeds, speeds and depths of cut to milling cutter type, geometry and diameter, determine the amount of metal removed and the rate of removal. Finish, type and size of chip are other factors that have a bearing on milling machine operation, and all are interrelated.

Speeds for face milling operations can be selected from a Milling Speeds and Feeds table like that in Table XX, *Appendix*. The material classification serves as a general indicator, as is the other data, and it will be necessary for the operator to determine what speeds, feeds and depths of cut are best within the ranges suggested and the capacity of the milling machine.

Cutting Fluids

Cutting fluids used in machining operations have the dual purpose of cooling the workpiece at the point where the chip is being formed, and of reducing the adhesion between the flowing chip and the tool face. The search for a cutting fluid with the "best" properties has led to the development and use of a variety of cutting fluids. The cutting fluids most generally used may be grouped as follows: (1) emulsions of water and other elements such as waxes, mineral oil containing soap or sulfur, and pastes, with anti-rust compounds added to the emulsions; (2) cutting oils, which may be straight mineral oils, animal and vegetable oils, or compounded oils having a sulfurized or sulfurized-chlorinated base; and (3) air supplied in the form of a jet.

Amount of Cutting Fluid to Use. The quantity as well as the kind of cutting fluid supplied to the cutting tool has considerable effect on the amount of heat removed and the cutting speed that can be used in the operation. If a large amount of cutting fluid is used and properly supplied to the tool, a considerable increase in cutting speed can be obtained.

Effect on Machining Accuracy. Although the average temperature of the workpiece will not be as high as that reached at the tool point, it will be sufficiently high to cause the workpiece to expand during the machining operation. This expansion fluid can result in the cutter removing slightly more metal than intended so that, when the part cools down to room temperature, the machined dimension will be slightly under the required dimension.

The effect of temperature on the dimensions of the workpiece is important when machining to close tolerances or machining thin sections. A plentiful supply of cutting may be instrumental in cooling both cutting tool and workpiece to maintain proper dimensions.

Chemical Effect of Cutting Fluids. In addition to cooling, a cutting fluid should minimize adhesion of the chip to the tool. This function is performed by cutting fluids which contain chemically active substances that form non-metallic films on the freshly-formed surfaces during formation of the chip. These films cling to the contacting surfaces and thus prevent adhesion between chip and tool. In addition, resistance to the movement of the chip over the tool face is considerably reduced due to the low shear strength of the non-metallic film. Under these conditions, the built-up edge is prevented from developing and a good surface finish is obtained.

Cooling Effect of Cutting Fluids. The cooling effect of any fluid is measured by its capacity to absorb heat. Water has the highest cooling effect of any cutting fluid, but is seldom used alone because it promotes rust. For this reason, chemicals with anti-corrosion properties are added to water or water-based emulsions.

Machinability Ratings

Recognizing the fact that the machinability of a material cannot be measured with an instrument as can hardness or tensile strength, attempts have been made to give various steels machinability ratings on a different basis. One method of specifying the machinability of tool steel is to use such ratings as: best, good, fair, and poor. Still another rating uses letters as follows: *A*, excellent; *B*, good; *C*, fair; *D*, poor.

A common practice today is to rate the machinability of tool steel by comparing its properties with those of 1.00% straight carbon tool steel, properly annealed, which is given a machinability rating of 100%. With this system the various low carbon and alloy steels can be rated as shown in Table I.

Machinability testing is an exact scientific comparison of the materials to be cut under accurate control conditions. Comparative machinability observations have been recorded with the result that information such as in Tables I and II can be used as guides. The figures given do not represent absolute values, but should serve only as a guide for use under average shop conditions.

Developing the Ratings

The machinability ratings given

TABLE I. RELATIVE MACHINABILITY OF STEELS

COLD-DRAWN AND HOT-ROLLED				HOT-ROLLED AND ANNEALED			
SAE Number	Condition	Approximate Brinell Hardness	Machinability Rating	SAE Number	Condition	Approximate Brinell Hardness	Machinability Rating
1010	cd	131–170	42	1095	annealed	190–220	45
1015	cd	135–170	50	1330	"	179–235	50
1020	cd	137–174	65	1340	"	179–235	45
1025	cd	160–200	65	2320	hr	175–220	50
1030	cd	170–212	65	2330	hr	179–235	45
1035	cd	175–217	60	2340	annealed	179–235	45
1040	cd	175–217	68	2350	hr	190–240	45
1112	cd	179–229	100	2515	hr	175–220	47
X1112	cd	179–229	135	3120	hr	140–160	50
1120	cd	179–229	80	3130	hr	185–220	45
X1315	cd	143–179	80	3140	annealed	187–229	55
2315	cd	174–220	55	X3140	"	187–229	55
3120	cd	163–206	60	3150	"	200–240	50
4615	cd	175–217	65	3250	"	195–230	44
5120	cd	117–212	65	X4130	"	187–229	65
6120	cd	180–218	50	4140	"	190–235	56
1015	hr	110–130	42	4150	"	187–235	50
1020	hr	130–150	48	X4340	"	220–245	58
X1020	hr	135–160	62	4620	hr	165–195	58
1025	hr	130–150	58	4640	annealed	187–235	55
1030	hr	135–150	60	4820	hr	190–220	55
1035	hr	160–180	55	5140	annealed	174–229	60
1045	hr	180–220	55	52100	"	183–229	30
1085	hr	185–220	48	6130	"	210–225	55
				6145	"	179–235	50

TABLE II. MACHINABILITY RATINGS

Type of Steel	Machinability Ratings		
Carbon tool steel with or without small additions of alloys.	Best	40 *	100 **
Carbon-vanadium tool steel.			95
Manganese oil-hardening steel.	Good	35	90
5% chrome air-hardening tool steel.	Fair		65
Chisel tool steel.			75
Low tungsten chromium tool steel.		25	65
Shock-resisting steel.	Fair		65
Tungsten alloy chisel steel.	Good	30	85
High carbon, high chromium steel.	Poor	15	50
Hot-work steel.	Fair		70
High speed steel.		20	50

* Based on SAE 1112 steel as 100.
** Based on carbon tool steel as 100.

in Table II are based on a 100% rating for SAE 1112 cold-drawn Bessemer steel, machined under normal cutting conditions and using a suitable cutting fluid. They take into consideration such factors as the type of machining, speed, power used, etc. When a rating of 50% appears with the material rated, it is an indication that the material has a general, overall machinability approximately one half that of SAE 1112. It means greater difficulty will be encountered in machining it, with a corresponding shorter tool life, poorer surface finish, and increased power consumption. Generally speaking, when machinability is lower the speed of the cutting action is decreased also.

Table II indicates that machinability ratings for alloy steels are lower than for SAE 1112 steel. Alloy steels in general are more difficult to machine, although not necessarily as much as the hot-rolled, low carbon and high carbon steels. For that reason low carbon and medium carbon steels are soft and difficult to machine. For best performance in machining, low carbon steels require tools ground with keener cutting edges than do alloy steels. In addition, tools used for machining alloy steels usually have larger lip angles and a smaller side rake.

Rating Tool Steel

Every toolmaker and mechanic should understand the essentials about the machinability of tool steel because of its economical importance in his work. It is commonly known that tool steel is too hard when it comes from the rolling mill and must be softened by annealing in order to machine satisfactorily. Tool steels are almost invariably annealed when they reach the user to assure the easiest practicable working quality. It must be remembered, however, that with the addition of alloying elements, some easy-machining quality must be sacrificed for the more desirable qualities the alloying elements give.

In purchasing tool steel it is usually good practice to buy standard sizes of standard brands that have been annealed for *average* machinability for most toolroom conditions. It is often necessary to experiment with different brands before deciding which annealing procedure will best suit the particular requirement. However, steel which has been annealed to give suitable machinability in a drilling operation will not necessarily give the best machinability in a milling operation.

Carbon Tool Steels. With a carbon content ranging from about 0.75% to 1.35% carbon tool steels are the oldest type of tool steel used for cutting and shaping materials. Steels with this carbon content are generally known as water-hardening tool steels. They are so widely used and accepted that even the development of various types of alloy steels have only slightly lessened their use.

Carbon tool steels in the thoroughly annealed condition are the *easiest* of the tool steels to machine. The 1% carbon tool steel is rated with a machinability of 100% on a basis of comparison with other tool steels.

Carbon-Vanadium Tool Steel. This type of steel is similar to straight carbon tool steel except for the addition of about 0.20% vanadium which gives the steel a finer grain structure after it is heat treated. Carbon-vanadium tool steels in a properly annealed condition are among the easiest of all tool steels to machine. A 1% carbon-vanadium tool steel is rated with a machinability of approximately 95%.

Oil-Hardening Tool Steel. This type of steel has a good combination of abrasion-resistance and toughness for a wide variety of tool-and-die applications. Oil-hardening tool steel when properly annealed is rated 90% in comparison with 1% carbon steel as 100%.

High Speed Tool Steels. When these steels contain 18% tungsten or more they are slightly harder to machine than the molybdenum types. Like all highly-alloyed steels, high speed tool steels machine with somewhat more difficulty than the lower-alloyed steels. Where a 1% carbon tool steel is rated at 100, high speed steel is given a rating of 60.

NOTE: Please see review questions at end of book.

Numerical Control:

Point-to-Point and
Continuous Path Systems

Numerical control (N/C) is an automated method used to operate general purpose machines from instructions stored on a roll of tape for future as well as present use. The method can be used with or without a computer.

Numerical control is just what the term implies—control by the numbers. The two words, "control" and "numbers", have brought about a revolution in manufacturing.

This control may be adapted to any kind of machine or process which must be directed by human intelligence. Assembly, welding equipment and drafting machines are just a few of the diverse operations which have been adapted to N/C. Metal working is a major field of application for this new control concept. Milling machines, drilling and boring machines, lathes and grinding machines—in fact all machine tools—can and are being used with N/C.

How an N/C System Operates

Numerical control systems fall into two categories: open-loop and closed-loop systems. An *open-loop* system, Fig. 1, is one in which a moving machine element, for example a machine tool worktable or spindle, is instructed to move to one or more precise locations. For example, the worktable of a numerically controlled drill press can be programmed to move to a number of different positions to present a workpiece to the drill spindle so that a series of holes are drilled at precise locations along the workpiece.

The open-loop system, however, incorporates no provision for checking whether the movable machine

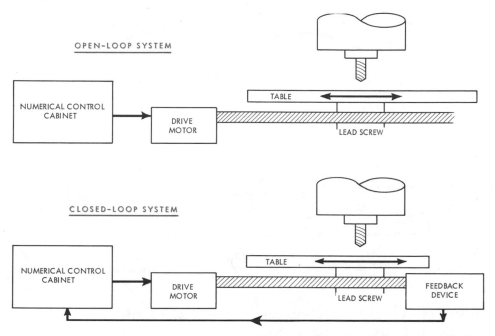

OPEN-LOOP SYSTEM

CLOSED-LOOP SYSTEM

Fig. 1. Open-loop and closed-loop numerical control systems. The closed-loop system includes a feedback device which monitors the position of the movable element (table) and automatically makes positioning adjustments.

element actually arrives at the required location so that, despite positioning errors or equipment malfunctions, the moving element continues through its programmed movements until the error or malfunction is discovered and corrected by the machine attendant.

If a machine is extremely accurate, and variable factors such as friction and general dynamics are adequately provided for, it is possible—although usually quite expensive—to achieve the desired accuracy with an open-loop system.

The majority of numerically controlled machine tools operate on

the servo-mechanism or *closed-loop* principle, Fig. 1. A closed-loop system incorporates a sensing device on the moving element of the machine to indicate the element's exact position as it moves. If a discrepancy arises between where the machine element *should be* and where it actually is, the sensing device "tells" the driving unit to make an adjustment to bring the movable element to the exact point required.

The closed-loop system employed in numerical control can be compared to a thermostatically-controlled home heating system, Fig. 2. When the room temperature drops

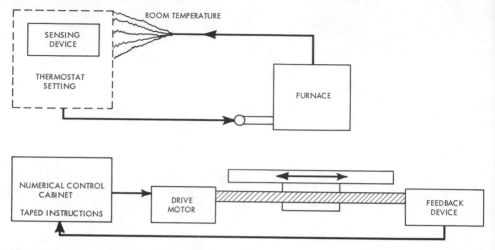

Fig. 2. A familiar application of the closed-loop control system is the thermostatically-controlled home heating system. Note the one-to-one relationship of the components in the two systems.

below the thermostat setting, the furnace is turned on by a sensing device. The furnace runs until room temperature again reaches the thermostat setting and the sensing device turns the furnace off.

In a closed-loop system, the taped instructions fed to the numerical control cabinet can be compared to the thermostat setting. The feedback device of the closed-loop system indicates the actual position of the movable machine element, just as the thermostat's sensing element measures actual room temperature. And the machine tool's drive motor and worktable correspond to the furnace and room temperature respectively since the drive motor actuates the table—just as the furnace raises the room temperature—to correct the unbalanced condition.

The Feedback Circuit

The field of electronics has made perhaps the most significant contribution to the successful development of numerical control, since it makes possible the high rate of information exchange required. Fig. 3 illustrates one of the more popular electronic closed-loop control systems. The coded information contained on the tape is converted to electronic pulses, each pulse being equivalent to a small incremental movement of the machine element (normally .0002 in.). Since it is possible that pulses can be "lost" during the servo process, they are converted to electric currents having a wave form similar to the common AC wave form.

The input wave pattern is com-

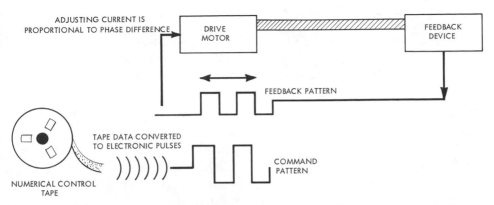

ADJUSTING CURRENT IS PROPORTIONAL TO PHASE DIFFERENCE

DRIVE MOTOR

FEEDBACK DEVICE

FEEDBACK PATTERN

TAPE DATA CONVERTED TO ELECTRONIC PULSES

COMMAND PATTERN

NUMERICAL CONTROL TAPE

Fig. 3. How the servo motor works: A phase difference between the command and feedback wave patterns creates a current which powers the motor which makes the positioning adjustment. When the wave forms coincide, power to the motor stops.

pared to an identical pattern generated by the feedback device. A phase difference between the two wave patterns — which indicates a positioning discrepancy—results in an adjusting current which activates the drive motor. As the motor adjusts the table position, the "command" and feedback wave patterns move closer together until the table reaches the correct position, the phases of the two wave patterns coincide, and the current to the drive motor caused by the phase difference stops.

Types of N/C

Numerical control systems are classified under two basic functional types—positioning and continuous path. The *positioning* or *point-to-point system* is used to control such machine tools as the drill press and jig borer which perform operations only at specified points on a workpiece. In drilling, for example, the drill spindle is positioned at a single specific point; the proper drill size, speed and feed are selected. The drill is advanced to cut a hole to the proper depth and is withdrawn when the operation is completed; then it is positioned to cut at the next workpoint, etc.

The *continuous path system* is used to control machine tools such as the lathe and milling machine which remove metal continuously from the surface of a workpiece. The task is to control continuously a cutting tool which requires frequent changes in movement with respect to two or more machine axes simultaneously and which is in constant contact with the workpiece. The continuous path system is more complex and requires a far greater input of detailed instructional information than a point-to-point positioning system. Therefore, the use of an electronic computer has become indispensable for the programmer preparing instructions for machine tools controlled by a continuous path system.

Measuring Basis for Numerical Control Systems

The basis for measuring for numerically controlled machine tools is the system of rectangular coordinates. In the system of rectangular coordinates, Fig. 4, sometimes called the Cartesian coordinate system, all point positions are described in terms of distances from a common point called the *origin* and measured along certain mutually perpendicular dimension lines called *axes*. Thus, to describe the geometry of

a part, it is only necessary to locate every point of the part within a framework of three such axes, called the X-, Y-, and Z-axes.

Although the angular orientation and the location of the origin can be chosen arbitrarily, the horizontal plane is customarily represented as containing the X- and Y-axes. In this plane, measurements taken to the right of the origin along the X-axis are considered to be in a plus x direction ($+x$), and measurements taken to the left are in a minus x direction ($-x$). In the same plane ($z = 0$), the Y-axis with its plus y and minus y directions is established exactly 90° from the X-axis. Perpendicular to both the X- and Y-axis is the third or Z-axis, with its plus z and minus z directions.

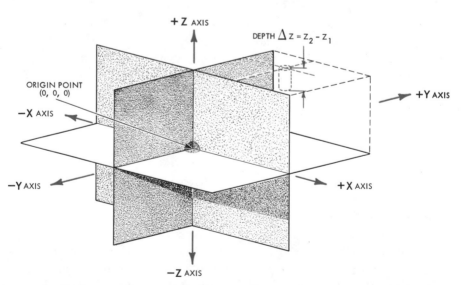

Fig. 4. The three-dimensional system of rectangular or Cartesian coordinates is used in most numerical control systems.

While the units of measurement along each of these three axes can be chosen arbitrarily, the decimal-inch system is commonly used in machine work throughout the United States. This coordinate system is used to designate machine axes, and all programming for numerical control is based on it.

Machine Axis Designation. For each machine tool, the machine axis designation is based on a rectangular coordinate system associated with the machine. The directions of motion indicated in Fig. 5 are characteristic of the usual motions of machine travel. The longest motion of which the machine is capable is generally designated as along, or parallel to, the X-axis. Thus, movement in one direction is considered as positive x ($+$ x), while movement in the opposite direction is considered as negative x ($-$ x). Lying horizontally from the point on the X-axis x $=$ 0, and at an angle of exactly 90° to the X-axis, is the Y-axis with its positive and

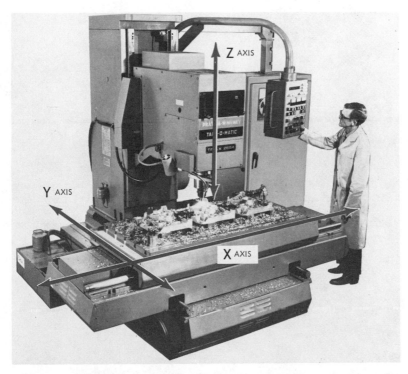

Fig. 5. Movement of the major movable elements of this 3-axis machining center can be described in terms of movement along the X, Y and Z axes of the rectangular coordinate system. (Pratt-Whitney-Colt Industries)

negative directions of motion, $(+y)$ and $(-y)$ respectively.

The motion of the table or head up or down (change in depth or Δz, that is) is designated as along, or parallel to, the Z-axis.

Figs. 6 to 10 illustrate the wide range of machine tools adapted to numerical control. In each case the major movements of the machine correspond to the coordinate axes principle.

Fig. 6. Numerically controlled 3-axis drilling and milling machine. (Superior Electric Co.)

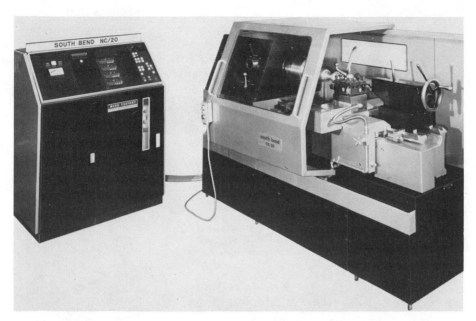

Fig. 7. Numerically controlled turning lathe with square turret. (South Bend Lathe, Inc.)

Fig. 8. Numerically controlled vertical bed turret lathe with a single spindle. (Monarch Machine Tool Co.)

Fig. 9. Numerically controlled eight-spindle machining center. (Burgmaster-Houdaille)

Fig. 10. Numerically controlled machining center featuring an automatic tool changer and tool storage. (Kearney & Trecker Corp.)

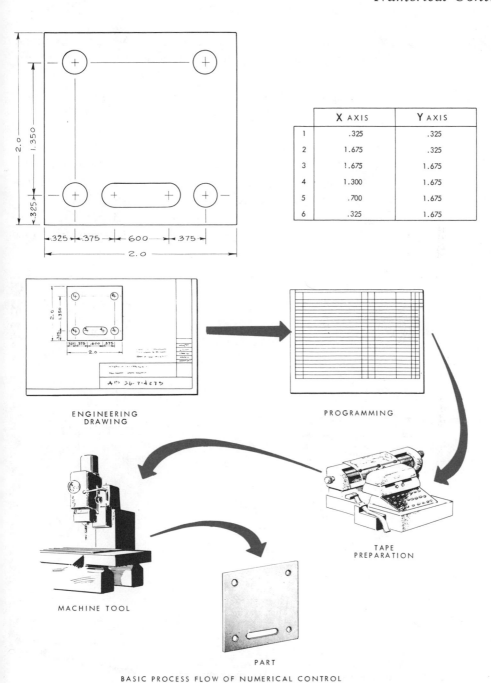

	X AXIS	Y AXIS
1	.325	.325
2	1.675	.325
3	1.675	1.675
4	1.300	1.675
5	.700	1.675
6	.325	1.675

ENGINEERING
DRAWING

PROGRAMMING

TAPE
PREPARATION

MACHINE TOOL

PART

BASIC PROCESS FLOW OF NUMERICAL CONTROL

Fig. 11. Before programming an operation, the part programmer studies the engineering drawing. Note that base line dimensioning is used on the drawing.

Production Steps—
Point-to-Point Control

Production steps in conventional machining—part design, methods planning, tooling, setup and machine tool operation—are present in N/C, but they are performed somewhat differently. Moreover there is a substantial shift in the burden to be borne by the various steps in the production process.

Part Design.   A part to be made on a point-to-point numerically controlled machine tool is designed by a design engineer who is familiar with manufacturing processes as well as the capabilities of N/C machine tools. After designing it is developed as an engineering drawing by a draftsman. Note the process sequence shown in Fig. 11.

Part Programming.   The engineering drawing is translated into

SHEET __1__ OF __1__

PROGRAM SHEET • HYDRA-POINT • BRIDGEPORT MILL

TAPE NO. __16__

PART NO. __SPACER__   REV. _____

PART NAME __MANIFOLD BLOCK__

PREPARED BY __KELLEY__ DATE _____

TYPED BY   bh   DATE

LOCATION NO	BLOCK NO 1·2·3	PREP FCT 4	X 5·6·7·8·9	Y 10·11·12·13·14	FEED FCT 15·16	TOOL NO 17	MISC FCT 18·19		SEQUENCE NO	PREP FCT	X	Y	FEED FCT	TOOL NO	MISC FCT
LOAD	000	5	19.000	1.000	00	1	02								
TOOL CHANGE	000	5	19.000	9.000	00	1	06								
1	001	5	2.325	2.325	00	1	08								
2	002	5	3.675	2.325	00	1	80								
3	003	5	3.675	3.675	00	1	80								
4	004	5	3.300	3.675	00	1	80								
5	005	6	2.700	3.675	00	1	80								
6	006	5	2.325	3.675	00	1	80								
LOAD	000	5	19.000	1.000	00	1	02								

Column group headers: ORIGINAL TAPE (left); REVISIONS (right).

Fig. 12. A manuscript program shows the sequence of operations to be performed in numerical control machine work. (Chicago Public Schools, Manpower Training Div.)

a manuscript program by a parts programmer. See Fig. 12. This step in production includes many steps that were previously done by the machinist or setup man at the machine tool. The programmer must consider the shape of the part, the material from which the part is to be made, the machine tools available to him and the capabilities of the machine tool for which he is programming. First the programmer studies the drawing and then, with all the foregoing factors in mind, he draws up the sequence of operations to be performed.

The simplest N/C machine tools have the X and Y distances controlled by perforated tape. Many machines have only a single tool holder. In these cases, part of the operation is performed by a machine tender, or operator, manually. Manual operations must be included on the manuscript program. These take the form of instructions to the operator. The operator will not change tools automatically without instructions. Once the sequence of operations and the necessary instructions to the operator have been determined, the programmer will make up his manuscript. All tool movements and operations, coolant control, speed and feeds, as well as tool selection and tool change must be listed on the manuscript. The programmer must anticipate all operations for he controls every phase of production .

Tape Preparation. When the programmer has completed the first draft of the manuscript, it is usually checked by a second programmer. When the manuscript is complete and correct, it is turned over to a typist for preparation of the tape. See Fig. 13. This is usually accomplished on a Friden Flexowriter or a similar machine.

A second tape is then made for verification; the first tape is inserted into the typewriter's reading unit and automatic retyping of the process sheet produces a second tape. This, in effect, provides mechanical proofreading as the second tape will not punch out if it differs at any point from the first tape. The process tape is now ready for insertion in the reader at the machine tool; the reader interprets the data and commands the machine tool.

For point-to-point N/C systems, as in contour N/C systems, the tape is then threaded on the control unit of the machine. A machine operator places the workpiece on the work table, changes tools and performs manual parts of the operation as instructed on the manuscript sheet.

Continuous Path or Contour Control

Contouring control, far more complex than point-to-point control, is used in milling, turning, or other operations where the entire path of the tool or workpiece must be de-

BINARY NUMBERS

ARABIC	BINARY	POWERS OF 2
0	0	
1	1	2^0
2	10	2^1
3	11	
4	100	2^2
5	101	
6	110	
7	111	
8	1000	2^3
9	1001	
10	1010	
11	1011	
12	1100	
13	1101	
14	1110	
15	1111	
16	10000	2^4
17	10001	
18	10010	
19	10011	
20	10100	
21	10101	
22	10110	
23	10111	
24	11000	
25	11001	
26	11010	
27	11011	
28	11100	
29	11101	
30	11110	
31	11111	
32	100000	2^5
64	1000000	2^6
128	10000000	2^7

Value of 182 expressed in
Arabic numbers

1×10^2 - 100
8×10^1 - 80
2×10^0 - 2

1 8 2 $\overline{182}$

Value of 182 expressed in
Binary numbers

1×2^7 - 128
0×2^6 - 0
1×2^5 - 32
1×2^4 - 16
0×2^3 - 0
1×2^2 - 4
1×2^1 - 2
0×2^0 - 0

1 0 1 1 0 1 1 0 $\overline{182}$

Remember: Any number raised to
the zero power equals 1.

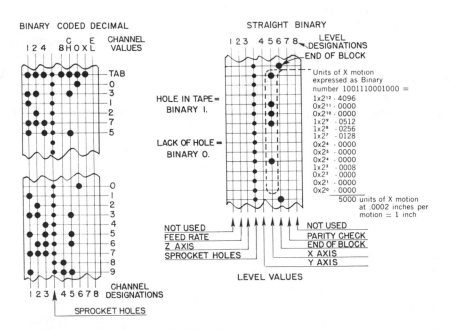

Fig. 13. After the manuscript program has been checked, a tape is prepared. For almost all point-to-point systems, the *BCD* form of numerical representation, shown here, is used. (Modern Machine Shop Magazine)

scribed and controlled. The steps required to program for continuous-path control depend on the particular system, so the description given here should be regarded only as typical. While manual programming may be practical for simple contours, the use of a computer makes programming faster and easier. Fig. 14 shows a tape-controlled engine lathe which uses a continuous path contour numerical control system.

As in point-to-point positioning, the programmer first prepares a manuscript from the engineering drawing. The data, including process and tooling information, are set up on punched cards. The cards are fed to the computer, which makes the necessary calculations. Punched cards from the computer are then converted to punched tape.

In preparing the process sheet, the programmer enters starting and end points for each major section of the contour, the type of curve, location of foci and other curve constants, cutter size, speeds, feeds, machining sequence, etc. All this information is converted on a keypunch to cards.

The computer, from card data, calculates the incremental points along the cutting path, time intervals, cutter offset and other information needed to direct the ma-

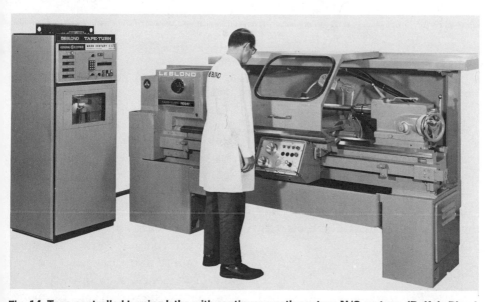

Fig. 14. Tape-controlled turning lathe with continuous path contour N/C system. (R. K. LeBlond Machine Tool Co.)

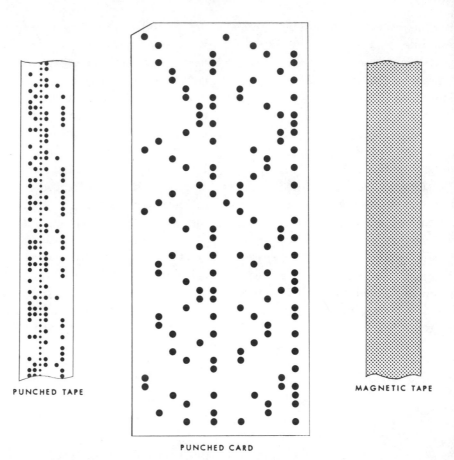

PUNCHED TAPE

PUNCHED CARD

MAGNETIC TAPE

Fig. 15. Machine tool instructions can be converted from one medium to another. The three media shown above are punched tape, punched card, and magnetic tape. (Modern Machine Shop Magazine)

chine control system. It reads out this information on a set of punched cards. The punched cards, when fed to a card-to-card tape converter, produce the punched process tape that controls the machine tools. See Fig. 15.

Opportunities in N/C

Parts Programmer. A skilled worker in this new occupation cre-

ated by the N/C technology is a key member of the new metal-working team. Parts programmers determine the detailed manufacturing steps and prepare instructions which are coded onto the control tape. These instructions direct the machine tool to perform automatically, and in sequence, such operations as positioning the work table, changing speeds and feeds, and even changing tools. To prepare optimum instructions, parts programmers must be familiar with general shop and machining procedures; they must know the functions and uses of cutting tools and fixtures; they must have a thorough knowledge of the various machine tools, their feeds and speeds; and they must be able to interpret engineering data, sketches, and blueprints. Parts programming incorporates many of the decision, judgment and setup functions of the highly skilled conventional machine tool operators as well as many of the duties of the toolmaker in making sure the work is measured, located, and guided correctly.

In addition to these basic skills, parts programmers need a thorough knowledge of the capabilities of N/C systems and procedures, and methods of programming. Several parts programmers may work as a team on one problem. A senior programmer may have overall responsibility. Junior programmers may be assigned to convert the senior programmer's completed manuscript into special programming languages or codes.

In the early days of N/C, a knowledge of college mathematics was considered a necessary programming background. Today, high school arithmetic, extraction of square roots, algebra, geometry, and trigonometry is reportedly sufficient in most instances. Many companies

rely on skilled machinists and tooling men, who are able to solve problems by a systematic approach, to assume parts programming duties.

Operator. N/C machine tools do not usually require the services of an experienced journeyman machinist. Operators are, however, required to know something about programming techniques and a basic knowledge of machine tools is helpful. See Fig. 16. The machinery which the operator handles is more expensive than conventional machinery and for this reason the pay for machine tool operators in an N/C shop is often comparable to the pay of conventional operators, in spite of the less rigorous training requirements.

Maintenance. Many employers have found that personnel servicing conventional machine tools can easily be shifted to service the numerically-controlled tool itself. With additional training, workers having a background of electrical or electronics experience have been reported to be best adaptable to servicing the N/C systems.

Designers and Toolmakers. The tool designer and tool and die maker must learn about new capabilities, functions, operations and programming necessary to use many of the various numerically-controlled machine tools. But there is no complete or general change from present occupational requirements. As numerically-controlled machines become

Fig. 16. The vocational students of today must be trained to prepare manuscripts for and operate the N/C machine tools of modern industry. (Chicago Public Schools, Manpower Training Div.)

more commonly used in the manufacture of tools, jigs and fixtures, however, many of the decisions, judgments, shop practices and precision-machinery functions presently required of these highly skilled craftsmen will be transferred to the planning and programming operations.

NOTE: Please see review questions at end of book.

Electrical Energy Processes:

Electro-Discharge, Electro-Chemical, Ultrasonic, Magnetic Pulse Forming, and Electrolytic Grinding

Developments in the technology of machining have made it possible to make parts of newer, harder metals, but which cannot be economically machined by conventional methods. Also, the more intricate designs and smaller tolerances demanded have led to difficulties in conventionally fabricating and machining such shapes.

The rapid developments and frequent use of high heat-resistant alloys has made it necessary for tool makers to reduce costs and improve tool making techniques. These new needs have resulted in the introduction of several new electrical machining concepts.

Electro-Discharge Machining (EDM)

This process removes metal through the action of an electrical discharge of very short duration and high current density between the electrode and the work. Material is removed from both the tool and the work. The proper selection of work-tool combinations can increase rates of removal of material from work and decrease tool wear. Both the tool and workpiece are submerged in a dielectric fluid which assists in confining the electrical conduction to the regions of discharge in the gap and washes out the debris. Note in Fig. 1, the basic electric circuit.

Electro-discharge machining is frequently used to machine unusual, or very small or large shapes which cannot be machined by conventional methods. Because of its low cost, EDM has become a key tool-and-die-making machine tool. Difficult-to-work materials such as hardened steels, stainless steels, carbides, and titanium are only a few of the materials that can be machined economically by the process. Intricate and precision parts and shapes which would require extensive investments to machine by conventional methods can be successfully machined by the EDM process. See Fig. 2.

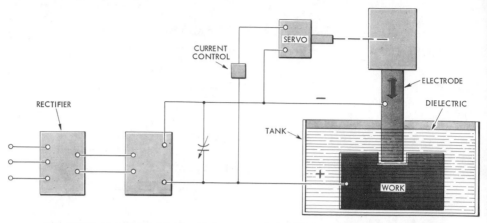

Fig. 1. Electro-discharge machining uses these basic elements in its circuit.

Fig. 2. Intricate, precision parts and shapes can be economically machined from difficult-to-work materials by EDM. *(Left)* a coining die; *(right)* a blanking die. (Charmilles Corp. of America)

EDM affords high degree of flexibility, control, and accuracy when drilling deep holes. Reworking of tools which otherwise could not be salvaged is another wide application for EDM.

The electro-discharge process can be used to generate almost any shape if a suitable electrode tool can be fabricated and brought into close proximity to the workpiece. The process has been used for:

1. cutting small-diameter holes.
2. machining molds and dies for blanking, extrusion and forging.

3. blanking parts for sheets.
4. cutting off rods of carbides or metals with poor machinability.
5. flat or form grinding.

Holes of almost any shape and size can be produced by the EDM process since neither the tool nor the workpiece rotates. Tolerances as low as 0.0005" can be held with slow rates of metal removal.

The rate of metal removal depends to a great extent on the average current in the discharge circuit. To a lesser extent, the rate also is influenced by other electrical parameters, the electrode characteristics and the nature of the dielectric fluid. In practice the machining rate is normally varied by altering the energy per discharge and/or the number of discharges per second. Rate is controlled in the power supply.

Electro-Discharge Machine Tools

A number of machines are commercially available and have proven their performance as tool-and-die-making machines. Typical units are shown in Fig. 3.

Principal Elements. The major elements of a typical electro-discharge machine tool are shown in Fig. 4.

1. The *power supply unit*, Fig. 4, supplies current to the machine tool while providing the signals for automatic control of the machining electrode, which is the cutting tool. Current can be either a pulsating or relaxation type. Circuits for each are shown in Fig. 5.

2. The *hydraulic servo unit*, Fig. 4, supports the electrode, is automatically fed in a vertical direction and maintains a constant gap between the cutting tool (negative electrode) and the workpiece (anode) being machined.

3. The *machine table*, Fig. 4, is adjustable for positioning the workpiece with respect to the X and Y axes. The dielectric tank is an integral part of the table for the purpose of containing the liquid dielectric since all machining must be done submerged in the fluid bath.

Electrode Materials. Electrodes in electrical discharge machining can be compared to cutting tools in conventional machining. As in conventional machining some materials will provide longer wearing qualities, some electrode materials in EDM will provide better machining qualities under certain conditions. Some electrode materials have very good wear rates, but are difficult to machine and are quite expensive.

Each machining application will dictate the selection of an electrode material. The following electrode materials have been widely used in industry: (1) tungsten carbide, (2) silver tungsten, (3) copper, (4) graphite, (5) brass, (6) copper and zinc alloys.

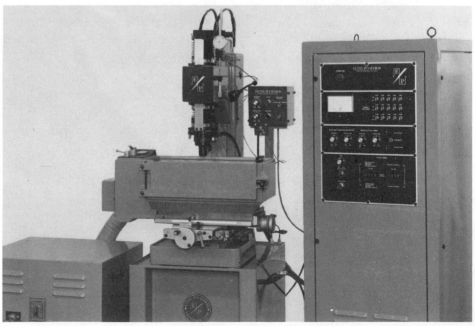

Fig. 3. Examples of currently available EDM machines: the 8-in. Ram Type EDM from Cincinnati Milacron *(top)* and the Single-Multilead 30 amp EDM from Eltee Pulsitron *(bottom)*.

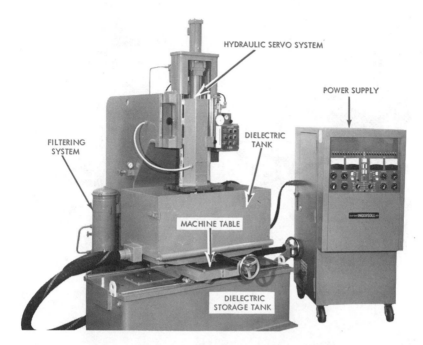

Fig. 4. Principal elements of the Electro-Discharge Machine (South Bend-Ingersol).

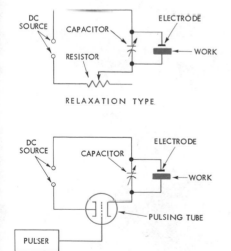

Fig. 5. Either of two circuit types is used in the EDM power supply unit, the relaxation type *(top)* or the pulse type circuit *(bottom)*.

Typical Tooling. The EDM process is largely used to machine the following types of tools: stamping dies, plastic molding dies, extrusion dies, die-cast dies, and forging dies. In machining these tools considerable savings in tool-and-die-making time can be realized, and difficult-to-machine shapes are obtainable much more easily with the EDM process. Fig. 6 illustrates forging dies, a motor lamination die and a plastic mold die, all produced economically and to close tolerances by the EDM process.

Electro-Chemical Machining (ECM)

Electro-chemical machining, or ECM, is essentially an electrolysis

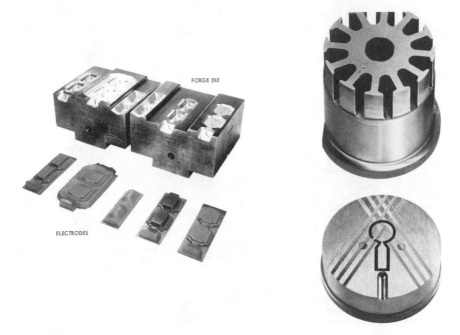

Fig. 6. Typical parts produced by the EDM process. *Left:* Forging dies and the electrodes used to produce them; *Right:* A motor lamination die and a plastic mold die. (Cincinnati Milacron)

Fig. 7. A view of the working area of Cincinnati's ECM. (Purdue University)

process where electro-chemical reaction dissolves metal from a workpiece into an electrolyte solution. A direct current is passed through an electrolyte solution between the electrode tool, which is negative (cathode), and the workpiece, which is positive (anode), causing metal removal ahead of the electrode tool as it is fed towards the workpiece. Chemical reaction, caused by the current in the electrolyte, dissolves the metal from the workpiece and hence the term, electro-chemical machining. See Fig. 7.

The Electrode. In ECM, the electrode tool is a carefully made insulated device of a particular size or shape through which the fluid electrolyte is carried to the machining area, Fig. 8. Regardless of the material used, the electrode must have these four characteristics: it must be machinable, rigid, a conductor of electricity and have resistance to corrosion. There are at least three good electrode materials for ECM applications: (1) copper, (2) brass, and (3) stainless steel. Copper is a proven excellent conductor of electricity but difficulties occur during its machining. On thin walled applications or deep holes where long electrodes are required copper is too soft. Brass can be used for electrode material with good results. While brass is normally considered a good conductor of electricity, it cannot compare favorably

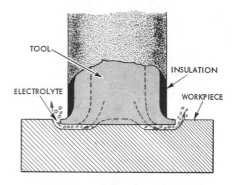

Fig. 8. Current from the electrode tool speeds up the chemical reaction between the electrolyte and the metal being machined.

with the conductivity of copper. This relatively greater loss in electrical transmission by brass may be offset by the metal's greater strength, ease in machining and lower initial cost. Where large volume flow and high pressures of electrolyte are required, the strength of stainless steel electrodes is excellent.

The electrolyte bath is not stationary in some large vat, as in plating operations, but is a swiftly flowing controlled stream. Machining currents reach very high levels, at times as high as 10,000 amperes per square inch of work material. All of these elements and the role they assume in the electrolytic system affect the success of an ECM application. A typical ECM system is diagrammed in Fig. 9.

The Machining Gap. In electrochemical machining, the "machining gap" is the physical distance between the tool and work. See Fig.

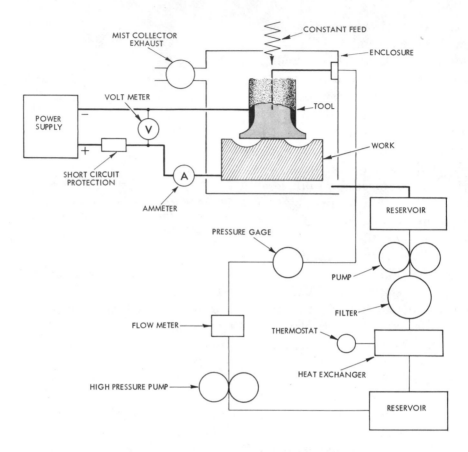

Fig. 9. This schematic diagram illustrates a typical ECM system.

10. The electrolyzing current must cross this gap to complete the electrolytic cell. The tool and work are positioned as close together as possible to encourage efficient localized electrical conduction of the electrolyte. The width of the gap, under average conditions, will vary from one-thousandth to three-thousandths of an inch. Any physical contact of the tool and work at ECM high current levels will result in arcing and serious damage to both members.

Machining Capabilities. The process will produce large complex parts from hard, tough and heat-sensitive materials. Because there is no physical contact with the workpiece, there is no surface hardening, heat distortion or burring. ECM will produce parts ten times faster than a conventional grinding operation, with high ECM penetrat-

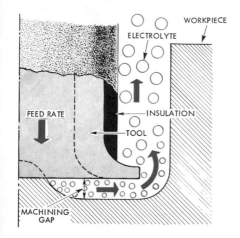

Fig. 10. A physical distance, called the *machining gap*, is necessary to the operation of an ECM system.

ing rates of .050 to .100 inch per minute.

Basically, five major configurations are considered in ECM operations: (1) round through holes, (2) square through holes, (3) round or square blind holes, (4) simple cavity operations, and (5) surface finishing. The penetration rates and speed of metal removal will vary for each of these shapes. Penetration rates are shown in the following tables.

Requirements for any machining operation are easily found. Just multiply the current density (CD value in Table I) by the current density multiplication factor in Table II. To find total current for the job, multiply this maximum current density by the surface area between tool and work. Once machining conditions have been established, and are held constant, dimensional accuracies are highly precise.

Ultrasonic Machining (USM)

Ultrasonic machining uses high frequency mechanical vibration transmitted through a shaped tool

TABLE I. CURRENT DENSITY OF METALS

MATERIAL	CD/0.100 IN./MIN. PENETRATION RATE
IRON (STEEL)	1100
NICKEL	1200
COBALT	1195
ALUMINUM	1190
TITANIUM	490
A-286	1100
RENE 41	1350
17-7PH	1180
HS-25	1425

TABLE II. TYPICAL ECM PENETRATION RATES

MACHINING OPERATION	PENETRATION RATE (IN./MIN.)	CD MULTIPLICATION FACTOR
ROUND THROUGH HOLES	0.5	5.0
SQUARE THROUGH HOLES	0.4	4.0
BLIND HOLES (ROUND AND SQUARE)	0.3	3.0
SIMPLE CAVITIES	0.25	2.5
PLANING	6.0	5.0
WIRE CUTTING	(DEPENDS UPON SIZE OF AREA MACHINED)	

1. COURTESY OF CINCINNATI MILLING MACHINE COMPANY

to an abrasive grit to remove material. The abrasive is suspended in water while being circulated. This process is often referred to as impact grinding because of the high energy of the abrasive particles used to cut into the work.

Major Elements. The major parts of the ultrasonic machine are:

1. The *electronic generator* which converts 110 volts single phase or 220 volts 3 phase A-C into high frequency power for a transducer. Fig. 11.

2. The *transducer* converts part of the electrical energy into mechanical energy which will cause the tool to vibrate at high frequencies.

3. The *tool,* which vibrates to

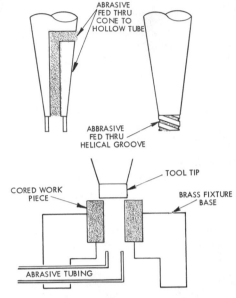

Fig. 12. The tool shape determines the shape of the finished workpiece.

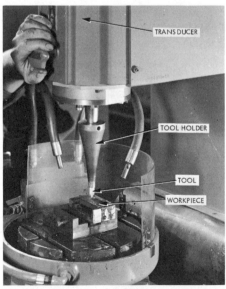

Fig. 11. The major working parts of an ultrasonic machine tool are the *transducer,* the *tool holder* and the *tool.* (Sheffield Corp.)

drive the abrasive grit into the workpiece, machines away minute particles of the work. Fig. 12.

Machining Concept. Ultrasonic machining involves the following concepts:

Abrasive. Abrasive grains are the cutting edges of the tool. They are carried, in granular form, in a liquid that flows between the workpiece and the tool and is recirculated for continued use.

The same abrasive types and grit sizes found in commercial grinding wheels are used in ultrasonic machining. Type and grit size are selected in accordance with the nature of the operation and surface finish desired. The most universally used

ABRASIVE

Fig. 13. Abrasive grains are the cutting edge of the ultrasonic machine tool.

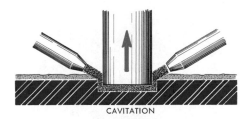

CAVITATION

Fig. 15. *Cavitation*, a partial vacuum in the liquid, creates the turbulence necessary to cutting action.

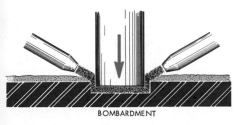

BOMBARDMENT

Fig. 14. Grains bombard the workpiece to produce a pattern.

abrasives are boron carbide, silicon carbide, and aluminum oxide.

Coarse grit is used for rapid cutting, the finer grits for finishing. See Fig. 13.

Bombardment. A tool stroke of a few thousandths of an inch and vibrations of approximately 20,000 times per second drive the abrasive grains. The frequency, as well as shape of the tool determine the precise bombardment pattern of the grit against the workpiece. See Fig. 14.

Cavitation. The extremely fast motion of the tool face produces cav-

itation (partial vacuum) and agitation of the abrasive liquid carrier necessary for material cutting. This turbulent action of the liquid acts as a pump for the abrasive grains and the minutely cut particles in the cutting area. See Fig. 15.

Major Areas of Application. Ultrasonic machining is used to cut and shape hard, brittle materials that are difficult to machine with conventional equipment. The process is capable of cutting holes, slots, intricate cavities, and complex shapes. Typical applications are the making of metal dies and molds for ceramic buttons and other nonmetallic products. Designs can be machined in gem stones and similar materials. Figs. 16 and 17.

The jewelry industry has found this process of value for engraving and drilling precious stones. Today the hard materials processing industry, which uses such substances as tungsten carbide and ceramics, is finding it invaluable.

537

Fig. 16. This typical tool used for ultrasonic machining is brazed to the toolholder shown.

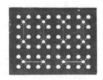

Fig. 17. Holes and slots of this workpiece were machined ultrasonically using the tool shown in Fig. 16.

Significant Advantages. Machining of hard and brittle materials is possible. Small externally applied forces are used which do not stress, distort, or significantly raise the temperature of the material being worked on. This method can offer considerable cost, time and quality improvements in its proper application.

Process Limitations. Ultrasonic machining should not be considered for removal of large amounts of material, but rather as a supplement to conventional machines extending their range of usefulness.

The abrasive grit vibrating wears away both the tool and the workpiece. Ratio of stock removal to tool wear varies from about 200 to 1 down to 1 to 1.

Fig. 18. The Magneform machine forms metal electro-magnetically. (Purdue University)

TABLE III. COMPARISON OF ELECTRO-DISCHARGE, ULTRASONIC, AND ELECTRO-CHEMICAL MACHINING CHARACTERISTICS

OPERATING CHARACTERISTICS	EDM	ULTRASONIC	ECM
METHOD OF REMOVAL	SPARK–ARC BOMBARDMENT	MECHANICAL IMPACT	ELECTROCHEMICAL
KIND OF WORK MATERIAL	CONDUCTIVE	ALL HARD MATERIAL	CONDUCTIVE
RATE OF REMOVAL	SLOW	SLOWEST	FASTEST
TYPE OF CUTS	ALL	ALL	ALL
THERMAL DAMAGE	YES	NO	NO
VISIBILITY OF WORK	SUBMERGED IN OIL	OPEN (WET)	OPEN (WET)
TYPE OF COOLANT	INSULATING OIL	WATER SLURRY	CONDUCTIVE ELECTROLYTE
POWER SUPPLY	A.C. PULSES	ULTRASONIC GENERATOR	D. C. RECTIFIER PLUS AUTOMATIC CONTROL

Practical machining areas vary from about three square inches down to one sq. in. This is a function of tool and material being cut. The average rate of material cutting using a ½″ diameter tool, ½″ deep varies from .02 IPM (inches per minute) to .10 IPM.

Table III compares the operating characteristics of the three electrical energy processes already discussed: electro-discharge, electro-chemical, and ultrasonic machining.

Magnetic Pulse Forming

Electro-magnetic metal forming machines are playing an important part in today's fabrication processes. The General Atomic Division of General Dynamics Corporation introduced the first commercial production machine that forms metal electro-magnetically. This machine is called the Magneform. See Fig. 18.

The Magneform machine operates with precisely controlled electro-magnetic forces and is capable of discharging magnetic pressures up to 50,000 pounds per square inch, in pulses of 10-20 ms (microseconds = millionths of a second) duration. The electrical energy is stored in a capacitor and discharged rapidly through a coil. See Fig. 19.

Basic Concept of System. The basic circuit consists of an energy storage capacitor, a switch, a coil, and a power supply that provides energy to charge a capacitor as shown in Fig. 19. In operation, the current through the coil produces a temporary magnetic field of high intensity between the coil and the workpiece. During this brief impulse, eddy currents in the workpiece restrict the magnetic field to

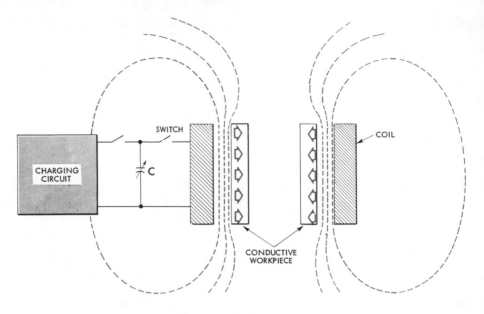

Fig. 19. The basic magnetic-pulse metalworking circuit of the Magneform.

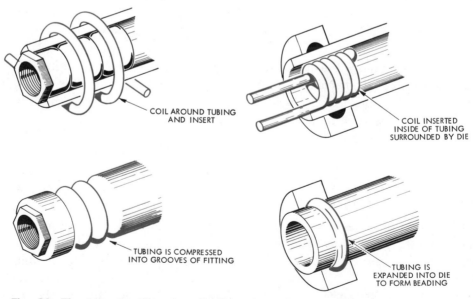

Fig. 20. The type A coil surrounds the work to create a compressive force which forms the metal.

Fig. 21. The type B coil is placed inside the work to produce an expansive force.

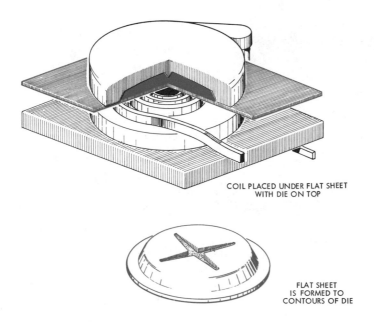

COIL PLACED UNDER FLAT SHEET
WITH DIE ON TOP

FLAT SHEET
IS FORMED TO
CONTOURS OF DIE

Fig. 22. The type C coil is used with a die to form flat stock.

Fig. 23. Electro-magnetic pulse forming can be used to make a variety of intricate shapes, such as medallions and coins. (General Dynamics Corp.)

the surface of the workpiece. The interaction of the magnetic field and the eddy currents will create a uniform central force.

Coils — the Forming Tools. There are three basic coils that must be employed with the machine. These coils transmit the energy pulse to form and shape the workpiece. The type *A coil* is designed for producing compressive forces. Fig. 20. The *B coil* is designed to produce expansive forces and the *C coil*, which is flat in design is used to form or pierce flat stock. Figs. 21 and 22.

Typical Applications. Many different applications are possible with the Magneform machine, Fig. 23.

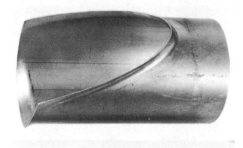

Fig. 24. Aluminum tubing can be flared *(top)*, or beaded, necked and formed *(bottom)* electro-magnetically. (General Dynamics Corp.)

The following are only a few that have proven themselves in practical production.

1. Shaping aluminum tubing into precise and difficult shapes which cannot be achieved by extrusion or other means. See Fig. 24.

2. Expanding tubing into bushings, hubs, and split dies.

3. Swaging inserts, fittings, terminals and collars onto many diverse parts, including rope, aircraft control rods, and other similar parts.

4. Easily and quickly assembling electrical components into a tightly bound package.

Electrolytic Grinding (EG)

Process Description. The electrolytic grinding process is a reverse or deplating method of grinding employing metal-bonded diamond wheels and electrolytic coolant supply. The action is approximately 90% deplating and 10% grinding, although the diamond particles actually act as scrubbers to remove the deplated carbide. Salts are added to water to make the solution conductive. The range of amperes available from the power pack is from 50-3000 amperes. The process is low voltage, 4-15 volts with variable power packs, depending on the job to be accomplished. Typical stock removal by

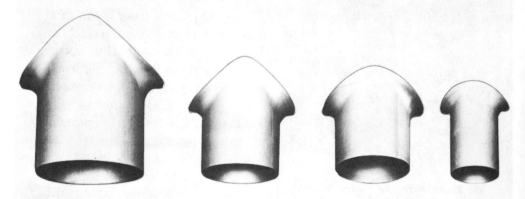

Fig. 25. These are some flared shapes possible with the Magneform. (General Dynamics Corp.)

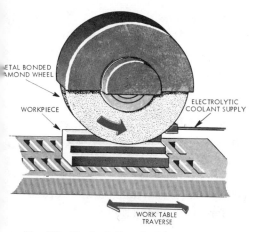

Fig. 26. Electrolytic grinding combines de-plating with grinding action.

this process is .031 cubic inches/min. which is considerably better than by any other method of carbide grinding (green grit wheel, diamond and electrical discharge machining process). See Fig. 26.

Electrolytic Grinders. The electrolytic grinding machine tool is basically a plain surface grinder, or cutter and tool grinder, equipped with a special wheel head (insulated from the wheel head housing and the rest of the machine) and an electrolyte power supply unit, Fig. 27.

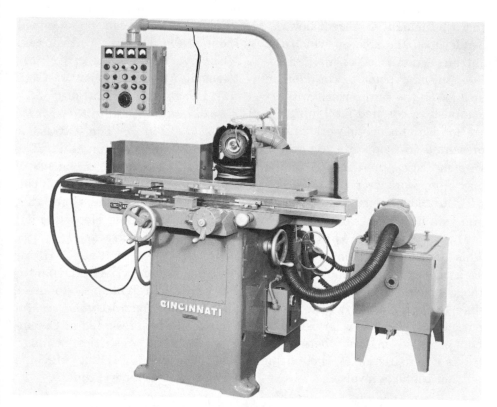

Fig. 27. An electrolytic grinding machine. (Cincinnati Milling Machine Co.)

Rate of Metal Removal. The presentation of the work to the wheel is the most important consideration when reviewing a job for electrolytic grinding. Conventional grinding practices must be ignored, especially cutter grinding practices.

The volumetric rate of electrolytic decomposition of metal occurs in proportion to the amount of current which flows between the work and the wheel. As a general "rule of thumb," stock can be removed at a rate of .010 cubic inches per minute for each 100 amperes of electrolytic grinding current used in a cut. Further, the amount of current flow depends upon the area on which the cutting action is occurring. Therefore, the first point to consider on any job is the setup which will accomplish the desired machining operation with maximum work surface presented just opposite the wheel. In some cases, presentation of the maximum work area must be sacrificed to reduce work handling time and gain higher rates of production.

Major Areas of Application. This process is used in tool rooms as a means of sharpening carbide tools faster, cheaper, and better than by previously available methods. It is also used for grinding honeycombed material because it leaves a burr-free edge—there being no tool contacts involved.

Significant Advantages. The removal of stock from high temperature materials and carbides can be accomplished with this process. Great savings in diamond wheels are possible. As a rule-of-thumb, the process will give 10 times greater wheel life and so reduces the consumption of diamond wheels. Better finish in regard to micro finish (RMS) is achievable, and there is no danger of heat cracks.

References

Additional information which may be of value regarding electrical machining methods can be obtained from the following references: *Magnetic Pulse Forming Notes* (General Atomic Division of General Dynamics Corporation); *Machining the Unmachinable with Sheffield-Cavitron*, The Ultrasonic Machine Tool (Dayton, Ohio: The Sheffield Corporation); *Elements of Electrolytic Grinding* (The Cincinnati Milling Machine Company, Electrical Machining Department); *Elements of Electrochemical Machining* (The Cincinnati Milling Machine Company, Meta-Dynamics Division); and Williams, Robin O. *Elements of Electro-Discharge Machining* (Paper presented at Cincinnati Technical Activities Seminar, The Cincinnati Milling Machine Company, Cincinnati, Ohio).

NOTE: Please see review questions at end of book.

Review Questions

Chapter 1—Machine Tools

1. Explain what is meant by the statement that tools are the basis of industry.

2. Explain what is meant by "mass production."

3. What is meant by the "new generation" of machine tools?

4. What are some of the causes of accidents and bodily injury that occur in machine shops?

Chapter 2—Measuring Tools

1. What are the fractional dimensions on a blueprint called?

2. Name the principal parts to the combination set.

3. Describe the two ways in which contact measurements can be made.

4. What is meant by the term "tolerance"?

5. What is meant by the term "decimal equivalent"?

6. What two ways can be used to check the accuracy of a micrometer caliper?

7. Show by a sketch or drawing some of the uses to be made of the dial indicator.

Chapter 3—Bench Tools

1. Name some of the uses to be made of a soft-headed hammer.

2. Draw the front and side views of the blade of a correctly ground screw driver.

3. Why is it sometimes necessary to place false jaws over the steel jaws of the machinist's vise when holding the workpiece?

4. Give several good reasons why files should never be used without a handle.

5. Why is it important that the wrench be of a size and type suited to the nut being tightened or loosened?

6. Distinguish between the three taps in the standard hand tap set.

7. Why is it important that the correct drill be used before tapping a hole?

8. What is the difference in appearance and use between a solid die and a split die?

9. What is the difference between a hand reamer and a machine reamer?

10. Why is it important that care be used in starting a new hack saw blade in a previously made saw cut?

11. Name some of the reasons why layout work is done on a workpiece to be machined.

Chapter 4—Power Saws

1. What are the principal elements of the vertical band machine?

2. What is the function of the butt welding system? What are its principal components?

3. Describe the steps involved in joining (welding) a saw band.

4. Describe the steps involved in installing the saw band on the machine.

5. What information is provided on the job selector dial?

6. In what respect does internal sawing differ from contour sawing? Describe the steps involved in an internal machining operation.

Chapter 5—Drill Press

1. Name the principal parts of the sensitive drill press and tell the function of each.

2. List all the cutting tools used in the drill press.

3. Make a sketch to show the difference between a countersink and a counterbore.

4. Distinguish between a cutter-holding device and a work-holding device.

5. List the operations that can be performed on the drill press.

6. Why is it sometimes necessary to "draw" a drill and how can this be done?

7. Distinguish between the counterboring and spot-facing operations.

Chapter 6—Engine Lathe

1. Identify the major types of engine lathes.

2. Name the principal parts of the standard engine lathe and briefly describe the function of each.

3. Identify and explain the function of the manual controls found on the lathe carriage.

4. Describe the steps involved in moving the tailstock into position when mounting the workpiece on the lathe.

5. Identify the four different types of chucks used on the engine lathe.

6. How is a lathe dog used to drive a workpiece on the lathe?

7. What is the function of the cross slide and compound rest? Name three types of tool posts used in lathe operations.

Chapter 7—Engine Lathe

1. Make a list of the various kinds of lathe cutting tools.

2. List some of the newer lathe cutting tool materials.

3. Before making a setup on the engine lathe, what are some of the things which must be taken into consideration?

4. List the operations that can be performed in the lathe on the external surface of the workpiece.

5. What is the difference between a roughing cut and a finish cut in turning work in the lathe?

6. List several reasons for knurling a piece of work after it has been turned in the lathe.

7. Describe two methods used in measuring external V-shaped threads.

8. List the three ways used to turn tapers in the engine lathe.

9. List the operations that can be performed in the lathe on the internal surface of the workpiece.

10. What are the three means by which the size of a bored hole can be measured?

11. How is a threaded hole meas-

ured after the thread has been cut in the lathe?

Chapter 8—Shapers

1. In making various setups on the horizontal shaper, what are some of the adjustments that usually have to be made?

2. List the operations that can be performed on the horizontal shaper.

3. List some of the precautions that must be taken to avoid injury when operating the shaper. Classify

4. Make a list of the tools, work-holding devices, and other things usually needed in making the setup on a shaper.

Chapter 9—Milling Machines

1. Show, by sketches carefully labeled, the difference between conventional and climb milling.

2. Identify the main parts of a horizontal knee-and-column type milling machine.

3. How does the horizontal knee-and-column milling machine differ from the vertical type?

4. How can one tell a plain knee-and-column type milling machine from a universal knee-and-column type milling machine?

5. Explain the difference between the style A, style B, and style C arbors. What are spacing collars? Which two arbor types use spacing collars?

6. Are arbor supports used in face milling operations? Explain.

7. Identify the milling machine manual table controls.

8. What is the purpose of a dividing head?

9. Identify the power controls found on the knee of the knee-and-column milling machine.

Chapter 10—Milling Machines

1. List by correct name the various cutters used in milling operations.

2. What is the major difference between a plain or slab mill and a face mill? A face mill and an end mill?

3. Make a list of the operations performed on the milling machine, using the correct title of the operation. Opposite each operation give the name of the cutter or cutters used.

4. Why do the manual controls of the milling machine have micrometer collars?

5. Why must the horsepower of the machine and the rigidity of the setup be carefully considered for each milling operation?

6. Using a string, pencil, and circular object, draw an involute curve.

7. Describe the steps necessary to position the milling machine table (workpiece) for the desired cut in a slab milling setup.

8. Are there any precautions to take into consideration when mounting the workpiece on the table? Explain.

Chapter 11—Production Turning

1. What are the main differences between the Ram-type and Saddle-type turret lathes?

2. What are the major tool-holding elements of the hand-operated turret lathe?

3. List five basic internal and external turret lathe machining operations.

4. Must internal machining operations be performed in a particular sequence? Why or why not?

5. What are the three general classes of automatic screw machines?

6. What are the major differences between the hand-operated turret lathe and the single-spindle automatic?

7. What are the chief operating features of the multiple-spindle automatic?

8. What are the chief operating features of the Swiss automatic?

Chapter 12—Grinding Machines

1. Describe just what happens when surface grinding is done.

2. List the major parts on the horizontal reciprocating-table surface grinder and tell the function of each.

3. Show by sketch the several operations that can be performed on the horizontal surface grinder.

4. Distinguish between a center-type cylindrical grinder and a centerless-type cylindrical grinder.

5. What is the purpose of dressing the grinding wheel and how is this done?

6. Write a paragraph to tell about "plunge-cut" grinding.

Chapter 13—Steels

1. Name the chief elements of cast iron.

2. What type of steel has gradually replaced wrought iron in commercial applications?

3. What elements occur in greater amounts in cast steel than in cast iron to insure thorough deoxidation and prevent gassing?

4. Name at least five reasons why alloying elements are added to ordinary steel.

5. List five kinds of tool steel used in industry today.

6. What does the prefix *B* denote in the AISI method of classification?

7. Explain what is meant by a spark test.

Chapter 14—Heat Treating

1. Name some of the advantages of controlling the grain size.

2. Describe the pack hardening process used in hardening carbon and low alloy steels.

3. What is surface hardening commonly called? Name two methods of surface hardening.

4. In tempering steel according to color, list the progression of colors that occur after a piece of steel is hardened and polished with emery cloth. What color is used for temper-

ing knives and chisels? For springs?

5. What are the principal differences between annealing and normalizing? What is the purpose of each operation?

6. Name three steels classified according to the medium in which they are quenched. Why is quenching necessary?

7. What dangers arise when metals are subjected to grinding? How are these dangers avoided?

8. Why is proper control of furnace atmosphere important? Name the different kinds of furnace atmospheres.

9. Name three instruments commonly used in industry to measure the hardness of metals. Explain how each is different from the others.

Chapter 15—Machinability

1. Name factors which commonly affect machinability.

2. In chip formation, what chip type travels smoothly up the tool face? What type is forced upward over the tool face and breaks into short segments? What type adheres to the face of the cutting tool?

3. Explain why cutting fluids may improve the accuracy of cutting operations. What other advantages are gained by using cutting fluids?

4. Why is it difficult to specify the machinability of metals?

5. Why are tool steels usually annealed before they reach the user?

6. What two types of steel are commonly used as a basis for rating the machinability of steels?

Chapter 16—Numerical Control

1. What measuring system is used to locate the center of a hole to be drilled through a workpiece mounted on a numerically controlled drill press? What is the name given to the center from which all distances are measured in N/C processes?

2. In a N/C machine, which axis is generally used to designate the direction(s) of the longest possible motion?

3. What are the two basic types of N/C control?

4. Describe the preparations necessary for machining a production part—such as the handle of a pencil sharpener, for example—by N/C.

5. Describe the duties of the parts programmer in a typical N/C machining operation.

Chapter 17—Electrical Energy Processes

1. What are the principal (and principle) differences between the EDM process and the ECM process?

2. What are the major parts of an ultrasonic machine? What actually removes the material during the ultrasonic machining process?

3. In general, what shapes can be produced with an electro-magnetic metal forming machine?

4. Describe an electrolytic grinding machine. What use is made of the process?

Appendix

TABLE I. UNIFIED AND AMERICAN STANDARD COARSE THREAD SERIES—BASIC DIMENSIONS*

Size	Basic Major Diam.	Thds. per Inch	Basic Pitch Diam.	Minor Diam. Ext. Thds.	Minor Diam. Int. Thds.	Lead Angle, Basic Pitch Diam.		Area, Minor Diam.	Stress Area,
	Inches		Inches	Inches	Inches	Deg.	Min.	Sq. In.	Sq. In.
1 (.073)	.0730	64	.0629	.0538	.0561	4°	31′	.0022	.0026
2 (.086)	.0860	56	.0744	.0641	.0667	4°	22′	.0031	.0036
3 (.099)	.0990	48	.0855	.0734	.0764	4°	26′	.0041	.0048
†4 (.112)	.1120	40	.0958	.0813	.0849	4°	45′	.0050	.0060
5 (.125)	.1250	40	.1088	.0943	.0979	4°	11′	.0067	.0079
†6 (.138)	.1380	32	.1177	.0997	.1042	4°	50′	.0075	.0090
†8 (.164)	.1640	32	.1437	.1257	.1302	3°	58′	.0120	.0139
†10 (.190)	.1900	24	.1629	.1389	.1449	4°	39′	.0145	.0174
12 (.216)	.2160	24	.1889	.1649	.1709	4°	1′	.0206	.0240
¼	.2500	20	.2175	.1887	.1959	4°	11′	.0269	.0317
⁵⁄₁₆	.3125	18	.2764	.2443	.2524	3°	40′	.0454	.0522
⅜	.3750	16	.3344	.2983	.3073	3°	24′	.0678	.0773
⁷⁄₁₆	.4375	14	.3911	.3499	.3602	3°	20′	.0933	.1060
½	.5000	13	.4500	.4056	.4167	3°	7′	.1257	.1416
⁹⁄₁₆	.5625	12	.5084	.4603	.4723	2°	59′	.1620	.1816
⅝	.6250	11	.5660	.5135	.5266	2°	56′	.2018	.2256
¾	.7500	10	.6850	.6273	.6417	2°	40′	.3020	.3340
⅞	.8750	9	.8028	.7387	.7547	2°	31′	.4193	.4612
1	1.0000	8	.9188	.8466	.8647	2°	29′	.5510	.6051
1⅛	1.1250	7	1.0322	.9497	.9704	2°	31′	.6931	.7627
1¼	1.2500	7	1.1572	1.0747	1.0954	2°	15′	.8898	.9684
1⅜	1.3750	6	1.2667	1.1705	1.1946	2°	24′	1.0541	1.1538
1½	1.5000	6	1.3917	1.2955	1.3196	2°	11′	1.2938	1.4041
1¾	1.7500	5	1.6201	1.5046	1.5335	2°	15′	1.7441	1.8983
2	2.0000	4½	1.8557	1.7274	1.7594	2°	11′	2.3001	2.4971
2¼	2.2500	4½	2.1057	1.9774	2.0094	1°	55′	3.0212	3.2464
2½	2.5000	4	2.3376	2.1933	2.2294	1°	57′	3.7161	3.9976
2¾	2.7500	4	2.5876	2.4433	2.4794	1°	46′	4.6194	4.9326
3	3.0000	4	2.8376	2.6933	2.7294	1°	36′	5.6209	5.9659
3¼	3.2500	4	3.0876	2.9433	2.9794	1°	29′	6.7205	7.0992
3½	3.5000	4	3.3376	3.1933	3.2294	1°	22′	7.9183	8.3268
3¾	3.7500	4	3.5876	3.4433	3.4794	1°	16′	9.2143	9.6546
4	4.0000	4	3.8376	3.6933	3.7294	1°	11′	0.6084	11.0805

Figures in bold type indicate Unified threads.

* Source: *Machinery's Handbook*

† These sizes were adopted in 1951 as Unified threads or certain limited applications, such as attachment of instruments to panels to facilitate interchangeability of American, British, and Canadian equipment.

TABLE II. UNIFIED AND AMERICAN STANDARD FINE AND EXTRA-FINE THREAD SERIES—BASIC DIMENSIONS*

Size	Basic Major Diam. Inches	Thds. per Inch	Basic Pitch Diam. Inches	Minor Diam. Ext. Thds. Inches	Minor Diam. Int. Thds. Inches	Lead Angle, Pitch Diam. Deg.	Min.	Area, Minor Diam. Sq. In.	Stress Area Sq. In.
				FINE THREAD SERIES					
0 (.060)	.0600	80	.0519	.0447	.0465	4°	23'	.0015	.0018
1 (.073)	.0730	72	.0640	.0560	.0580	3°	57'	.0024	.0027
2 (.086)	.0860	64	.0759	.0668	.0691	3°	45'	.0034	.0039
3 (.099)	.0990	56	.0874	.0771	.0797	3°	43'	.0045	.0052
4 (.112)	.1120	48	.0985	.0864	.0894	3°	51'	.0057	.0065
5 (.125)	.1250	44	.1102	.0971	.1004	3°	45'	.0072	.0082
6 (.138)	.1380	40	.1218	.1073	.1109	3°	44'	.0087	.0101
8 (.164)	.1640	36	.1460	.1299	.1339	3°	28'	.0128	.0146
†10 (.190)	.1900	32	.1697	.1517	.1562	3°	21'	.0175	.0199
12 (.216)	.2160	28	.1928	.1722	.1773	3°	22'	.0226	.0257
1/4	.2500	28	.2268	.2062	.2113	2°	52'	.0326	.0362
5/16	.3125	24	.2854	.2614	.2674	2°	40'	.0524	.0579
3/8	.3750	24	.3479	.3239	.3299	2°	11'	.0809	.0876
7/16	.4375	20	.4050	.3762	.3834	2°	15'	.1090	.1185
1/2	.5000	20	.4675	.4387	.4459	1°	57'	.1486	.1597
9/16	.5625	18	.5264	.4943	.5024	1°	55'	.1888	.2026
5/8	.6250	18	.5889	.5568	.5649	1°	43'	.2400	.2555
3/4	.7500	16	.7094	.6733	.6823	1°	36'	.3513	.3724
7/8	.8750	14	.8286	.7874	.7977	1°	34'	.4805	.5088
1	1.0000	14	.9536	.9124	.9227	1°	22'	.6464	.6791
1	1.0000	12	.9459	.8978	.9098	1°	36'	.6245	.6624
1 1/8	1.1250	12	1.0709	1.0228	1.0348	1°	25'	.8118	.8549
1 1/4	1.2500	12	1.1959	1.1478	1.1598	1°	16'	1.0237	1.0721
1 3/8	1.3750	12	1.3209	1.2728	1.2848	1°	9'	1.2602	1.3137
1 1/2	1.5000	12	1.4459	1.3978	1.4098	1°	3'	1.5212	1.5799
				EXTRA-FINE THREAD SERIES					
12 (.216)	.2160	32	.1957	.1777	.1822	2°	55'	.0242	.0269
1/4	.2500	32	.2297	.2117	.2162	2°	29'	.0344	.0377
5/16	.3125	32	.2922	.2742	.2787	1°	57'	.0581	.0622
3/8	.3750	32	.3547	.3367	.3412	1°	36'	.0878	.0929
7/16	.4375	28	.4143	.3937	.3988	1°	34'	.1201	.1270
1/2	.5000	28	.4768	.4562	.4613	1°	22'	.1616	.1695
9/16	.5625	24	.5354	.5114	.5174	1°	25'	.2030	.2134
5/8	.6250	24	.5979	.5739	.5799	1°	16'	.2560	.2676
1 1/16	.6875	24	.6604	.6364	.6424	1°	9'	.3151	.3280
3/4	.7500	20	.7175	.6887	.6959	1°	16'	.3685	.3855
1 3/16	.8125	20	.7800	.7512	.7584	1°	10'	.4388	.4573
7/8	.8750	20	.8425	.8137	.8209	1°	5'	.5153	.5352
15/16	.9375	20	.9050	.8762	.8834	1°	0'	.5979	.6194
1	1.0000	20	.9675	.9387	.9459	0	57'	.6866	.7095
1 1/16	1.0625	18	1.0264	.9943	1.0024	0	59'	.7702	.7973
1 1/8	1.1250	18	1.0889	1.0568	1.0649	0	56'	.8705	.8993
1 3/16	1.1875	18	1.1514	1.1193	1.1274	0	53'	.9770	1.0074
1 1/4	1.2500	18	1.2139	1.1818	1.1899	0	50'	1.0895	1.1216
1 5/16	1.3125	18	1.2764	1.2443	1.2524	0	48'	1.2082	1.2420
1 3/8	1.3750	18	1.3389	1.3068	1.3149	0	45'	1.3330	1.3684
1 7/16	1.4375	18	1.4014	1.3693	1.3774	0	43'	1.4640	1.5010
1 1/2	1.5000	18	1.4639	1.4318	1.4399	0	42'	1.6011	1.6397
1 9/16	1.5625	18	1.5264	1.4943	1.5024	0	40'	1.7444	1.7846
1 5/8	1.6250	18	1.5889	1.5568	1.5649	0	38'	1.8937	1.9357
1 11/16	1.6875	18	1.6514	1.6193	1.6274	0	37'	2.0493	2.0929
1 3/4	1.7500	16	1.7094	1.6733	1.6823	0	40'	2.1873	2.2382
2	2.0000	16	1.9594	1.9233	1.9323	0	35'	2.8917	2.9501

Figures in bold type indicate Unified threads.

* Source: *Machinery's Handbook.*

† This size was adopted in 1951 as a Unified thread for certain limited applications, such as attachment of instruments to panels to facilitate interchangeability of American, British, and Canadian equipment.

TABLE III. UNIFIED AND AMERICAN STANDARD SCREW THREAD PITCHES AND RECOMMENDED TAP DRILL SIZES[1]

	AMERICAN NATIONAL COARSE STANDARD THREAD (N.C.) Formerly U.S. Standard				AMERICAN NATIONAL FINE STANDARD THREAD (N.F.) Formerly S.A.E. Thread				
Sizes	Threads per Inch	Outside Diameter of Screw	Tap Drill Sizes	Decimal Equivalent of Drill	Sizes	Threads per Inch	Outside Diameter of Screw	Tap Drill Sizes	Decimal Equivalent of Drill
1......	64	.073	53	.0595	0.....	80	.060	$\frac{3}{64}$	.0469
2......	56	.086	50	.0700	1.....	72	.073	53	.0595
3......	48	.099	47	.0785	2.....	64	.086	50	.0700
4......	40	.112	43	.0890	3.....	56	.099	45	.0820
5......	40	.125	38	.1015	4.....	48	.112	42	.0935
6......	32	.138	36	.1065	5.....	44	.125	37	.1040
8......	32	.164	29	.1360	6.....	40	.138	33	.1130
10......	24	.190	25	.1495	8.....	36	.164	29	.1360
12......	24	.216	16	.1770	10.....	32	.190	21	.1590
$\frac{1}{4}$.....	20	.250	7	.2010	12.....	28	.216	14	.1820
$\frac{5}{16}$....	18	.3125	F	.2570	$\frac{1}{4}$....	28	.250	3	.2130
$\frac{3}{8}$....	16	.375	$\frac{5}{16}$	.3125	$\frac{5}{16}$....	24	.3125	I	.2720
$\frac{7}{16}$....	14	.4375	U	.3680	$\frac{3}{8}$....	24	.375	Q	.3320
$\frac{1}{2}$....	13	.500	$\frac{27}{64}$	.4219	$\frac{7}{16}$...	20	.4375	$\frac{25}{64}$	.3906
$\frac{9}{16}$....	12	.5625	$\frac{31}{64}$	.4843	$\frac{1}{2}$....	20	.500	$\frac{29}{64}$	.4531
$\frac{5}{8}$....	11	.625	$\frac{17}{32}$	.5312	$\frac{9}{16}$...	18	.5625	.5062	.5062
$\frac{3}{4}$....	10	.750	$\frac{21}{32}$	.6562	$\frac{5}{8}$...	18	.625	.5687	.5687
$\frac{7}{8}$....	9	.875	$\frac{49}{64}$	.7656	$\frac{3}{4}$...	16	.750	$\frac{11}{16}$	.6875
1......	8	1.000	$\frac{7}{8}$	.875	$\frac{7}{8}$...	14	.875	.8020	.8020
$1\frac{1}{8}$....	7	1.125	$\frac{63}{64}$	.9843	1....	14	1.000	.9274	.9274
$1\frac{1}{4}$....	7	1.250	$1\frac{7}{64}$	1.1093	$1\frac{1}{8}$....	12	1.125	$1\frac{3}{64}$	1.0468
					$1\frac{1}{4}$....	12	1.250	$1\frac{11}{64}$	1.1718

[1]Courtesy of South Bend Lathe Works. Table is based on 75% thread depth.

TABLE IV. RECOMMENDED DRILL GEOMETRY FOR VARIOUS MATERIALS[1]

Cast Iron

The machinability of alloyed gray iron castings is not readily measurable. While Brinell hardness is an indication of the structure, within the same hardness range the machinability varies inversely proportional to the tensile strength and changes caused by alloying.

Cast Iron (Soft) under 150 Brinell
Less than 30,000 T.S.

WEB THINNING

Recommendations:

Cutting Speed	100-150 feet per min.
Feed Rate	5-7 inches per min.
Coolant	Dry or air jet
Drill	Standard twist drill modified as shown
Point Angle	100°
Lip Relief	12-15°
Web	Thinned at chisel edge only

TABLE IV. RECOMMENDED DRILL GEOMETRY FOR VARIOUS MATERIALS (*Cont.*)

Cast Iron (Medium) up to 175 Brinell
30,000-35,000 T.S.

Recommendations:

Cutting Speed	90-120 feet per min.
Feed Rate	3-4 inches per min.
Coolant	Dry or air jet
Drill	Standard twist drill modified as shown
Point Angle	100°
Lip Relief	12-15°
Web	Thinned at chisel edge only

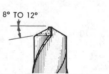

WEB THINNING

Cast Iron (Hard) 175-250 Brinell
30,000-40,000 T.S.

Recommendations:

Cutting Speed	70-100 feet per min.
Feed Rate	2½-3½ inches per min.
Coolant	Dry or air jet
Drill	Standard twist drill as shown
Point Angle	118°
Lip Relief	8-12°

Steel (Mild)—Hot Rolled, Cold Drawn, Normalized or Annealed—Not Heat Treated for Hardness

SAE #1010-1030, 1112-1115, 2015-2115, 2315-2340, 2512, 2515, 3115-3130, 4615-4620, 5120-5140, 5210, 6115-6130.

Recommendations:

Cutting Speed	70-90 feet per min.
Feed Rate	3-5 inches per min.
Coolant	Soluble or sulphurized oil
Drill	Standard twist drill as shown
Point Angle	118°
Lip Relief	7-9°

Steel (Medium Soft)—Hot Rolled, Cold Drawn, Annealed or Normalized—Not Heat Treated for Hardness

SAE #1035-1060, 2345, 2350, 3140-3150, 3240-3250, 3435, 4135-4150

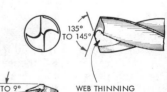

Recommendations:

Cutting Speed	50-70 feet per min.
Feed Rate	2-3 inches per min.
Coolant	Sulphurized oil
Drill	Standard twist drill modified as shown
Point Angle	135-145°
Lip Relief	7-9°
Web	Thinned at chisel edge only

WEB THINNING

TABLE IV. RECOMMENDED DRILL GEOMETRY FOR VARIOUS MATERIALS (*Cont.*)

Steel (Medium Hard)—Hot Rolled, Cold Drawn, Annealed or Normalized—Not Heat Treated for Hardness

SAE #1060-1092, 3250, 3312-3340, 3430-3450, 6150

Recommendations:

Cutting Speed	35-50 feet per min.
Feed Rate	1½-2 inches per min.
Coolant	Sulphurized oil
Drill	Standard twist drill with heavy web and modified as shown
Point Angle	145°
Lip Relief	7°
Web	Thinned at chisel edge only

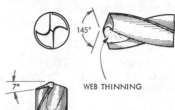

WEB THINNING

**Stainless Steel
(For Tough Machining Grades)**

Recommendations:

Drill	Standard twist drill modified as shown
Point Angle	135°
Lip Relief	6-8°
Web	Thinned at chisel edge only

WEB THINNING

**Stainless Steel
(For Free Machining and Chrome Nickel Grades)**

Recommendations:

Drill	Standard twist drill modified as shown
Point Angle	118°
Lip Relief	8-12°
Web	Thinned at chisel edge only

WEB THINNING

**Stainless Steel
(For Free Machining and Chrome Nickel Grades)**

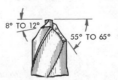

Recommendations:

Drill	Standard twist drill modified as shown
Point Angle	118°
Lip Angle	8-12°
Point	Crankshaft point for deep holes

TABLE IV. RECOMMENDED DRILL GEOMETRY FOR VARIOUS MATERIALS (*Cont.*)

Stainless Steel
(For Tough Machining Grades)

Recommendations:

Drill	Standard twist drill modified as shown
Point Angle	135°
Lip Angle	6-8°
Point	Crankshaft point for deep holes

Aluminum

No general recommendation to cover all grades of aluminum and aluminum alloys can be given. Aluminum not alloyed is usually free cutting and easily drilled.

Recommendations for aluminum not alloyed:

Cutting Speed	200-300 feet per min.
Feed Rate	4-6 inches per min.
Coolant	Soluble oil, kerosene and lard oil compounds or neutral oils
Drill	Standard twist drill for shallow holes (3 times diam. deep) with wide polished flutes modified as shown
Point Angle	118°
Lip Relief	15-18°
Cutting Edge Rake Angle	10°

Drill	Standard twist drill for deep holes (over 3 times diam. deep) with wide polished flutes modified as shown
Point Angle	118°
Lip Relief	15°
Cutting Edge Rake Angle	10°
Helix Angle	45° wide polished flutes

TABLE IV. RECOMMENDED DRILL GEOMETRY FOR VARIOUS MATERIALS (*Cont.*)

Bronze
SAE #40, 41, 58, 62-67, 620-622

Recommendations:

Cutting Speed	150-300 feet per min.
Feed Rate	2-6 inches per min.
Coolant	Mineral oil with 5-15% lard oil
Drill	Standard twist drill modified as shown
Point Angle	118°
Lip Relief	12-15°
Cutting Edge	
Rake Angle	0°
Helix Angle	10-15°
	Polished Flutes

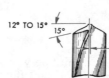

Bronze
SAE #68A, 68B

Recommendations:

Cutting Speed	75-150 feet per min.
Feed Rate	1½-4 inches per min.
Drill	Standard twist drill modified as shown
Point Angle	100-110°
Lip Relief	12-15°
Rake and Helix Angles	28°
	Polished Flutes

Copper and Brass
(Soft and Medium Hard)

Recommendations:

Cutting Speed	150-300 feet per min.
Feed Rate	2-6 inches per min.
Coolant	Dry or mineral oil
Drill	Standard twist drill modified as shown
Point Angle	118°
Lip Relief	15°
Rake Angle	0°
Helix Angle	18°
	Wide Polished Flutes

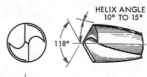

TABLE IV. RECOMMENDED DRILL GEOMETRY FOR VARIOUS MATERIALS (*Cont.*)

Plastics
(Hot Set Plastics, Hard Rubber and Fiber)

Recommendations:

Cutting Speed	100-300 feet per min.
Feed Rate	1-6 inches per min.
Coolant	Dry with air jet wherever possible
Drill	Standard twist drill modified as shown
Point Angle	60°
Lip Relief	10-12°
Rake and Helix Angles	10°
Web	Thinned at point only

Extra wide polished flutes and double standard margin relief.

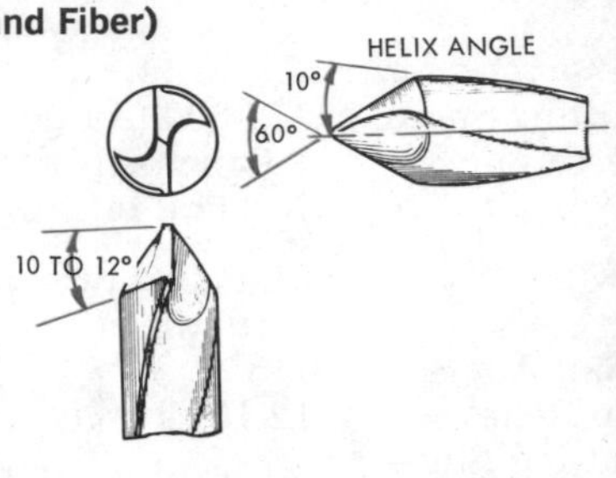

Plastics
(Cold Set Plastics such as Plexiglas, Lucite, etc.)

Recommendations:

Cutting Speed	100-300 feet per min.
Feed Rate	1-6 inches per min.
Coolant	Soapy water or a mixture of 3 parts kerosene to 1 part carbon tetrachloride
Drill	Standard twist drill modified as shown
Point Angle	118-135°
Lip Relief	20°
Rake Angle	10°
Helix Angle	28°

Wide polished flutes with double standard margin clearance.

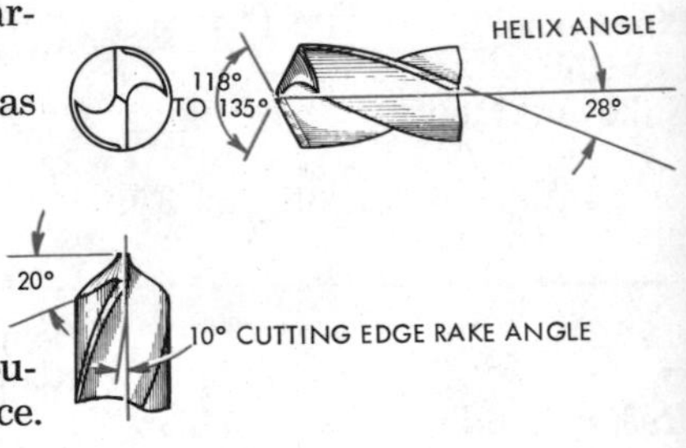

1. Courtesy of National Automatic Tool Company, Inc.

TABLE V. SUGGESTED DRILL SPEEDS FOR VARIOUS MATERIALS[1]

Material to be Drilled	Cutting Speed[2] (Surface Feet per Minute)
Aluminum and its alloys	200–300
Bakelite	100–150
Brass and bronze, soft	200–300
Bronze, high tensile	70–150
Carbon, pure (carbide drills)	100
Cast iron, soft	100–150
Cast iron, hard	70–100
Cast iron, chilled	30–40
Copper graphite alloy (carbide drills)	60–70
Glass (carbide drills)	20–30
Magnesium and its alloys	250–400
Malleable iron	80–90
Marble	15–25
Marble (carbide drills)	60–80
Nickel and monel	40–60
Slate	15–25
Slate (carbide drills)	40
Steel, machinery (0.2–0.3 c)	80–110
Steel, annealed (0.4–0.5 c)	70–80
Steel, tool (1.2 c)	50–60
Steel, forged	50–60
Steel, alloy (300 to 400 Brinnel)	20–30
Steel, stainless, free machining	30–40
Steel, stainless, hard	30–40
Steel, manganese	15
Stone	15–25
Stone (carbide drills)	30
Wood	300–400

[1] Source: *American Machinist.*
[2] Except as noted, speeds given above are for high-speed steel drills. Carbon steel drills should be run at from 40 to 50 per cent of those for high-speed steel drills. These speeds are given as starting points; the best speed in each case must be based on the prevailing conditions, material, set-up, etc.

TABLE VI. R.P.M. SPEEDS FOR FRACTION SIZE DRILLS[1]

Feet per Min.	30	40	50	60	70	80	90	100	110	120	130	140	150
Diam. (Inches)	Revolutions per Minute												
$\frac{1}{16}$	1833	2445	3056	3667	4278	4889	5500	6111	6722	7334	7945	8556	9167
$\frac{1}{8}$	917	1222	1528	1833	2139	2445	2750	3056	3361	3667	3973	4278	4584
$\frac{3}{16}$	611	815	1019	1222	1426	1630	1833	2037	2241	2445	2648	2852	3056
$\frac{1}{4}$	458	611	764	917	1070	1222	1375	1528	1681	1833	1986	2139	2292
$\frac{5}{16}$	367	489	611	733	856	978	1100	1222	1345	1467	1589	1711	1833
$\frac{3}{8}$	306	407	509	611	713	815	917	1019	1120	1222	1324	1426	1528
$\frac{7}{16}$	262	349	437	524	611	698	786	873	960	1048	1135	1222	1310
$\frac{1}{2}$	229	306	382	458	535	611	688	764	840	917	993	1070	1146
$\frac{5}{8}$	183	244	306	367	428	489	550	611	672	733	794	856	917
$\frac{3}{4}$	153	203	255	306	357	407	458	509	560	611	662	713	764
$\frac{7}{8}$	131	175	218	262	306	349	393	436	480	524	568	611	655
1	115	153	191	229	267	306	344	382	420	458	497	535	573
$1\frac{1}{8}$	102	136	170	204	238	272	306	340	373	407	441	475	509
$1\frac{1}{4}$	92	122	153	183	214	244	275	306	336	367	397	428	458
$1\frac{3}{8}$	83	111	139	167	194	222	250	278	306	333	361	389	417
$1\frac{1}{2}$	76	102	127	153	178	204	229	255	280	306	331	357	382
$1\frac{5}{8}$	70	94	117	141	165	188	212	235	259	282	306	329	353
$1\frac{3}{4}$	65	87	109	131	153	175	196	218	240	262	284	306	327
$1\frac{7}{8}$	61	81	102	122	143	163	183	204	224	244	265	285	306
2	57	76	95	115	134	153	172	191	210	229	248	267	287
$2\frac{1}{4}$	51	68	85	102	119	136	153	170	187	204	221	238	255
$2\frac{1}{2}$	46	61	76	92	107	122	137	153	168	183	199	214	229
$2\frac{3}{4}$	42	56	69	83	97	111	125	139	153	167	181	194	208
3	38	51	64	76	89	102	115	127	140	153	166	178	191

[1] Courtesy of Cleveland Twist Drill Co.

TABLE VII. R.P.M. SPEEDS FOR NUMBER SIZE DRILLS[1]

Feet per Min.	30	40	50	60	70	80	90	100	110	120	130	Decimal Equivalents
No. Size					Revolutions per Minute							
1	503	670	838	1005	1173	1340	1508	1675	1843	2010	2179	.2280
2	518	691	864	1037	1210	1382	1555	1728	1901	2074	2247	.2210
3	538	717	897	1076	1255	1434	1614	1793	1974	2152	2331	.2130
4	548	731	914	1097	1280	1462	1645	1828	2010	2193	2376	.2090
5	558	744	930	1115	1301	1487	1673	1859	2045	2230	2416	.2055
6	562	749	936	1123	1310	1498	1685	1872	2060	2247	2434	.2040
7	570	760	950	1140	1330	1520	1710	1900	2090	2281	2470	.2010
8	576	768	960	1151	1343	1535	1727	1919	2111	2303	2495	.1990
9	585	780	975	1169	1364	1559	1754	1949	2144	2339	2534	.1960
10	592	790	987	1184	1382	1579	1777	1974	2171	2369	2566	.1935
11	600	800	1000	1200	1400	1600	1800	2000	2200	2400	2600	.1910
12	606	808	1010	1213	1415	1617	1819	2021	2223	2425	2627	.1890
13	620	826	1032	1239	1450	1652	1859	2065	2271	2479	2684	.1850
14	630	840	1050	1259	1469	1679	1889	2099	2309	2518	2728	.1820
15	638	851	1064	1276	1489	1702	1914	2127	2334	2546	2759	.1800
16	647	863	1079	1295	1511	1726	1942	2158	2374	2590	2806	.1770
17	662	883	1104	1325	1546	1766	1987	2208	2429	2650	2870	.1730
18	678	904	1130	1356	1582	1808	2034	2260	2479	2704	2930	.1695
19	690	920	1151	1381	1611	1841	2071	2301	2531	2761	2991	.1660
20	712	949	1186	1423	1660	1898	2135	2372	2610	2847	3084	.1610
21	721	961	1201	1441	1681	1922	2162	2402	2644	2883	3123	.1590
22	730	973	1217	1460	1703	1946	2190	2433	2676	2920	3164	.1570
23	744	992	1240	1488	1736	1984	2232	2480	2728	2976	3224	.1540
24	754	1005	1257	1508	1759	2010	2262	2513	2764	3016	3267	.1520
25	767	1022	1276	1533	1789	2044	2300	2555	2810	3066	3322	.1495
26	779	1039	1299	1559	1819	2078	2338	2598	2858	3118	3378	.1470
27	796	1061	1327	1592	1857	2122	2388	2653	2919	3183	3448	.1440
28	816	1088	1360	1631	1903	2175	2447	2719	2990	3262	3534	.1405
29	843	1124	1405	1685	1966	2247	2528	2809	3090	3370	3651	.1360
30	892	1189	1487	1784	2081	2378	2676	2973	3270	3567	3864	.1285
31	955	1273	1592	1910	2228	2546	2865	3183	3501	3821	4138	.1200
32	988	1317	1647	1976	2305	2634	2964	3293	3622	3951	4281	.1160
33	1014	1352	1690	2028	2366	2704	3042	3380	3718	4056	4394	.1130
34	1032	1376	1721	2065	2409	2753	3097	3442	3785	4129	4474	.1110
35	1042	1389	1736	2083	2430	2778	3125	3472	3821	4167	4514	.1100
36	1076	1435	1794	2152	2511	2870	3228	3587	3945	4304	4663	.1065
37	1102	1469	1837	2204	2571	2938	3306	3673	4040	4407	4775	.1040
38	1129	1505	1882	2258	2634	3010	3387	3763	4140	4516	4892	.1015
39	1152	1536	1920	2303	2687	3071	3455	3839	4222	4007	4991	.0005
40	1169	1559	1949	2339	2729	3118	3508	3898	4287	4677	5067	.0980
41	1194	1592	1990	2387	2785	3183	3581	3979	4377	4775	5172	.0960
42	1226	1634	2043	2451	2860	3268	3677	4085	4494	4902	5311	.0935
43	1288	1717	2146	2575	3004	3434	3863	4292	4721	5150	5579	.0890
44	1333	1777	2221	2665	3109	3554	3999	4442	4886	5330	5774	.0860
45	1397	1863	2329	2795	3261	3726	4192	4658	5124	5590	6056	.0820

[1] Courtesy of Cleveland Twist Drill Co.

Table VII. R.P.M. Speeds for Number Size Drills (Continued)

Feet per Min.	30	40	50	60	70	80	90	100	110	120	130	Decimal Equivalents
No. Size						Revolutions per Minute						
46	1415	1886	2358	2830	3301	3773	4244	4716	5187	5659	6130	.0810
47	1460	1946	2433	2920	3406	3893	4379	4866	5352	5839	6326	.0785
48	1508	2010	2513	3016	3518	4021	4523	5026	5528	6031	6534	.0760
49	1570	2093	2617	3140	3663	4186	4710	5233	5756	6279	6808	.0730
50	1637	2183	2729	3274	3820	4366	4911	5457	6002	6548	7094	.0700
51	1710	2280	2851	3421	3991	4561	5131	5701	6271	6841	7413	.0670
52	1805	2406	3008	3609	4211	4812	5414	6015	6619	7218	7820	.0635
53	1924	2566	3207	3848	4490	5131	5773	6414	7062	7704	8346	.0595
54	2084	2778	3473	4167	4862	5556	6251	6945	7639	8334	9028	.0550
55	2204	2938	3673	4408	5142	5877	6611	7346	8080	8815	9549	.0520
56	2465	3286	4108	4929	5751	6572	7394	8215	9036	9857	10678	.0465
57	2671	3561	4452	5342	6232	7122	8013	8903	9771	10660	11548	.0430
58	2729	3637	4547	5456	6367	7275	8186	9095	10004	10913	11823	.0420
59	2795	3726	4658	5590	6521	7453	8388	9316	10248	11180	12111	.0410
60	2865	3820	4775	5729	6684	7639	8594	9549	10504	11459	12414	.0400
61	2938	3918	4897	5876	6856	7835	8815	9794	10774	11753	12732	.0390
62	3015	4020	5025	6030	7035	8040	9045	10050	11057	12060	13068	.0380
63	3096	4128	5160	6192	7224	8256	9288	10320	11366	12398	13421	.0370
64	3183	4244	5305	6366	7427	8488	9549	10610	11671	12732	13793	.0360
65	3273	4364	5455	6546	7637	8728	9819	10910	12005	13096	14187	.0350
66	3474	4632	5790	6948	8106	9264	10422	11580	12732	13890	15047	.0330
67	3582	4776	5970	7164	8358	9552	10746	11940	13130	14324	15517	.0320
68	3696	4928	6160	7392	8624	9856	11088	12320	13554	14786	16018	.0310
69	3918	5224	6530	7836	9142	10488	11754	13060	14389	15697	17006	.0292
70	4091	5456	6820	8184	9548	10912	12276	13640	15006	16370	17734	.0280
71	4419	5892	7365	8838	10311	11784	13257	14730	16160	17629	19099	.0260
72	4584	6112	7640	9168	10696	12224	13752	15280	16807	18335	19863	.0250
73	4776	6368	7960	9552	11144	12736	14328	15920	17507	19099	20690	.0240
74	5106	6808	8510	10212	11914	13616	15318	17020	18674	20372	22069	.0225
75	5457	7276	9095	10914	12733	14552	16371	18190	20008	21827	23646	.0210
76	5730	7640	9550	11460	13370	15280	17190	19100	21008	22918	24828	.0200
77	6366	8488	10610	12732	14854	16976	19098	21220	23343	25465	27587	.0180
78	7161	9548	11935	14322	16709	19096	21483	23870	26260	28648	31035	.0160
79	7902	10536	13170	15804	18438	21072	23706	26340	28988	31611	34246	.0145
80	8490	11320	14150	16980	19810	22640	25470	28300	31123	33953	36782	.0135

¹ Courtesy of Cleveland Twist Drill Co.

TABLE VIII. R.P.M. SPEEDS FOR LETTER SIZE DRILLS[1]

Feet per Min.	30	40	50	60	70	80	90	100	110	120	130	Decimal Equivalents
Letter Size					Revolutions per Minute							
A	491	654	818	982	1145	1309	1472	1636	1796	1959	2122	.234
B	482	642	803	963	1124	1284	1445	1605	1765	1926	2086	.238
C	473	631	789	947	1105	1262	1420	1578	1736	1894	2052	.242
D	467	622	778	934	1089	1245	1400	1556	1708	1863	2018	.246
E	458	611	764	917	1070	1222	1375	1528	1681	1834	1968	.250
F	446	594	743	892	1040	1189	1337	1486	1635	1784	1932	.257
G	440	585	732	878	1024	1170	1317	1463	1610	1756	1903	.261
H	430	574	718	862	1005	1149	1292	1436	1580	1723	1867	.266
I	421	562	702	842	983	1123	1264	1404	1545	1685	1826	.272
J	414	552	690	827	965	1103	1241	1379	1517	1655	1793	.277
K	408	544	680	815	951	1087	1223	1359	1495	1631	1767	.281
L	395	527	659	790	922	1054	1185	1317	1449	1581	1712	.290
M	389	518	648	777	907	1036	1166	1295	1424	1554	1683	.295
N	380	506	633	759	886	1012	1139	1265	1391	1518	1644	.302
O	363	484	605	725	846	967	1088	1209	1330	1450	1571	.316
P	355	473	592	710	828	946	1065	1183	1301	1419	1537	.323
Q	345	460	575	690	805	920	1035	1150	1266	1384	1496	.332
R	338	451	564	676	789	902	1014	1127	1239	1355	1465	.339
S	329	439	549	659	769	878	988	1098	1207	1317	1427	.348
T	320	426	533	640	746	853	959	1066	1173	1280	1387	.358
U	311	415	519	623	727	830	934	1038	1142	1246	1349	.368
V	304	405	507	608	709	810	912	1013	1114	1219	1317	.377
W	297	396	495	594	693	792	891	989	1088	1188	1286	.386
X	289	385	481	576	672	769	865	962	1058	1155	1251	.397
Y	284	378	473	567	662	756	851	945	1040	1135	1229	.404
Z	277	370	462	555	647	740	832	925	1017	1110	1202	.413

[1] Courtesy of Cleveland Twist Drill Co.

TABLE IX. FEEDS FOR DRILLING[1]

Diameter of Drill (Inches)	Feed[2] (Inches per Rev.)
Under ⅛	.001 to .002
⅛ to ¼	.002 to .004
¼ to ½	.004 to .007
½ to 1	.007 to .015
1 inch and over	.015 to .025

[1] Source: *American Machinist.*
[2] It is best to start with a moderate speed and feed, increasing either one, or both, after observing action and condition of drill.

TABLE X. DRILLS AND DRILLING SPEEDS AND FEEDS

R.P.M.= No. of revolutions per minute.
F/R   = Feed rate in inches per revolution.
Dia.   = Diameter of drill.
S.F.M.= Surface or peripheral speed in feet per minute.
F/M   = Feed rate in inches per minute.

KNOWN	TO FIND	
Dia. and S.F.M.	R.P.M.	$= \dfrac{\text{S.F.M.} \times 12}{\text{Dia.} \times 3.1416}$
Dia. and R.P.M.	S.F.M.	$= \dfrac{\text{Dia.} \times \text{R.P.M.} \times 3.1416}{12}$
Dia. S.F.M. and F/M	F/R	$= \dfrac{\text{Dia.} \times \text{F/M} \times 3.1416}{\text{S.F.M.} \times 12}$
R.P.M. and F/M	F/R	$= \dfrac{\text{F/M}}{\text{R.P.M.}}$
F/R and R.P.M.	F/M	$= \text{R.P.M.} \times \text{F/R}$

TABLE XI. CUTTING FLUIDS FOR DRILL PRESS WORK[1]

Metal	Drilling	Reaming	Tapping
Machine steel (hot and cold rolled)	Soluble oil Mineral lard oil	Mineral lard oil	Soluble oil Lard oil
Tool steel (carbon and high speed)	Soluble oil Lard oil with sulphur	Lard oil	Sulphur-base oil Mineral lard oil
Alloy steel	Soluble oil Mineral lard oil	Lard oil	Sulphur-base oil Mineral lard oil
Brass and bronze	Dry Lard oil Kerosene mixture	Soluble oil	Dry Soluble oil Lard oil
Copper	Soluble oil	Soluble oil	Soluble oil Lard oil
Aluminum	Kerosene Lard oil	Mineral lard oil	Soluble oil Mineral lard oil
Monel metal	Lard oil Sulphur-base oil	Mineral lard oil Sulphur-base oil	Mineral lard oil Sulphur-base oil
Malleable iron	Soluble oil	Soluble oil	Soluble oil
Cast iron	Dry	Dry	Small amount of mineral lard oil

[1] Source: *American Machinist.*

TABLE XII. PRINCIPAL DIMENSIONS OF MORSE STANDARD TAPERS (INCHES)

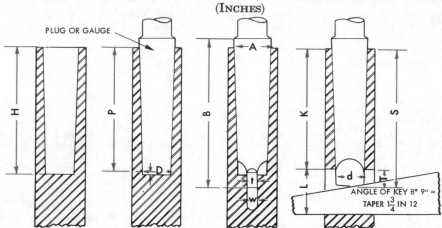

Chart Showing Principal Dimensions of Morse Standard Tapers Which Are Listed in Tabulation Below

Number of Taper	Diam. of Plug at Small End	Diam. at End of Socket	Shank		Depth of Hole	Standard Plug Depth	Tongue		Keyway		End of Socket to Keyway	Taper per Foot
			Whole Length	Depth			Thickness	Length	Width	Length		
	D	A	B	S	H	P	t	T	W	L	K	
0	0.252	0.3561	$2\frac{11}{32}$	$2\frac{7}{32}$	$2\frac{1}{32}$	2	$\frac{5}{32}$	$\frac{1}{4}$	0.160	$\frac{9}{16}$	$1\frac{15}{16}$	.6246
1	0.369	0.475	$2\frac{9}{16}$	$2\frac{7}{16}$	$2\frac{3}{16}$	$2\frac{1}{8}$	$\frac{13}{64}$	$\frac{3}{8}$	0.213	$\frac{3}{4}$	$2\frac{1}{16}$	.5986
2	0.572	0.700	$3\frac{1}{8}$	$2\frac{15}{16}$	$2\frac{5}{8}$	$2\frac{9}{16}$	$\frac{1}{4}$	$\frac{7}{16}$	0.260	$\frac{7}{8}$	$2\frac{1}{2}$	.5994
3	0.778	0.938	$3\frac{7}{8}$	$3\frac{11}{16}$	$3\frac{1}{4}$	$3\frac{3}{16}$	$\frac{5}{16}$	$\frac{9}{16}$	0.322	$1\frac{3}{16}$	$3\frac{1}{16}$	.6023
4	1.020	1.231	$4\frac{7}{8}$	$4\frac{5}{8}$	$4\frac{1}{8}$	$4\frac{1}{16}$	$\frac{15}{32}$	$\frac{5}{8}$	0.478	$1\frac{1}{4}$	$3\frac{7}{8}$	.6233
5	1.475	1.748	$6\frac{1}{8}$	$5\frac{7}{8}$	$5\frac{1}{4}$	$5\frac{3}{16}$	$\frac{5}{8}$	$\frac{3}{4}$	0.635	$1\frac{1}{2}$	$4\frac{15}{16}$	.6315
6	2.116	2.494	$8\frac{9}{16}$	$8\frac{1}{4}$	$7\frac{3}{8}$	$7\frac{1}{4}$	$\frac{3}{4}$	$1\frac{1}{8}$	0.760	$1\frac{3}{4}$	7	.6256
7	2.750	3.270	$11\frac{5}{8}$	$11\frac{1}{4}$	$10\frac{3}{8}$	10	$1\frac{1}{8}$	$1\frac{3}{8}$	1.135	$2\frac{5}{8}$	$9\frac{1}{2}$	.6240

The figures in the "Taper per Foot" column have been revised to conform with the standard end diameters and lengths.

TABLE XIII. LATHE CUTTING SPEEDS[1]

Feet per Min.	40	50	60	70	80	90	100	120	140	160	180
Diam. Inches	REVOLUTIONS PER MINUTE										
1/4	611	764	917	1070	1222	1375	1528	1833	2140	2444	2750
3/8	408	509	611	713	815	916	1018	1222	1426	1630	1832
1/2	306	382	458	535	611	688	764	916	1070	1222	1376
5/8	244	306	367	428	489	550	611	733	856	978	1100
3/4	204	254	306	357	407	458	509	611	714	814	916
7/8	175	218	262	306	349	393	436	523	612	698	786
1	153	191	229	267	306	344	382	458	534	612	688
1 1/8	136	170	204	238	272	305	339	407	476	544	610
1 1/4	122	153	183	214	244	275	305	366	428	488	550
1 3/8	111	139	166	194	222	249	277	332	388	444	498
1 1/2	102	127	153	178	204	229	254	305	356	408	458
1 3/4	87	109	131	153	175	196	218	262	306	350	392
2	76	95	114	133	153	172	191	229	266	306	344
2 1/4	68	85	102	119	136	153	170	204	238	272	306
2 1/2	61	76	92	107	122	137	153	183	214	244	274
2 3/4	55	69	83	97	111	125	139	166	194	222	250
3	51	64	76	89	102	115	127	153	178	204	230
3 1/4	47	59	70	82	94	106	117	141	164	188	212
3 1/2	44	54	65	76	87	98	109	131	152	174	196
3 3/4	41	51	61	71	81	92	102	122	142	162	184
4	38	48	57	67	76	86	95	114	134	152	172
4 1/2	34	42	51	59	68	76	85	102	118	136	152
5	30	38	46	53	61	69	76	92	106	122	138
5 1/2	28	35	42	49	55	62	69	83	98	110	124
6	25	32	38	44	51	57	64	76	88	102	114
6 1/2	23	29	35	41	47	53	59	70	82	94	106
7	22	27	33	38	44	49	54	65	76	88	98
7 1/2	20.4	25	31	36	41	46	51	61	72	82	92
8	19.1	24	29	33	38	43	48	57	66	76	86
8 1/2	18	22	27	31	36	40	45	54	62	72	80
9	17.0	21.2	25	30	34	38	42	51	60	68	76
9 1/2	16.1	20.1	24	28	32	36	40	48	56	64	72
10	15.3	19.1	23	27	31	34	38	46	54	62	68
11	13.9	17.4	20.8	24	28	31	35	41	48	56	62
12	12.7	15.9	19.1	22	25	29	32	38	44	50	58
13	11.8	14.7	17.6	20.6	23	26	29	35	41	46	52
14	10.9	13.6	16.4	19.1	22	24	27	33	38	44	48
15	10.2	12.7	15.3	17.8	20.4	23	25	30	35	41	46
16	9.5	11.9	14.3	16.7	19.1	21.4	24	29	33	38	43
17	9.0	11.2	13.5	15.7	18.0	20.2	22	27	31	36	40
18	8.5	10.6	12.7	14.8	17.0	19.1	21	25	30	34	38

[1] Source: *Turret Lathe Operator's Manual*, The Warner & Swasey Co.

TABLE XIV. MILLING MACHINE R.P.M. NECESSARY TO GIVE A DESIRED CUTTING SPEED[1]

DIAMETER (INCHES)	CUTTING SPEEDS IN FEET PER MINUTE						
	40	50	60	70	80	90	100
	Revolutions per Minute						
¼	611	764	917	1,070	1,222	1,375	1,528
⁵⁄₁₆	489	611	733	856	978	1,100	1,222
⅜	407	509	611	713	815	917	1,019
⁷⁄₁₆	349	437	524	611	698	786	873
½	306	382	458	535	611	688	764
⅝	244	306	367	428	489	550	611
¾	204	255	306	357	407	458	509
⅞	175	218	262	306	349	393	437
1	153	191	229	267	306	344	382
1⅛	136	170	204	238	272	306	340
1¼	122	153	183	214	244	275	306
1⅜	111	139	167	194	222	250	278
1½	102	127	153	178	204	229	255
1⅝	94	117	141	165	188	212	235
1¾	87	109	131	153	175	196	218
1⅞	81	102	122	143	163	183	204
2	76	95	115	134	153	172	191
2¼	68	85	102	119	136	153	170
2½	61	76	92	107	122	137	152
2¾	56	69	83	97	111	125	139
3	51	64	76	89	102	115	127
3½	44	55	65	76	87	98	108
4	38	48	57	67	76	86	95
4½	34	42	51	59	68	77	85
5	31	38	46	54	61	69	76
5½	28	35	42	49	56	63	70
6	25	32	28	45	51	57	64
7	22	27	33	38	44	49	55
8	19	24	29	33	38	43	48
9	17	21	25	30	34	38	42
10	15	19	23	27	31	34	38
11	14	17	21	24	28	31	35
12	13	16	19	22	25	29	32
13	12	15	18	21	24	27	29
16	10	12	14	17	19	22	24
18	8	11	13	15	17	19	21

[1] Source: *American Machinist.*

TABLE XV. CUTTING SPEEDS IN SURFACE FEET PER MINUTE[1,2]

| WORKPIECE MATERIAL | CUTTER MATERIAL | | | | Coolant |
| | High-Speed Steel | | Carbide-Tipped | | |
	Rough	Finish	Rough	Finish	
Cast iron......	50–60	80–110	180–200	350–400	Dry
Semisteel.....	40–50	65–90	140–160	250–300	Dry
Malleable iron.	80–100	110–130	250–300	400–500	Soluble, sulphurized, or mineral oil
Cast steel.....	45–60	70–90	150–180	200–250	Soluble, sulphurized, mineral, or mineral lard oil
Copper.......	100–150	150–200	600	1,000	Soluble, sulphurized, or mineral lard oil
Brass.........	200–300	200–300	600–1,000	600–1,000	Dry
Bronze.......	100–150	150–180	600	1,000	Soluble, sulphurized, or mineral lard oil
Aluminum....	400	700	800	1,000	Soluble or sulphurized oil, mineral oil and kerosene
Magnesium...	600–800	1,000–1,500	1,000–1,500	1,000–5,000	Dry, kerosene, mineral lard oil
SAE steels 1020 (coarse feed).....	60–80	60–80	300	300	Soluble, sulphurized, mineral, or mineral lard oil
1020 (fine feed).....	100–120	100–120	450	450	" " "
1035.......	75–90	90–120	250	250	" " "
X-1315.....	175–200	175–200	400–500	400–500	" " "
1050......	60–80	100	200	200	" " "
2315.......	90–110	90–110	300	300	" " "
3150.......	50–60	70–90	200	200	" " "
4340.......	40–50	60–70	200	200	Sulphurized and mineral oils
Stainless steel..	100–120	100–120	240–300	240–300	" " "

[1] Feeds should be as much as the work and equipment will stand, provided a satisfactory surface finish is obtained.
[2] Source: *American Machinist.*

Machine Shop Operations and Setups

TABLE XVI. PERMISSIBLE FEED PER TOOTH (INCHES) FOR HIGH-SPEED STEEL MILLING CUTTERS[1]

Work Material	Face Mills	Spiral Mills	Slotting & Side Mills	End Mills	Form Cutters	Saws
Aluminum and soft bronze...	.022	.017	.013	.011	.006	.005
Medium bronze and soft cast iron...............	.018	.014	.011	.009	.005	.004
Malleable iron and medium cast iron...............	.015	.012	.009	.008	.005	.004
SAE X-1112 steel and hard cast iron...............	.013	.010	.008	.006	.004	.003
SAE 1020 steel and SAE X-1335 steel........	.011	.009	.007	.005	.004	.003
SAE 1045 steel and cast steel.	.009	.007	.006	.005	.003	.003
Alloy steel, medium........	.008	.006	.005	.004	.003	.002
Alloy steel, tough..........	.007	.005	.004	.004	.002	.002
Alloy steel, 250 to 300 Brinell.	.006	.005	.004	.003	.002	.0015
Alloy steel, hard—300 to 360 Brinell.................	.005	.004	.003	.003	.002	.0015

[1] Source: *American Machinist.*

TABLE XVII. R.P.M. FOR VARIOUS DIAMETERS OF GRINDING WHEELS TO GIVE PERIPHERAL SPEEDS IN S.F.P.M.[1,2]

Diameter of Wheel in Inches	Peripheral Speed in Feet per Minute					
	4000	4500	5000	5500	6000	6500
	Revolutions per Minute					
1	15,279	17,189	19,098	21,008	22,918	24,828
2	7,639	8,594	9,549	10,504	11,459	12,414
3	5,093	5,729	6,366	7,003	7,639	8,276
4	3,820	4,297	4,775	5,252	5,729	6,207
5	3,056	3,438	3,820	4,202	4,584	4,966
6	2,546	2,865	3,183	3,501	3,820	4,138
7	2,183	2,455	2,728	3,001	3,274	3,547
8	1,910	2,148	2,387	2,626	2,865	3,103
10	1,528	1,719	1,910	2,101	2,292	2,483
12	1,273	1,432	1,591	1,751	1,910	2,069
14	1,091	1,228	1,364	1,500	1,637	1,773
16	955	1,074	1,194	1,313	1,432	1,552
18	849	955	1,061	1,167	1,273	1,379
20	764	859	955-	1,050	1,146	1,241
22	694	781	868	955	1,042	1,128
24	637	716	796	875	955	1,034
26	588	661	734	808	881	955
28	546	614	682	750	818	887
30	509	573	637	700	764	828
32	477	537	597	656	716	776
34	449	505	562	618	674	730
36	424	477	530	583	637	690

[1] Formula for finding s.f.p.m. of wheel:

$$\text{S.f.p.m. of wheel} = \frac{\text{diam. of wheel (inches)} \times 3.1416 \times \text{r.p.m. of spindle}}{12}$$

[2] Source: The Norton Company, Worcester, Mass.

TABLE XVIII. GUIDE FOR CARBIDE GRADE SELECTION

	APPLICATION		Newcomer Products	Adamas	Carmet	Carboloy	Sandvik	Howmet	Firthite	Kennametal	Talide	Unimet	Valenite	V-R Wesson	Walmet	Willey
CHIP REMOVAL	Cast Irons	Roughing Cuts	C-1	N10	B	CA3	44A	H20	FA-5	H	K1	C89	U-10	VC-1	VR54 2A68	WA-1	E8
	Non-Ferrous, Non-Metallic, Hi-Temp. Alloys	General Purpose	C-2	N20 N22	A AM	CA4	860 883	H20 H1P	FA-6	HA	K6 K68	C91	U-20	VC-2 VC-28	2A5 VR54	WA-2	E6
		Light Finishing	C-3	N30	AA	CA7	905	H1P H05	FA-7	HE	K8	C93	U-30	VC-3	2A7	WA-3	E5
	200 & 300 Series Stainless	Precision Boring	C-4	N40	AAA	CA8	999	H05	FA-8	HF	K11	C95	U-40	VC-4	VR52	WA-4	E3
	Carbon Steels	Roughing Cuts	C-5	N50	GG 434	CA740	370	S6 S4	FT-3	TO4	K21 K42	S-880	U-53	VC-55 VC-125	VR89 VR77	WA-5	945 10A
		General Purpose	C-6	N60	D 6X	CA610 CA720	78B	S4 S2	FT-4 FT-5	NTA TXH	K25	S-901	U-60	VC-6	VR75	WA-6	8A
	Alloy Steels	Finishing Cuts	C-7	N70 N72	C548 TI-80	CA606 CA711	78 350	S1P F02	FT-62 FT-6	T22 TXL	K4H K45	S-92	U-70 U-73	VC-7	VR73	WA-7	6A 606
		Precision Boring	C-8	N80 N93	CC TI-80	CA704	320 210	F02	FT-7	T31 WF	K7H K165	S-94	U-80	VC-8 VC-83	VR71 VR65	WA-8	509
	400 Series Stainless	Hi-Velocity Cuts	C-80	N95			0-30				CO6				VR97		

TABLE XIX. GUIDE TO CUTTING SPEEDS AND FEEDS FOR TURNING WITH TITANIUM CARBIDE

MATERIAL	CONDITION	BRINELL HARDNESS NUMBER	SPEED F.P.M.	FEED I.P.R.
Grey Cast Iron				
Class 20 thru 25	Annealed	100/150	350/1200	.008/.020
Class 30	Cast	140/190	350/1000	.008/.020
Class 35 thru 40	Cast	180/220	300/900	.008/.020
Class 45 thru 60	Cast	210/270	150/900	.007/.015
Nodular Cast Iron				
SAE 60-40-10	Annealed	120/200	350/1500	.008/.020
SAE 80-60-03	Cast	200/270	250/900	.008/.020
SAE 100-70-03	Norm. & Tempered	240/300	150/800	.007/.015
SAE 120-90-02	Tempered	250/360	125/600	.007/.015
Malleable Cast Iron				
Grade 32510	Malleablized	100/150	400/1200	.008/.020
Grade 43010	Malleablized	150/200	325/1000	.008/.020
Grade 60003	Malleablized	200/240	250/800	.007/.015
Cast Steel Plain Carbon, Plain Carbon and Alloy Steels	Normalized	140/180	400/900	.010/.015
	Quenched & Tempered	190/240	400/800	.010/.015
Low Carbon Free Machining	Cold Drawn	150/240	500/1000	.005/.024
	Quenched & Tempered	250/300	300/900	.002/.015
Medium Carbon Free Machining	Cold Drawn	190/240	400/1000	.005/.020
	Quenched & Tempered	240/310	350/900	.005/.015
Plain Medium Carbon (C .40 thru .50)	Annealed	110/190	400/900	.005/.015
Plain Medium Carbon	Quenched & Tempered	200/260	350/700	.005/.015
	Quenched & Tempered	260/325	300/650	.005/.015
Plain High Carbon (C .55 thru .59)	Annealed	150/210	375/800	.005/.015
Plain Carbon Steel	Quenched & Tempered	300/375	250/700	.005/.015
Leaded Alloy	Annealed	130/190	600/1500	.005/.020
	Normalized	250/300	400/1000	.005/.020

MATERIAL	CONDITION	BRINELL HARDNESS NUMBER	SPEED F.P.M.	FEED I.P.R.
Alloy Steels	Annealed	150/250	300/1000	.005/.020
	Normalized or Quenched & Tempered	230/325	300/1000	.005/.015
	Quenched & Tempered	300/380	300/800	.005/.015
	Quenched & Tempered	390/450	250/750	.005/.012
	Quenched & Tempered	460/510	220/600	.005/.010
	Quenched & Tempered	520/600	150/500	.003/.008
Tool Steels High-Carbon, High-Chrome		150/200	225/600	.010/.015
High-Speed Steels		210/285	200/500	.010/.015
Hot-Work Die Steels		160/220	350/900	.010/.015
		340/375	250/600	.010/.015
		515/560	75/200	.008/.015
Stainless Steels 405, 430, 430F, 442 & 446	Annealed	125/200	350/900	.005/.020
403, 410, 420 & 431	Annealed	150/230	400/1000	.005/.020
Stainless Steel 440 Series	Annealed	200/260	300/800	.004/.015
	Quenched & Tempered	250/375	200/800	.004/.015
	Quenched & Tempered	380/450	150/650	.004/.012
Precipitation-Hardening	Annealed	160/180	300/600	.010/.015
	Quenched & Tempered	380/440	200/600	.008/.015
Aluminum Alloys Non-Heat-treatable Cast Alloys	Cast	50/70*	850 min.	.010/.015
Heat-treatable Cast Alloys	Solution-Treated and Aged	70/105*	700 min.	.010/.015
Non-Heat-treatable Wrought Alloys	Cold Drawn	40/70*	700 min.	.010/.015
Heat-treatable Wrought Alloys	Solution-Treated and Aged	65/105*	700 min.	.010/.015

*500-Kg Load

TABLE XX. SPEEDS AND FEEDS FOR FACE MILLING

Material To Be Machined	Operation	Grade		Cutting Speed Range–SFPM	Feed– In./Tooth
		Throw–Away	Inserted Blade		
Steels: Free Machining Including Most Hot Rolled or Cold Finished Low Carbon and Medium Carbon Steels	Roughing	WS, VR–77, HR	WS	400–800	.010–.025
	Finishing	WM, VR–75	WM	500–1000	.010–.025
Soft Alloy	Roughing	WS, VR–77, HR	WS	300–600	.008–.020
	Finishing	WM, VR–75	WM	400–800	.008–.020
Medium Hard Alloy	Roughing	WS, VR–77, HR	WS	200–400	.008–.015
	Finishing	WM, VR–75	WM	200–400	.008–.015
Hard Alloy	Rouhging	WS, VR–77, HR	WS	80–200	.004–.010
	Finishing	WM, VR–75	WM	100–300	.004–.010
Cast	Roughing	WS, VR–77, HR	WS	150–300	.008–.015
	Finishing	WM, VR–75	WM	250–450	.008–.015
Cast Iron: Soft	Roughing			440–600	.015–.040
	Finishing			500–600	.015–.040
Medium	Roughing	2A5, GI, VR–54	2A5, GI, VR–54	220–375	.010–.025
	Finishing			300–500	.010–.025
Hard	Roughing			125–275	.005–.015
	Finishing			175–400	.005–.015
Brass	Roughing	2A5, GI, VR–54	2A5, GI, VR–54	500–2000	.010–.020
	Finishing			1000–3000	.010–.020
Bronze	Roughing	2A5, GI, VR–54	2A5, GI, VR–54	300–1000	.010–.020
	Finishing			500–1000	.010–.020
Aluminum	Roughing	2A5, GI, VR–54	2A5, GI, VR–54	3000–15,000	.010–.025
	Finishing			5000 and up	.010–.025
Magnesium	Roughing	2A5, GI, VR–54	2A5, GI, VR–54	5000–15,000	.010–.025
	Finishing			5000 and up	.010–.025

Note: Depth of cut for roughing operations, .250 in. or less; for finishing operations, .050 in. or less.